PARTICLE ACCELERATION AND TRAPPING IN SOLAR FLARES

Science and history joined together in the Stuart's Castle in Aubigny-sur-Nère (France) where the second CESRA Workshop on 'Particle Acceleration and Trapping in Solar Flares' was held (June 23–26, 1986).

A view of the Nançay Radioheliograph showing a portion of the east–west antenna array, one antenna of the north–south array and the laboratory where the receivers and digital facilities are located. Nançay Observatory is about twenty kilometers distant from Aubigny.

PARTICLE ACCELERATION AND TRAPPING IN SOLAR FLARES

Selected Contributions to the Workshop held at Aubigny-sur-Nère (Bourges), France, June 23–26, 1986

Edited by

GÉRARD TROTTET and MONIQUE PICK (Eds.)
Observatoire de Paris, Meudon, France

Reprinted from Solar Physics, Vol. 111, No. 1 (1987)

D. Reidel Publishing Company
Dordrecht / Boston

Library of Congress Cataloging-in-Publication Data

CIP

ISBN 90-277-2609-4

DATA APPEAR ON SEPARATE CARD

Published by D. Reidel Publishing Company,
P.O. Box 17, 3300 AA Dordrecht, Holland.

Sold and distributed in the U.S.A. and Canada
by Kluwer Academic Publishers,
101 Philip Drive, Norwell, MA 02061, U.S.A.

In all other countries, sold and distributed
by Kluwer Academic Publishers Group,
P.O. Box 322, 3300 AH Dordrecht, Holland.

Printed in Belgium

TABLE OF CONTENTS

(*Particle Acceleration and Trapping in Solar Flares*)

SESSION V: CYCLOTRON MASER, SPIKES

SESSION VI: HARD X-RAY BURSTS

PREFACE

These are the Proceedings of the Second CESRA Workshop on Particle Acceleration and Trapping in Solar Flares. The Workshop was held on June 23–26, 1986, at the city hall of Aubigny-sur-Nère (France), near Bourges and near the Nançay Radio Observatory. The scientific organizing committee included M. Pick (chairperson), A. O. Benz, A. Magun, G. Trottet, and P. Zlobec. During this meeting the main objective was to review the different acceleration processes and interpretations of continuum radio emission and fine structure. This volume contains the reviews and a selection of contributions presented at the workshop.

We wish to thank the local organizers, in particular N. Vilmer (chairperson), Y. Avignon, G. Cordeille, B. Darchy, P. Collin, and V. Etiève. Financial support was provided by Observatoire de Paris, the Centre National de la Recherche Scientifique (C.N.R.S.), the Centre National d'Études Spatiales (C.N.E.S.), the Institut National des Sciences de l'Univers (I.N.S.U.), the Direction des Recherches, Études et Techniques (D.R.E.T.), and the European Space Agency (E.S.A.).

We would like to express our gratitude to the Mayor of Aubigny-sur-Nère for his kind hospitality.

GÉRARD TROTTET MONIQUE PICK

LIST OF PARTICIPANTS

Aschwanden, M.	Institut für Astronomie ETH, Zürich, Switzerland
Benz, A. O.	Institut für Astronomie ETH, Zürich, Switzerland
Boischot, A.	Observatoire de Paris, Meudon, France
Bougeret, J. L.	Observatoire de Paris, Meudon, France
Bradu, P.	Dir. Rech. Etu. Tech., Paris, France
Carbone, V.	Universita della Calabria, Roges, Italy
Chiuderi, F.	Observatoire de Paris, Meudon, France
Dennis, B.	Goddard Space Flight Center, Greenbelt, U.S.A.
Dulk, G.	University of Colorado, Boulder, U.S.A.
Gary, D.	CALTECH, Pasedena, U.S.A.
Güdel, M.	Institut für Astronomie ETH, Zürich, Switzerland
Hurford, G.	CALTECH, Pasedena, U.S.A.
Kai, K.	Tokyo Astron. Observatory, Tokyo, Japan
Kaufmann, P.	INPE, São Paulo, Brazil
Klein, K.-L.	Observatoire de Paris, Meudon, France
Kundu, M. R.	University of Maryland, College Park, U.S.A.
Lang, K. R.	Tufts University, Medford, U.S.A.
Le Quéau, D.	CNET–CRPE, Issy les Moulineaux, France
Li, H. W.	Trieste Astron. Observatory, Trieste, Italy
Louarn, P.	CNET–CRPE, Issy les Moulineaux, France
Magun, A.	IAP University of Bern, Bern, Switzerland
Mangeney, A.	Observatoire de Paris, Meudon, France
Mercier, C.	Observatoire de Paris, Meudon, France
Messeroti, M.	SISSA/ISAS and Trieste Astron. Observatory, Italy
Pick, M.	Observatoire de Paris, Meudon, France
Raoult, A.	Observatoire de Paris, Meudon, France
Ruždjak, V.	Faculty of Geodesy, University of Zagreb, Zagreb, Yugoslavia
Sawant, H.	INPE, São Paulo, Brazil
Schanda, E.	IAP, University of Bern, Bern, Switzerland
Simnett, G.	University of Birmingham, Birmingham, U.K.
Slottje, C.	Netherlands Foundation for Radioastronomy, Dwingeloo, The Netherlands
Spicer, D.	Goddard Space Flight Center, Greenbelt, U.S.A.
Stähli, M.	IAP University of Bern, Bern, Switzerland
Thejappa, G.	Indian Inst. of Astrophysics, Bangalore, India
Trevisan, R.	Univesidade Londrina, Londrina, Brazil
Trottet, G.	Observatoire de Paris, Meudon, France
Urbarz, H.	University of Tübingen, Weissenau, Rasthalde, F.R.G.
Veltri, P.	Universita della Calabria, Roges, Italy
Vilmer, N.	Observatoire de Paris, Meudon, France
Vlahos, L.	University of Thessaloniki, Thessaloniki, Greece
Vršnak, B.	Faculty of Geodesy, University of Zagreb, Zagreb, Yugoslavia
Zlobec, P.	Trieste Astron. Observatory, Trieste, Italy

ACCELERATION AND ENERGIZATION BY CURRENTS AND ELECTRIC FIELDS*

(Invited Review)

A. O. BENZ

Institute of Astronomy, ETH, Zürich, Switzerland

Abstract. The acceleration of the influential $\lesssim 100$ keV electrons in flares observed in hard X-rays and several radio emissions is unknown. Shock-waves and MHD turbulence, successfully applied to interprete interplanetary energetic particles, have recently been called in question concerning energetic flare electrons and ions. Other possible mechanisms are considered which are closely related to the primary flare energy release. In particular, runaway acceleration by the electric field of the reconnection current sheet, bulk heating by microturbulence, and cross-field ion currents due to bulk motion as a primary result of reconnection are reviewed. All three are likely to occur in some way. Their relative importance cannot be definitively assessed due to the lack of information on non-thermal, low energy protons.

1. Introduction

The object of the present review is to give an account of the emerging new concepts of particle acceleration in solar flares and to expose some problems of the not well known grey zone between magnetohydrodynamic (MHD) models and kinetic (i.e. microscopic) theory. MHD flare models based on reconnection theory (for reviews cf. Vasyliunas, 1975; van Hoven, 1981; Priest, 1985, 1986, etc.) have received much attention during the past cycle. They may give a global view of the flare phenomenon and provide ideas on the background environment within which particle acceleration occurs. There is a subtle interaction between MHD and kinetic theory in the current layers and shocks. These places coincide with the regions where particles are heated and/or accelerated. In fact, these regions are generally treated as 'Pandora's box' containing a host of fickle physics which is considered better not to open. However, these regions control the global evolution by their resistivity and heat conduction. Most flare models use enhanced resistivity above single particle (Coulomb) friction thus implying turbulence of some sort in the current layer. Most energy in reconnection is released by Ohmic dissipation due to such resistivity and sub-Alfvénic ion motions (jets). Before instabilities occur leading to such turbulence, electric fields in current layers may accelerate particles thus draining energy from the field.

An extensive review on the theory of acceleration has been written by Heyvaerts (1981). Since then considerable progress has been achieved in both observations of acceleration related phenomena and theory. Recent observations and a general discussion have also been collected by Vlahos *et al.* (1984). Not only the knowledge on energetic flare particles and their various signatures has tremendously increased, but interesting inferences on the acceleration process itself have become possible. Some of

* Proceedings of the Second CESRA Workshop on Particle Acceleration and Trapping in Solar Flares, held at Aubigny-sur-Nère (France), 23–26 June, 1986.

Solar Physics **111** (1987) 1–18.

the major breakthroughs are mentioned here in Section 2. In the rest we concentrate on processes intimately related to electric currents and the primary energy release.

It has long been recognized that acceleration occurs in two phases (Wild *et al.*, 1963; de Jager, 1969). The first phase coincides with the main flare energy release which supposedly dumps a considerable amount of the total flare energy into $\lesssim 100$ keV electrons. The second phase is connected to the appearance of interplanetary energetic particles (ions and electrons) on time scales $\gtrsim 100$ s after the peak of the impulsive phase. Second phase acceleration is generally attributed to Fermi acceleration by shock waves, a subject not covered in this review. Here we concentrate on the first phase and in particular on acceleration processes due to runaway and microturbulence of the initial current sheet and cross-field ion currents due to primary bulk motion. All these processes are very likely to occur but their relative importance will remain unclear.

2. New Observations on Acceleration

It is widely accepted that energies up to 10^{32} erg can be released in solar flares presumably within a region with $\lesssim 10^4$ km diameter near the top of a magnetic loop or arcade. However, it is unclear whether this energy first appears in electrons of roughly 20 keV (corresponding to 2×10^8 K), or whether particles of such energies and higher are a negligible fraction of the total energy at all times. The first view results if the observed hard X-ray (HXR) flux is produced by electron beams impinging on a relatively cold background (non-thermal model). Their energy is then comparable to or greater than the total energy required for the post-impulsive phase phenomena, such as thermal soft X-ray (SXR) emission, mass motions, second phase particle acceleration, etc. A far smaller number of energetic electrons is required in a well contained source where energetic particles more efficiently emit bremsstrahlung by collisions among themselves. (This model is known under the misnomer 'thermal'.) Bipolar hard X-ray sources simultaneously lighting up have been observed by Duijveman *et al.* (1982) and support a partially non-thermal model. Directivity effects of > 300 keV electron bremsstrahlung yield further evidence in favor of vertically descending beams (Vestrand *et al.*, 1986). On the other hand, Brown *et al.* (1983), from the analysis of stereoscopic observations, and similarly Machado and Lerner (1984) from a limb flare, find that the hard X-ray source distribution in height is not compatible with a purely non-thermal interpretation. The latter authors conclude that more than half of the HXR emission is thermally emitted by a plasma of $> 5 \times 10^7$ K. The absence of any radio signature of electron beams in about 15% of major HXR flares led Simnett and Benz (1986) to conclude that the containment of the energetic electrons may sometimes by remarkably good. The picture that is emerging thus requires partial trapping with a variable degree of containment. It seems nevertheless plausible that a large fraction of the dissipated flare energy (at least 20% according to Duijveman *et al.*, 1982) transitorily resides in energetic electrons. It would seem rather contrived to accelerate such a fraction through a heat engine driven by the 10^7 K SXR emitting plasma. These electrons seem to originate from the primary energy release.

Where is the acceleration site? Since we cannot measure the magnetic topology, we may try to answer the more humble questions of the density and temperature of the site, important parameters in all acceleration processes. Strong *et al.* (1984) and Webb (1985) have measured a preflare density of 4–10 × 10^9 cm^{-3} by SXR lines of moderate coronal temperatures. From the propagation length of electron beams Duijveman *et al.* (1982) find an upper limit of 2 × 10^{10} cm^{-3}. Densities between 10^{10} and 7 × 10^{11} cm^{-3} have been suggested by Antonucci *et al.* (1986) who identify regions of high turbulence observed in SXR lines with acceleration. A lower limit can be derived from the starting frequency of normal type III bursts. The start of reversed slope type III events yields an upper limit. In cases events showing both kinds of type III's, such as shown in Figure 1, a limiting density range of the acceleration site can be determined. In the case of Figure 1 it corresponds to 750–800 MHz or, if the emission is at the plasma frequency, 7–8 × 10^9 cm^{-3}. There are, however, mavericks of reversed bursts at lower frequency and density. Note that the acceleration of type III causing electrons may not coincide with the main energy release (Sprangle and Vlahos, 1983). Associating the sources of millisecond radio spikes with acceleration sites, the band in which spikes have been reported suggests a density range of 10^9–10^{11} cm^{-3} (Benz, 1986). This is today – for the various reasons given – the mosr widely assumed range of acceleration site density.

Since the discovery of energetic flare particles in type III bursts it has been clear that their acceleration is impulsive and intermittent. De Jager and de Jonge (1978) have proposed 'elementary flares' to interprete HXR peaks of half-power durations of 4–24 s. With better sensitivity Kiplinger *et al.* (1983a, 1984) have found time variations in the HXR emission of the order of 0.1 s and less. Peaks separated by a few seconds have been interpreted as sequentially firing flare elements (Kiplinger *et al.*, 1983b; Kane *et al.*, 1983; Kouveliotou *et al.*, 1984). This is confirmed by spatially resolved UV observations (Cheng *et al.*, 1981) and type III bursts (Raoult and Pick, 1980), which both show sources at varying positions in the transition region and the upper corona, respectively. The highest fragmentation of flare radiation is observed in millisecond radio spikes. Hundreds of spikes can be associated with one type III of HXR peak. Radio spikes of 1–50 ms duration observed in the impulsive phase of flares have an extremely narrow bandwidth of about 1.5% of their center frequency. This frequency, however, varies chaotically from spike to spike up to a factor of two. It indicates that the spike sources must be very small regions in a larger volume. Benz (1985) has suggested that the spike phenomenon is caused by a myriad of exciters due to a fragmentation of the primary energy release into more than 10^4 elements in some flares.

Extensive observations in γ-rays have revealed a wealth of new results on protons and highly relativistic electrons in flares. Until recently it was thought that these phenomena were part of the second phase. However, improved sensitivity and time resolution have shown that the γ-rays sometimes peak simultenaously with the 20 keV electrons within less than 2 s (Chupp *et al.*, 1981, Figure 2). This prompt acceleration obviously makes them part of the first phase. Since the number of interplanetary protons is independent of these γ-ray-producing first phase protons (Cliver *et al.*, 1983), it is generally agreed

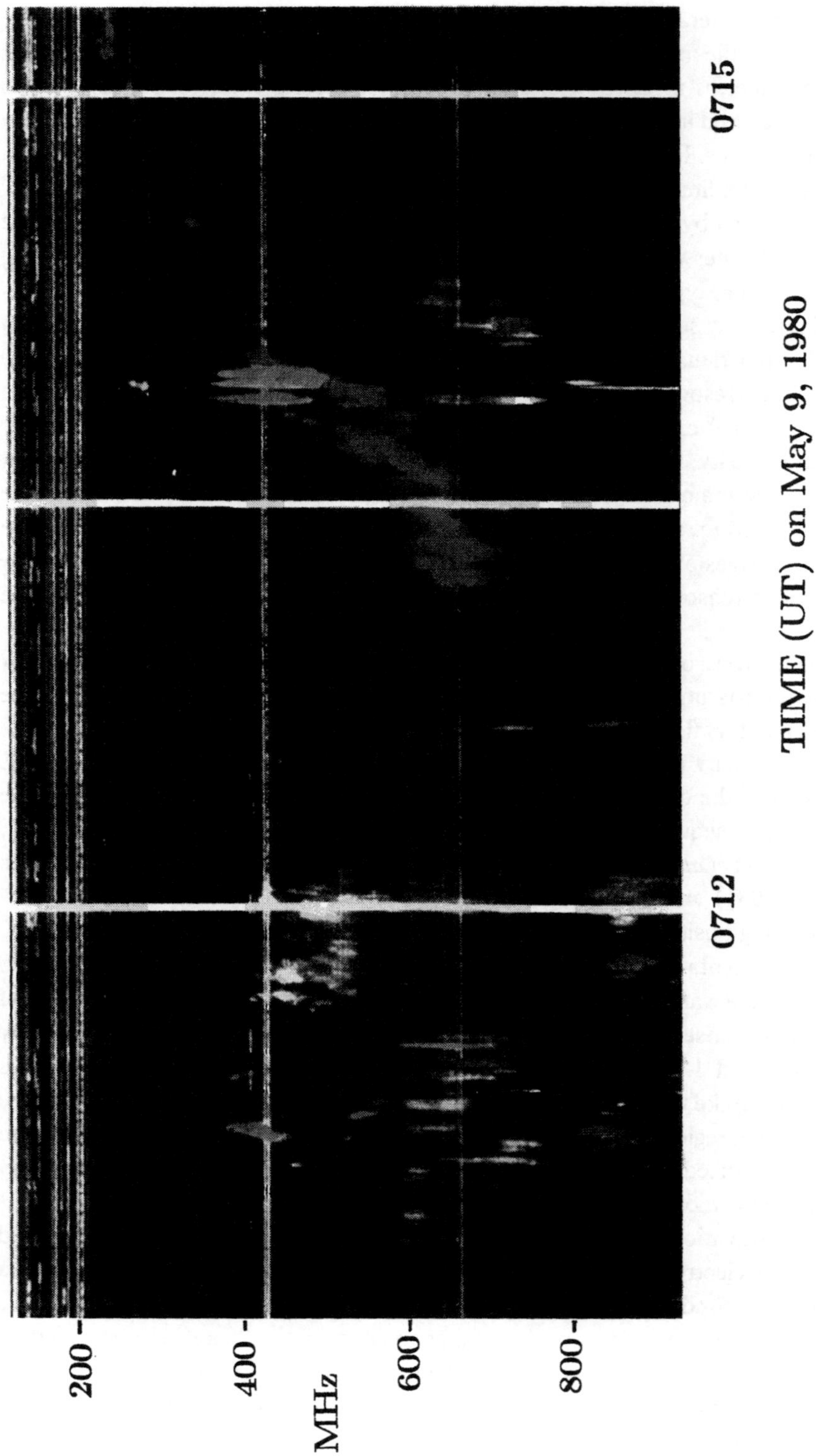

Fig. 1. Decimetric radio spectrogram recorded by the Zürich (Bleien) analog spectrograph (DAEDALUS). Enhanced emission is bright. *Yellow*: normal drift (upward moving) type III bursts or blips; *red*: reversed drift (downward moving) type III bursts; *green*: U-bursts; *blue*: type II (shockwave) emission; *white*: unidentified or terrestrial interference.

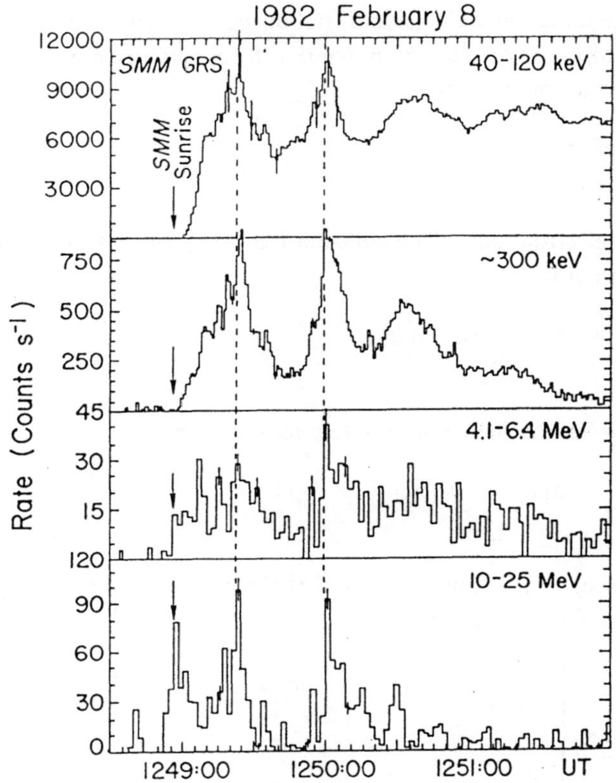

Fig. 2. The time history of HXR and γ-ray flare emission. The low-energy photons early in the flare have been occulted by the Earth's atmosphere. The 4.1–6.4 MeV band contains γ-ray lines due to accelerated protons (from Chubb, 1984).

that they represent independent acceleration processes. Delays of a few seconds of $\gtrsim 100$ keV electrons (as observed in hard X-rays and microwaves) and protons have also been reported for a small but significant fraction of flares (Hoyng *et al.*, 1976; Benz, 1977; Hudson *et al.*, 1980; Vilmer *et al.*, 1982; Bai and Dennis, 1985, and references therein). They may indicate that the various species of first phase energetic particles are accelerated by different mechanisms.

3. Runaway Acceleration

The energy released in solar flares is generally understood to derive from energy stored in non-potential magnetic fields and associated currents. Field annihilation and currents produce slowly varying electric fields in most flare models. If the electric field is parallel to the magnetic field, it provides a very simple acceleration mechanism and a direct means of energy transfer from fields to particles.

The parallel component of the electric field accelerates electrons and ions with a force

eE into opposite directions. Collisional friction between the two species introduces a drag of force of $m_e v \nu_c$ on an electron moving with velocity v relatively to the ions. There is an equilibrium at the velocity

$$v = \frac{eE_{\parallel}}{m\nu_c} , \tag{1}$$

where m_e is the electron mass. The collision frequency (inverse slowing-down time) is a function of velocity

$$\nu_c = \frac{8\pi Z^2 e^4 n \ln \Lambda \, G(v/v_e)}{m^2 v v_e^2} , \tag{2}$$

where ions have been considered infinitely massive, and

$$G(x) = \frac{\Phi(x) - x\Phi'(x)}{2x^2} \tag{3}$$

(Spitzer, 1962, see, e.g., Schmidt, 1979). Φ is the error function, and v_e is the electron thermal velocity. We will use

$$G(x) \approx \begin{cases} \dfrac{2x}{3\sqrt{\pi}} & x \ll 1 , \quad \text{(4a)} \\[2ex] \dfrac{1}{2x^2} & x \gg 1 . \quad \text{(4b)} \end{cases}$$

For

$$E < E_{\mathrm{D}} \equiv \frac{e}{\lambda_{\mathrm{D}}^2} \ln \Lambda \tag{5}$$

(where λ_{D} is the Debye length, end E_{D} the Dreicer field), Equation (1) has two solutions. The first solution is at $v_1 < v_e$ (electron thermal velocity). In the limit $E \ll E_{\mathrm{D}}$

$$v_1 \approx \left(\frac{E}{E_{\mathrm{D}}}\right) v_e \tag{6a}$$

is in a velocity range of *increasing* friction with speed. Slightly faster electrons are decelerated and *vice versa*. Thus v_1 is a stable equilibrium and corresponds roughly to the drift velocity of the electric current, $v_{\mathrm{D}} \approx v_1$ (cf. Figure 4). The second solution at $v_2 > v_e$ is in a velocity range of *decreasing* friction and is, therefore, an unstable equilibrium. Faster electrons are more and more accelerated until friction is negligible. For $E \ll E_{\mathrm{D}}$ this velocity above which runaway occurs is (using Equations (1), (2), (4a), and (5))

$$v_2 \approx \left(\frac{E_{\mathrm{D}}}{E}\right)^{1/2} v_e . \tag{6b}$$

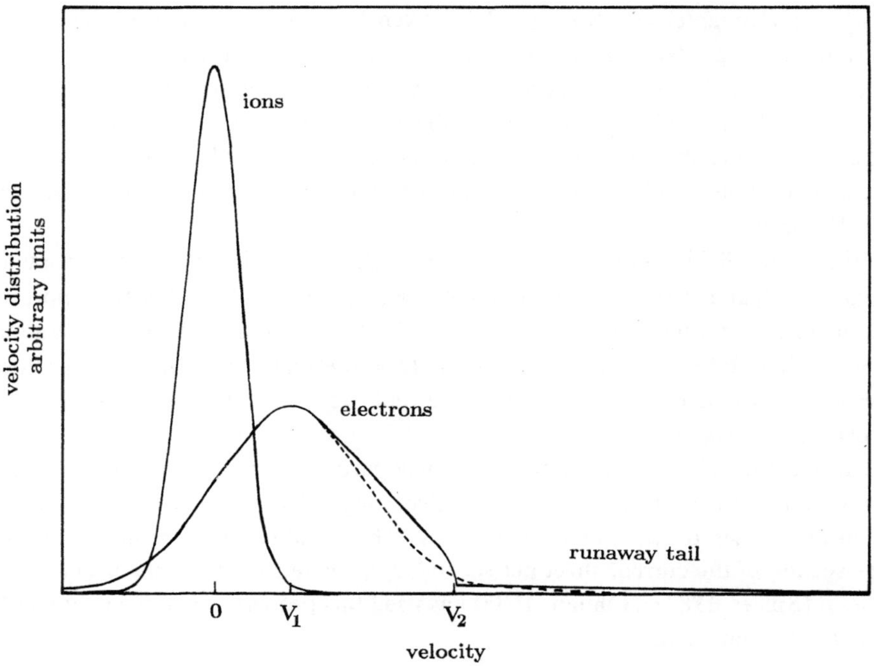

Fig. 3. Schematic drawing of parallel velocity distributions in the presence of a non-vanishing parallel component of the electric field in the rest frame of the ions.

For $E = E_D$, $v_1 = v_2 = v_e$ and all the electrons are in the runaway regime. Electron runaway acceleration has been reviewed by Knoepfel and Spong (1979) with particular reference to tokamak theory and experiments. The process has been proposed for flare energetic particles by Syrovatskii (1966), Coppi and Friedland (1971), Coppi (1975), Kuperus (1976) and others. These considerations have been made before quantitative estimates of the number of fast electrons involved in HXR bursts were available. The initial fraction of electrons that runaway is given by the condition $v > v_2$, i.e., for $v_2^2 \gg v_e^2$

$$\frac{n_{ra}}{n} \approx \exp\left[-\frac{1}{2}\left(\frac{v_2}{v_e}\right)^2\right] \approx \exp\left[-\frac{v_e}{2v_1}\right]. \tag{7}$$

For equal ion and electron temperatures a drift velocity of the order of the thermal ion velocity (i.e., $v_1 = v_i$) yields a ratio $n_{ra}/n \approx 5 \times 10^{-10}$. Collisions accelerate electrons beyond v_2 with about the collision rate at v_2. This constant particle supply into the runaway regime yields a runaway acceleration rate of

$$\frac{\dot{n}}{n} \approx 0.35\, \nu_c(v_e) \left(\frac{v_2}{v_e}\right)^{3/4} \exp\left[-2^{1/2}\left(\frac{v_2}{v_e}\right) - \frac{1}{4}\left(\frac{v_2}{v_e}\right)^2\right] \quad \mathrm{s}^{-1}, \tag{8}$$

for $v_2^2 \gg v_e^2$ (Knoepfel and Spong, 1979). Even for possibly increased densities and collision rates in preflare current sheets (say $\nu_c(v_e) \approx 10^3$ Hz, cf. Heyvaerts *et al.*, 1977), Equation (8) describes an acceleration mechanism which is limited to $\lesssim 1$ electron cm^{-3}. This is clearly insufficient for even the most conservative estimates of the flare volume and the number of electrons observed in HXR. Kuijpers *et al.* (1981) have proposed this mechanism for the less demanding number of electrons observed in type III bursts.

If the electric field approaches the Dreicer limit, both the initial runaway population (Equation (7)) and the runaway production rate (Equation (8)) may become appreciable and sufficient for a flare. Recently, Tsuneta (1985) has used runaway acceleration to explain HXR flare emission in the beam (non-thermal) scenario. He was forced, however, to use $E/E_D \approx 0.3$ causing a current very close to the threshold for the ion-cyclotron instability.

If the electric field rises, the flux of runaway particles carries more and more of the sheet current. However, the total current given by $j = (c/4\pi)\nabla \times B$ cannot increase in the current sheet by the runaway process. The number of electrons traversing a cross-section of the current sheet per second, I/e, where I is the total current, remains constant (Spicer, 1983). Holman (1985) has used this property to derive an upper limit of the total runaway rate

$$\dot{N} = \frac{cwB}{2\pi e} = 10^{30}\left(\frac{w}{10^9\ \mathrm{cm}}\right)\left(\frac{B}{100\ \mathrm{G}}\right)\ \text{electrons s}^{-1}, \tag{9}$$

where w is the width of the sheet (parallel to B in a neutral sheet). He finds that the runaway production rate (given in Equation (8) for $E \ll E_D$) is not the most restrictive time-scale, but the limited electron flux in the current:

$$\tau \equiv \frac{N}{\dot{N}} \approx 100\left(\frac{N}{10^{32}}\right)\left(\frac{w}{10^9\ \mathrm{cm}}\right)^{-1}\left(\frac{B}{100\ \mathrm{G}}\right)^{-1}\ \mathrm{s}, \tag{10}$$

given by Equation (9). Equations (9) and (10) suggest that even a very small electron beam containing a number of particles far below the threshold of detection with current HXR telescopes would have a formation time comparable to the duration of typical impulsive flares. A possible way out, first mentioned by Hoyng (1977), is to fold a current sheet of much larger dimension of many sheets into the observed volume and effectively increasing w in Equations (9) and (10). Uchida and Sakurai (1977) have considered the interchange instability to increase the length of the current sheet and speed up reconnection. More recently, filamentary current structure has been found in simulations of fast, high magnetic Reynolds number reconnection (Matthaeus and Lamkin, 1986, and references therein).

Runaway also occurs when turbulence due to kinetic current instabilities increases the resistivity. Since the current density remains constant within the inductive time-scale, and if the anomalous collision frequency has the same velocity dependence as

Equation (2) combined with (4b) (Kaplan and Tsytovich, 1973), the ratio E_D/E and, therefore, the runaway velocity v_2 (Equation (6)) remain constant (Papadopoulos, 1977). A sudden change in local resistivity does not produce a burst of runaway particles. In fact, if the temperature increases due to Ohmic heating by anomalous resistivity, the fraction of runaway particles and their production rate decrease (cf. Equations (7) and (8)). The only possibility to increase runaway acceleration is to create small current circuits with little inductance, such as apparently occurring by island formation and current filamentation in reconnection.

Particle runaway (both ions and electrons) seems to occur in every electric field and current parallel to the magnetic field. These particles take up energy, decrease resistivity by their lower friction and possibly change the properties of current instabilities in the build-up phase of current sheets. The dynamic role of the runaway process in flares should be better investigated. It is unclear, however, how it can be reconciled with HXR rise times of 30–50 ms (cf. Section 2) suggesting similar or shorter acceleration times.

4. Stochastic Resonance Acceleration

Stochastic acceleration involves the diffusion of particles in velocity space by the action of random phase turbulence. Here we consider resonant particle-wave interactions. Contrary to first order Fermi acceleration by MHD waves and shocks only the fraction of particles in resonance participates in the statistical energy exchange. The result is a diffusion that can produce general heating of a particle species, suprathermal tails in the velocity distribution, or further accelerate nonthermal particles depending on the velocity of resonance. The acceleration is then described by the equation of continuity of the particle velocity distribution f

$$\frac{\mathrm{d}f}{\mathrm{d}t} = \sum_{i,j} \frac{\partial}{\partial v_i} D_{ij} \frac{\partial}{\partial v_j} f + \left(\frac{\partial f}{\partial t}\right)_{\mathrm{coll}}, \tag{11}$$

where D is a diffusion coefficient depending on the nature and the level of the wave turbulence; the last term is due to collisions. The acceleration properties of resonant interactions have been investigated by Kaplan and Tsytovich (1973), Pikel'ner and Tsytovich (1976). Detailed accounts of this process are available in standard text books such as Melrose (1980). Effects of strong turbulence may considerably change the resulting energy distribution. In particular, parametric decay to smaller k vectors accelerates suprathermal electron tails to much higher velocity. Detailed simulations have been done for laser fusion (e.g., Hubbars and Joyce, 1976, and references therein). Smith (1977a, b) has applied the idea of strongly turbulent Langmuir waves to the flare electrons starting from some previously accelerated nonrelativistic electrons.

The most important point on this acceleration is the origin of the wave turbulence. If it is to accelerate a major constituent of the flare in terms of energy, it must be itself a major receiver of the primary energy. This could come about in several ways:

(i) First, turbulence can be directly connected to the main energy release process in the form of anomalous resistivity caused by unstable current densities. It is widely believed that the reconnection rate depends on the resistivity in the reconnection region and it seems most plausible that a flare is triggered by the sudden development of anomalous resistivity. Since the energy deposition of a constant current increases with resistivity, heating is strongly enhanced. Such an energy transfer of electromagnetic energy into thermal energy is sometimes also referred to as 'anomalous Ohmic dissipation'. The difference to ordinary Ohmic heating by Coulomb collisions between electrons and ions is that the randomization of energy by wave turbulence can affect electrons, ions, and various velocity regimes in very different ways. The end product may be unequal temperatures for electrons and ions and/or nonthermal velocity distributions. Since the resulting velocity distribution is not necessarily a Maxwellian, we will use the terms acceleration or energyzation rather than heating. The process may release a fair fraction of the energy of reconnecting current sheets. Most of the rest is taken up by ion bulk outflow from the sheet. The ratio between these two processes depends on resistivity and other model parameters, in particular on the length of the current sheet. Current estimates of the ratio of Ohmic dissipation to bulk kinetic energy are still very crude and for solar flares of the order of 1 (Priest, 1987).

A direct link between anomalous resistivity and stochastic resonance acceleration has been first proposed by Benz (1977) to explain the observed correlation of HXR spectral hardness and flux (cf. more recently Brown and Loran, 1985). For current sheet densities exceeding 10^{11} cm^{-3} and acceleration times of > 30 s, Holman (1985) finds that ordinary Ohmic heating would be adequate to power the observed HXR flux in the purely thermal model. The observations reviewed in Section 2, however, strongly indicate that these 3 assumptions generally do not simultaneously hold. Whatever the resistivity, Holman (1985) has shown that Ohmic dissipation is always stronger than the power put into runaway electrons. In the currents become unstable, the energy in runaway particles becomes negligible and kinetic wave turbulence in a major route of the flow of the energy from a magnetic to a kinetic or therexistamal form.

(ii) In some reconnection models (Petschek–Sonnerup mode) the central current sheet bifurcates at each end to form two pairs of stationary slow-mode magnetic shocks (cf. Figure 4). They convert $\frac{2}{5}$ of the incoming magnetic energy to Ohmic heat and the rest to kinetic bulk motion. They generally release more energy than the reconnection layer. Hoyng (1977) has suggested that the shock associated currents may become unstable and be a second source of wave turbulence and stochastic resonance absorption. Hoyng *et al.* (1980) have further developed the idea to include a multitude of shock produced current layers. The existence of the slow-mode shocks, however, is highly controversial. Biskamp (1966) does not find them in his numerical models with small resistivity.

(iii) A third relevant source of accelerating wave turbulence are ion motions due to outflowing material from the central reconnection layer (jets, cf. Figure 4). In these jets ions may have a bulk motion perpendicular to the magnetic field susceptible to cross-field current instabilities. Smith (1985) has proposed this as the cause for electron heating and, in the case of a ratio of thermal to magnetic energy density $\beta < 0.3$, for the

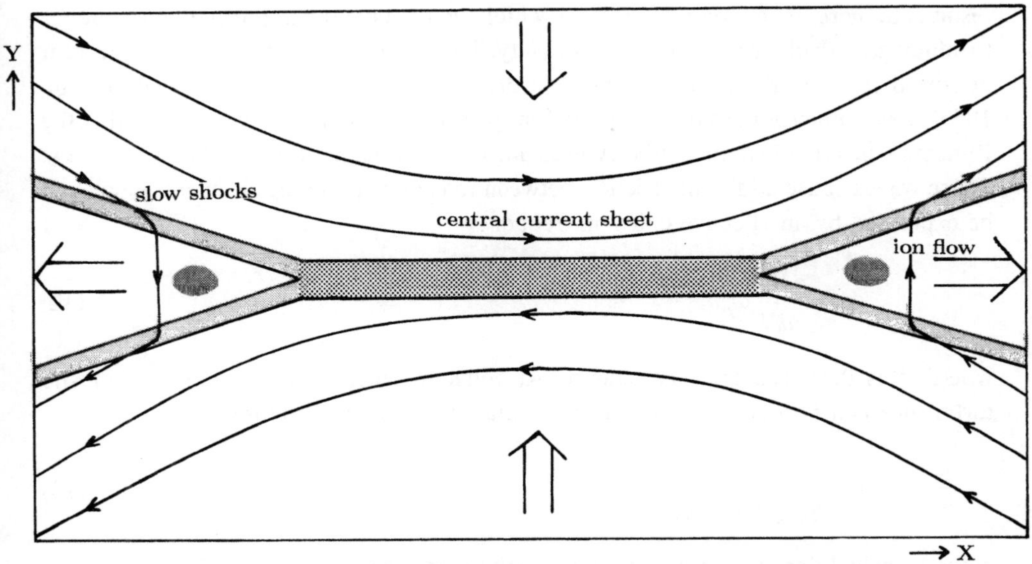

Fig. 4. Schematic magnetic field configuration in a typical reconnection scenario illustrating the features discussed in the text, which may not all exist in reality. Hatched and stippled are possible regions of enhanced turbulence and anomalous resistivity.

formation of suprathermal electron tails. The jet flow is asymptotically sub-Alfvénic, but near the end points of the central current sheet it can be highly super-Alfvénic (in terms of the local Alfvén speed) and be sensitive to fast-mode shock formation, another source of turbulence and particle acceleration.

(iv) Finally, if the current sheet is very long and thin, it becomes tearing unstable. The tearing mode leads to magnetic islands, which produce current sheets of their own (cf. review by Priest, 1986). This could further complicate the picture schematically drawn in Figure 4, and be an additional source of turbulence.

4.1. Parallel Currents

Which wave modes are excited in the sources of turbulence identified above? Which waves are, furthermore, most likely to energyze electrons and ions? The answers depend mainly on the direction of the current relative to the magnetic field. If the magnetic field in the reconnection region is aligned like the steps of a spiral staircase, the current is largest at the 'widest step' (largest $\nabla \times \boldsymbol{B}$). In this layer (labelled 'central current sheet' in Figure 4) it flows essentially parallel to the magnetic field. The presence of a magnetic field permits cyclotron modes. The instabilities of field aligned currents have been reevaluated by Morrison and Ionson (1982) for distributions skewed by the electric field. For equal electron and ion temperatures the threshold for instability is lowest for growing electrostatic ion cyclotron waves. It is generally agreed and confirmed by experiments that the resulting ion-cyclotron turbulence saturates at a low level and that its anomalous resistivity is small if not negligible (Papadopoulos, 1977). The main result is to increase the ion temperature, which would have the important consequence to inhibit growth of the ion-acoustic instability. The next instability for increasing current density then is the Buneman instability at drift velocities $v_2 \gtrsim 1.7\,(v_e + v_i)$ (Buneman, 1959). Two-dimensional simulations by Lampe *et al.* (1974), however, revealed that the Buneman instability nonlinearly evolves into an ion-acoustic instability if $\Omega_e < \omega_e$. These waves cause additional friction between ions and streaming electrons which can be expressed by an effective collision frequency

$$\nu_{\text{eff}} = \frac{W}{nkT}\,\omega_{pe}\,, \tag{12}$$

where W is the turbulent wave energy. At constant current density the ion-acoustic turbulence can produce a steady-state anomalous resistivity amounting to

$$\eta = \frac{1}{8\omega_{pe}}\left(\frac{T_i}{T_e}\right)\left(\frac{v_e}{v_{\text{D}}}\right) \tag{13}$$

in the so-called Sagdeev approximation (weak turbulence theory and $T_i/T_e \ll 1$). A very important result is the preferential heating of electrons for the large drift velocities considered here:

$$\frac{\dot{T}_e}{\dot{T}_i} = \frac{T_e}{T_i}\,\frac{v_{\text{D}}}{c_s}\,(1 - c_s/v_{\text{D}}) > 1 \tag{14}$$

(Papadopoulos, 1977), where the temperatures are effective values and c_s is the ion sound velocity,

$$c_s \equiv \left(\frac{m_e}{m_i}\right)^{1/2} v_e\,. \tag{15}$$

The actual turbulence level and spectrum of the ion-acoustic instability at saturation is not well established for high drift velocities giving possibly rise to strong turbulence. Experiments in this regime have been made by Gekelman and Stenzel (1978) and confirmed the previous picture in general. They observed an electron distribution which deviates considerably from a shifted Maxwellian having a flatter slope in electron drift direction. Ion tails have also been reported and interpreted within the theory of the ion-acoustic instability (Manheimer and Flynn, 1974). Stochastic acceleration (respectively, anomalous heating) continues after wave saturation until the criterion for instability is not satisfied anymore. For constant j (large induction time) this roughly happens when c_s reaches v_{D} allowing the electron temperature to increase by a factor of $1.7 m_i/m_e$.

The first scenarios of electric stochastic acceleration in flares involved Langmuir wave turbulence (Benz, 1977; Hoyng, 1977; Hoyng *et al.*, 1980, and others). These models were based on a process proposed by Tsytovich *et al.* (1975) that could amplify Langmuir waves by ion-acoustic turbulence. The mechanism has been shown to be impossible by DuBois and Pesme (1984). We do not consider Langmuir waves here, since they seem to be excited only by electrons moving already faster than v_e and since a high level of Langmuir turbulence would produce a radio signature far exceeding observed impulsive phase bursts.

4.2. Perpendicular currents

The region of purely parallel currents is probably very thin in reality. The perpendicular component of the current increases both in the direction out of the current sheet (y) and along the x-direction (cf. Figure 4). All currents are perpendicular in the reconnection geometry of antiparallel fields (neutral sheets). In the Petchek–Sonnerup model (if appropriate at all) most turbulence energy is released in the slow-mode shocks having perpendicular currents. Very strong perpendicular currents may also arise in the jets and their associated fast-mode shocks. The ratio of energy released by the $j_\perp$ and $j_\parallel$ components of the current depends on the model and the detailed geometry of the magnetic field in the current sheet. The perpendicular component may well be more effective since its predominant mode of instability is different having a threshold nearly two orders of magnitude below the threshold of the parallel component.

The anomalous resistivity of cross-field currents has been reviewed by Davidson and Krall (1977) with emphasis on theta-pinch implosions. Most relevant are the lower-hybrid-drift (LHD) instability (with gradients driving waves at $\boldsymbol{kB}_0 = 0$ in the presence of inhomogeneities in density, temperature and magnetic field) and the modified-two-stream (MTS) instability (convectively driven at $\boldsymbol{kB}_0 \approx 0$). Hsia *et al.* (1979) have extended the LDH theory to include a finite $k_\parallel$ bridging the LDH and MTS instabilities by a generalized lower-hybrid instability. Both produce lower-hybrid waves characterized by strongly magnetized electrons and unmagnetized ions, i.e., the wave frequency ω is between the gyrofrequencies of ions and electrons:

$$\Omega_i \ll \omega \ll \Omega_e \,. \tag{16}$$

For perpendicular motion the electrons are thus strongly tied to the magnetic field (B_0) and the ions can freely move allowing hydrodynamic double streams. The two instabilities give rise to similar electrostatic waves with frequency ω near the lower-hybrid frequency ω_{Lh}:

$$\omega^2 = \omega_{Lh}^2 \left(1 + \left(\frac{k_\parallel}{k}\right)^2 \frac{m_i}{m_e}\right) \tag{17}$$

(Krall and Liewer, 1971), where

$$\omega_{Lh} = \omega_{pi}(1 + \omega_{pe}^2/\Omega_e^2)^{-1/2} \,. \tag{18}$$

In both cases the threshold requirement is

$$v_{d\perp} \gtrsim v_i \,, \tag{19}$$

where $v_{d\perp}$ is the perpendicular component of the electron-ion drift. The growth rate is not very sensitive to T_e/T_i but decreases for $T_e \gg T_i$ (McBride *et al.*, 1972, Hsia *et al.*, 1979, and others).

Huba *et al.* (1978) have interpreted observed electrostatic noise in the magnetotail as lower-hybrid-wave turbulence. Winske (1981) has investigated the curren-driven lower-hybrid instability in neutral sheets and found it to occur away from the plane of field reversal, at the place where the density gradients are steepest. The resulting anomalous resistivity at the center of the sheet is effectively enhanced by wave propagation despite the stabilizing effect of a sheared magnetic configuration (Huba *et al.*, 1982). An analytical study by Drake *et al.* (1981) revealed that the LHD instability occurring in the region of maximum density gradient flattens the density distribution across a neutral sheet. Since pressure balance requires that magnetic flux is transported into the sheet, the current density dramatically increases. The built up magnetic flux can then be explosively released when the central current exceeds a threshold for instability.

Simulations by McBride *et al.* (1972) have shown that lower-hybrid turbulence heats ions in perpendicular velocity and electrons parallel to the field to comparable temperature ($T_{i\perp} \approx T_{e\parallel}$). At saturation the wave energy W_{Lh} reaches a value of

$$\frac{W_{Lh}}{nKT} \approx \frac{0.05}{1 + \omega_{pe}^2/\Omega_e^2} \,. \qquad (20)$$

More detailed simulations for reversed field geometry by Winske (1981) show that saturation in the null region is due to electron orbit modification, and the wave energy scales with v_{D}^2. For $v_{\mathrm{D}}/v_i = 3.5$ he finds an anomalous resistivity of $\eta \approx 0.05\,\omega_{pe}^{-1}$, about an order of magnitude below the value in the most unstable region, but many orders of magnitude above Spitzer resistivity.

4.3. Electron acceleration by lower-hybrid waves

Electron acceleration by lower-hybrid turbulence has been proposed for the discrete aurora by Bingham *et al.* (1984). The observed wave energy density – possibly supplied by ion streams in the plasma sheet boundary layer – is found sufficient to stochastically accelerate electrons to the measured energies and numbers. Stochastic electron energyzation by lower-hybrid waves has been proposed for the sources of solar type I radio bursts (Spicer *et al.*, 1981; Wentzel, 1981). Independently, Smith (1983, 1985) and Benz (1984) have qualitatively considered lower-hybrid turbulence as the main driver of flare electron acceleration. More detailed investigations by Benz and Smith (1987) suggest a model with very fragmented and small sized regions of enhanced lower-hybrid turbulence (as suggested by the radio spikes) leaving open the question of the source of turbulence and the global organization (flare trigger).

The electron accelerating potential of lower-hybrid waves stems from the Čerenkov resonance between waves and particles

$$\frac{\omega}{k_\parallel} = v_\parallel \,, \qquad (21)$$

where ω is given by Equation (17). Since $k_\parallel$ is generally small (for the MTS instability $k_\parallel / k_\perp \lesssim 10(m_e/m_i)^{1/2}$), the resonance condition parallel to the field can easily be fulfilled by thermal as well as non-thermal electrons. The case $k_\parallel = 0$, however, arising from the LHD instability, is not suitable for electron acceleration. The diffusion equation (11) can be reduced to parallel direction:

$$\frac{\partial f_e}{\partial t} + \frac{\partial}{\partial t} v_\parallel f_e = \frac{\partial}{\partial v_\parallel} D \frac{\partial}{\partial v_\parallel} f_e + \left(\frac{\partial f_e}{\partial t} \right)_{\text{coll}}, \tag{22}$$

where the diffusion coefficient D is determined by the wave spectral distribution. Complete and realistic simulation of the nonlinear evolution of the MTS instability have not run long enough to reach a steady state and do not limit the choice of the wave spectrum. A remarkable tail in the parallel electron distribution extending up to 6–10v_e in both directions has emerged in detailed simulations of the MTS instability by Tanaka and Papadopoulos (1983). It is the result of a nonlinear shift of the wave frequency occurring after ion trapping and increasing the parallel phase velocity of the waves. It can, therefore, carry electrons to higher energies. Tail formation only occurs for $v_D \lesssim 4.2 v_i$ and $\beta \lesssim 0.3$. The tail is generally made of the electrons having initial parallel velocities in excess of the most unstable mode and amounts to typically 10% of the total electron population. A very complete study by Winske *et al.* (1985) on lower-hybrid turbulence at the Earth's bowshock shows that about 50% of the initial drift energy may end up the electron tail.

The reduced diffusion equation (22) has been solved by Benz and Smith (1987) for model spectra of the power-law form

$$W(\mathbf{k}) = \delta(k_y) \Psi(k_y) A_\alpha k_x^{-\alpha}, \tag{23}$$

where $k_{\text{min}} < k_x < k_{\text{max}}$, $\Psi(k_y)$ does not need to be specified, and A_x is a normalization constant such that

$$W_{Lh} = \int W(\mathbf{k}) \frac{\mathrm{d}^3 k}{(2\pi)^3}. \tag{24}$$

The diffusion coefficient for lower-hybrid waves is

$$D = \frac{2\pi \omega_{pe}^2 v_d^2 W_{Lh} A_\alpha}{m_e n_e |v_x|^3} \left| \frac{k_{\text{min}} v_x}{\omega} \right|^\alpha \tag{25}$$

(Wu *et al.*, 1982) and defines a stochastic acceleration time

$$t_a \equiv \frac{v_x^2}{D(v_x)}. \tag{26}$$

Appreciable acceleration is only possible if t_a is much shorter than the time an electron interacts with the waves, which is either limited by the duration of the turbulence Δt (for low $v_{\parallel}$ particles) or escape of the particle (high $v_{\parallel}$ particles). The two limits on acceleration refer to two regimes in velocity space separated by $v_c = L/\Delta t$, where L is the length of the acceleration region parallel to the magnetic field. Benz and Smith (1987) have proposed this division of velocity space to be the cause of double power-law spectras frequently observed in flare HXR. The break is about at the observed energy if a length $L \approx 200$ km, as suggested by radio spikes, is assumed.

5. Conclusions

There have recently been pointed out several difficulties with runaway acceleration by electric fields (current limitation by induction, particle number, etc.). It is certainly inappropriate to completely dismiss the process at this stage. There remain several open questions such as:

(1) Why would the current and the associated electric field increase to a considerable fraction of the Dreicer limit (to produce enough runaway particle flux), but stop before exceeding the threshold of some instability?

(2) We have not discussed here the possibility of double layers, localized potential drops possibly observed in the laboratory and the magnetosphere (cf. review by Fälthammar, 1983). The role of double layers has remained obscure even in these relatively well accessible places of large parallel currents.

On the other hand, anomalous resistivity is a very multifarious process which can both release considerable energy and even accelerate electrons into a high energy tail. Parallel currents at $T_e = T_i$ have a high threshold and dump their energy into ion-acoustic turbulence. This can energize preferentially electrons and provide a sufficient number of 20–100 keV electrons. Perpendicular currents are even more attractive having a lower threshold and, in certain circumstances, produce an electron tail. The relevant waves are near the lower-hybrid frequency and are well known in laboratory and magnetospheric physics. The ratios $j_{\perp}/j_{\parallel}$ and T_e/T_i are the important parameters in the question of which instability will occur. Both ion acoustic and lower-hybrid instabilities considerably heat the ions, which in the latter case can take up half of the turbulence energy. Particles beyond 300 keV may be accelerated by a different mechanism. Runaway and shocks are possible candidates.

Observations of the energy ratio in ions and electrons, and the low (< 1 MeV) energy ion distribution would be of particular interest. As pointed out by Melrose (1983) ion abundancy anomalies could be important indicators of primary energy transfer processes.

Radio observations at frequencies corresponding to the plasma or electron cyclotron frequencies (or their harmonics) could be valuable. Low-frequency turbulence (such as discussed in Section 4) would interact with Langmuir waves (if present) to produce observable radio emission. In addition, beam radiation (via the bump-in-tail instability of Langmuir waves or maser emission) could yield signatures of newly accelerated particles. A continued study of spikes and their interpretation as well as other coherent emissions in the 0.3–5 GHz range is of great importance.

Acknowledgements

I thank E. R. Priest and D. F. Smith for helpful discussions and criticism, and acknowledge the Swiss National Science Foundation for partial financial support of the radio astronomy work at ETH.

References

Antonucci, E., Rosner, R., and Tsinganos, K.: 1986, *Astrophys. J.* **301**, 975.
Bai, T., and Dennis, B. R.: 1985, *Astrophys. J.* **292**, 699.
Benz, A. O.: 1977, *Astrophys. J.* **211**, 270.
Benz, A. O.: 1984, 'Trends in Physics', in J. Janta and J. Pantoflicek (eds.), *Proc. 6th General Conference of EPS* **1**, 156.
Benz, A. O.: 1985, *Solar Phys.* **96**, 357.
Benz, A. O.: 1986, *Solar Phys.* **104**, 99.
Benz, A. O. and Smith, D. F.: 1987, *Solar Phys.* **107**, 299.
Bingham, R., Bryant, D. A., and Hall, D. S.: 1984, *Geophys. Res. Letters* **11**, 327.
Biskamp, D.: 1986, *Phys. Fluids* **29**, 1520.
Brown, J. C. and Loran, J. M.: 1985, *Monthly Notices Roy. Astron. Soc.* **212**, 245.
Brown, J. C., Carlaw, V. A., Cromwell, D., and Kane, S. R.: 1983, *Solar Phys.* **88**, 281.
Buneman, O.: 1959, *Phys. Rev.* **115**, 503.
Cheng, C.-C., Tandberg-Hanssen, E., Bruner, E. C., Orwig, L., Frost, K. J., Kenny, P. J., Woodgate, B. E., and Shine, R. A.: 1981, *Astrophys. J.* **248**, L39.
Chupp, E. L., Forrest, D. J., Ryan, J. M., Cherry, M. L., Reppin, C., Kanbach, G., Rieger, E., Pinkau, K., Share, G. M., Kinzer, R. L., Strickman, M. S., Johnson, W. N., and Kurfess, J. D.: 1981, *Astrophys. J.* **244**, L171.
Chupp, E. L.: 1984, *Am. Rev. Astron. Astrophys.* **22**, 359.
Cliver, E. W., Kahler, S. W., and McIntosh, P. S.: 1983, *Astrophys. J.* **264**, 699.
Coppi, B.: 1975, *Astrophys. J.* **195**, 545.
Coppi, B. and Friedland, A. B.: 1971, *Astrophys. J.* **169**, 379.
Davidson, R. C. and Krall, N. A.: 1977, *Nuclear Fusion* **17**, 1313.
De Jager, C.: 1969, in C. de Jager and Z. Švestka (eds.), *Solar Flares and Space Research*, D. Reidel Publ. Co., Dordrecht, Holland, p. 1.
De Jager, C. and de Jonge, G.: 1978, *Solar Phys.* **58**, 127.
Drake, J. F., Gladd, N. T., and Huba, J. D.: 1981, *Phys. Fluids* **24**, 78.
DuBois, D. F., and Pesme, D.: 1984, *Phys. Fluids* **27**, 218.
Duijveman, A., Hoyng, P., and Machado, M. E.: 1982, *Solar Phys.* **81**, 137.
Fälthammar, C. G.: 1983, *ESA J.* **7**, 385.
Gekelman, W. and Stenzel, R. L.: 1978, *Phys. Fluids* **21**, 2014.
Heyvaerts, J.: 1981, in E. R. Priest (ed.), *Solar Flare Magnetohydrodynamics*, Gordon and Breach, New York, p. 429.
Heyvaerts, J., Priest, E. R., and Rust, D. M.: 1977, *Astrophys. J.* **216**, 123.
Holman, G. D.: 1985, *Astrophys. J.* **293**, 584.
Hoyng, P.: 1977, *Astron. Astrophys.* **55**, 23.
Hoyng, P., Brown, J. C., and v. Beek, H. F.: 1976, *Solar Phys.* **48**, 197.
Hoyng, P., Duijveman, A., van Grunsven, T. F. J., and Nicholson, D. R.: 1980, *Astron. Astrophys.* **91**, 17.
Hsia, J. B., Chin, S. M., Hsia, M. F., Chou, R. L., and Wu, C. S.: 1979, *Phys. Fluids* **22**, 1737.
Huba, J. D., Gladd, N. T., and Papadopoulos, K.: 1978, *J. Geophys. Res.* **83**, 5217.
Huba, J. D., Gladd, N. T., and Drake, J. F.: 1982, *Geophys. Res.* **87**, 1697.
Hubbard, R. F. and Joyce, G.: 1976, *Plasma Phys.* **18**, 681.
Hudson, H. S., Bai, T., Gruber, D. G., Metteson, J. L., Nolan, P. L., and Peterson, L. E.: 1980, *Astrophys. J.* **236**, L91.
Kane, S. R., Kai, K., Kosugi, T., Landecker, P. B., and McKenzie, D. L.: 1983, *Astrophys. J.* **271**, 376.
Kaplan, S. A. and Tsytovich, V. N.: 1973, *Plasma Astrophysics*, Pergamon Press, New York, p. 134.
Kiplinger, A. L., Dennis, B. R., Emslie, A. G., Frost, K. J., and Orwig, L. E.: 1983a, *Astrophys. J.* **265**, L99.

Kiplinger, A. L., Dennis, B. R., Frost, K. J., and Orwig, L. E.: 1983b, *Astrophys. J.* **273**, 783.
Kiplinger, A. L., Dennis, B. R., Frost, K. J., and Orwig, L. E.: 1984, *Astrophys. J.* **287**, L105.
Knoepfel, H. and Spong, D. A.: 1979, *Nucl. Fusion* **19**, 785.
Kouveliotou, C., Treumann, R. A., and Scholer, M.: 1984, *Astrophys. J.* **278**, 420.
Krall, N. A. and Liewer, P. C.: 1971, *Phys. Rev.* **A4**, 2094.
Kuijpers, J., van der Post, P., and Slottje, C.: 1981, *Astron. Astrophys.* **103**, 331.
Kuperus, M.: 1976, *Solar Phys.* **47**, 79.
Lampe, M., Haber, I., Orens, J. H., and Boris, J. P.: 1974, *Phys. Fluids* **17**, 428.
Machado, M. E. and Lerner, G.: 1984, *Adv. Space Res.* **4**, 239.
Mannheimer, W. M. and Flynn, R. W.: 1974, *Phys. Fluids.* **17**, 409.
Matthaeus, W. H. and Lamkin, S. L.: 1986, *Phys. Fluids.* **29**, 2513.
McBride, J. B., Ott, E., Boris, J. P., and Orens, J. H.: 1972, *Phys. Fluids* **15**, 2367.
Melrose, D. B.: 1980, *Plasma Astrophysics*, Vol. 2, Gordon and Breach, New York.
Melrose, D. B.: 1983, *Solar Phys.* **89**, 149.
Morrison, P. J. and Ionson, J. A.: 1982, *Phys. Fluids* **25**, 1183.
Papadopoulos, K.: 1977, *Rev. Geophys. Space Phys.* **15**, 113.
Pickel'ner, S. B. and Tsytovich, V. N.: 1976, *Soviet Astron.* **19**, 450.
Priest. E. R.: 1985, *Rep. Prog. Phys.* **48**, 955.
Priest, E. R.: 1986, *Solar Phys.* **104**, 1.
Priest, E. R.: 1987, personal communication.
Raoult, A. and Pick, M.: 1980, *Astron. Astrophys.* **87**, 63.
Schmidt, G.: 1979, *Physics of High Temperature Plasmas*, Academic Press, New York.
Simnett, G. M. and Benz, A. O.: 1986, *Astron. Astrophys.* **165**, 227.
Smith, D. F.: 1977a, *Astrophys. J.* **212**, 891.
Smith, D. F.: 1977b, *Astrophys. J.* **217**, 644.
Smith, D. F.: 1980, *Solar Phys.* **66**, 135.
Smith, D. F.: 1983, in M. R. Kundu and G. D. Holman (eds.), 'Unstable Current Systems', *IAU Symp.* **107**, 509.
Smith, D. F.: 1985, *Astrophys. J.* **288**, 801.
Spicer, D. S.: 1983, *Adv. Space Res.* **2**, 135.
Spicer, D. S., Benz, A. O., and Huba, J. D.: 1981, *Astron. Astrophys.* **105**, 221.
Spitzer, L.: 1962, *Physics of Fully Ionized Gases*, Interscience, New York.
Sprangle, P. and Vlahos, L.: 1983, *Astrophys. J.* **273**, L95.
Strong, K. T., Benz, A. O., Dennis, B. R., Leibacher, J. W., Mewe, R., Poland, A. I., Schrijver, J., Simnett, G. M., Smith, J. B., Jr., and Sylwester, J.: 1984, *Solar Phys.* **91**, 325.
Syrovatskii, S. I.: 1966, *Soviet Astron. AJ* **10**, 270.
Tanaka, M. and Papadopoulos, K.: 1983, *Phys. Fluids* **26**, 1697.
Tsytovich, V. N., Stenflo, C., and Wilhelmsson, H.: 1975, *Phys. Scripta* **11**, 251.
Tsuneta, S.: 1985, *Astrophys. J.* **290**, 353.
Uchida, Y. and Sakurai, K.: 1977, *Solar Phys.* **51**, 413.
Van Hoven, G.: 1981, in E. R. Priest (ed.), *Solar Flare Magnetohydrodynamics*, Gordon and Breach, New York, p. 217.
Vasyliunas, V. M.: 1975, *Rev. Geophys. Space Phys.* **13**, 303.
Vestrand, W. T., Forrest, D. J., Chupp, E. L., Rieger, E., and Share, G. H.: 1986, *Adv. Space Res.*, in press.
Vilmer, N., Kane, S. R., and Trottet, G.: 1982, *Astron. Astrophys.* **108**, 306.
Vlahos, L. *et al.*: 1984, in M. R. Kundu and B. Woodgate (eds.), *Energetic Phenomena in Solar Flares*, Proc. of SMM Workshop, Ch. 2.
Wenn, D. F.: 1985, *Solar Phys.* **97**, 321.
Wentzel, D. G.: 1981, *Astron. Astrophys.* **100**, 20.
Wild, J. P., Smerd, S. F., and Weiss, A. A.: 1963, *Ann. Rev. Astron. Astrophys.* **1**, 291.
Winske, D.: 1981, *Phys. Fluids* **24**, 1069.
Winske, D., Tanaka, M., Wu, C. S., and Quest, K. B.: 1985, *J. Geophys. Res.* **90**, 123.
Wu, C. S., Gaffey, J. D. Jr., and Liberman, B.: 1981, *J. Plasma Phys.* **25**, 391.

ELECTRON ACCELERATION WITHIN CORONAL LOOPS: A WAVE-PARTICLE PROCESS?

D. LE QUÉAU and A. ROUX

CRPE/CNET-CNRS, Issy-lès Moulineaux, France

Abstract. A simple model as given which aims at explaining how electrons could be accelerated within solar coronal loops due to their trapping by quasi-monochromatic whistler waves.

1. Introduction

The presence of a significant population of electrons having suprathermal energies ranging about a few tens of keV is an essential ingredient of any coronal loops plasma model. Such a population must indeed be responsible of the X-ray emissions observed near the Sun surface, particularly during flaring events.

Motivated by these observational facts we here propose a model for accelerating electrons up to such energies within magnetic solar loops. Quasi-monochromatic whistler waves (hereafter QMW) are supposed to be generated near the top of the loop, due to the presence of an anisotropic electron population. These waves are then likely to propagate back and forth along the coronal loop and to become electrostatic whistlers which are able to trap thermal electrons at the top of the loop. Due to the inhomogeneity of the B-field, the phase velocity of the wave in increased. Thus the trapped electrons can be accelerated before they reach the feet of the coronal loop where they are likely to be detrapped.

This scenario closely follows the mechanism invoked by Roux *et al.* (1984) for explaining suprathermal electron tails observed in the earth's magnetosphere in correlation with low frequency Alfven waves, strongly modified by the presence of heavy ions, which have propagative properties similar to those of whistler waves (Norris *et al.*, 1983).

2. Determination of the Energy Gain

The magnetic field intensity along the model loop is supposed to vary quadratically:

$$B = B_{\text{top}}(1 + x^2), \tag{1}$$

where $x = (z/L)(R_m - 1)^{1/2} = z/L_0$, z is the distance along the field lines, taken from the top of the loop, L is the half length of the loop ($L = 2 \times 10^7$ m); R_m is the magnetic field mirror ratio between the top of the loop ($B_{\text{top}} \approx 50$ G) and its root ($R_m \approx 2$ to 10).

* Proceedings of the Second CESRA Workshop on Particle Acceleration and Trapping in Solar Flares, held at Aubigny-sur-Nère (France), 23–26 June, 1986.

Solar Physics **111** (1987) 19–22.

Inside the loop, the electronic density will be taken to be $n_e \approx 3 \times 10^{10}$ cm^{-3} and the temperature of the order of $T_e \approx 2 \times 10^7$ K (2 keV). Depending on the altitude over the chromosphere the ratio Ω_{pe}/Ω_{ce} will change, but it is always thought to be large (≈ 5 to 10).

Let us now assume that electromagnetic QMW with $k_\perp \approx 0$ are produced in the vicinity of the top ($x = 0$) by an anisotropic electronic population having a typical energy W_r. Then, using both the gyroresonance condition and the whistler dispersion equation it is a simple matter to relate W_r to the quantity $t = \Omega_{ce}/\Omega_{|\text{top}}$ where Ω_{ce} is the gyrofrequency at the top and Ω the wave frequency:

$$W_r/W_{cr} = f(t) = (t-1)^3/t^2\,, \tag{2}$$

where $W_{cr} = (mc^2/2)\,(\Omega_{ce}/\Omega_{pe})^2_{|\text{top}}$.

After a few back and forth travel, the QMW is thought to approach its resonance cone, by increasing its perpendicular wave vector $k_\perp$ and thus becomes electrostatic, exhibiting a significant parallel electric field. The dispersion equation now writes:

$$(k_\parallel/k_\perp)^2 = (\Omega_{ce}^2 - \Omega^2)\,, \tag{3}$$

where Ω_{LH} is the lower hybrid frequency $\Omega_{\text{LH}} \approx (m_e/m_{\text{H}_+})^{1/2}\,\Omega_{ce}$. From a previous ray tracing study (Rauch and Roux, 1982), concerning waves whose refractive index surface are very similar to those of whistlers, it is expected that, at each pass, the parallel wave number at the loop's top roughly recovers its initial value. Moreover, the perpendicular wave number, after a few back and forth paths, only slightly changes during a travel between the loop's top and the point at which the wave is reflected. To be brief, the ray path is expected to closely follow the magnetic field lines and the wave to be reflected at the point where locally $\Omega \approx \Omega_{\text{LH}}(k_\parallel \approx 0)$. This last set of hypothesis forms the strongest unprooved assumptions of the model, and will have to be verified by further ray tracing studies.

Now, if the electrostatic whistlers have a sufficient amplitude they will be able to trap electrons in the potential troughs linked to their parallel electric field. This occurs more easily at the top of the loop, where the B-fields inhomogeneity is weaker. Conversely, a detrapping is expected, at a lower altitude (x_d), when the detrapping force due to the inhomogeneity will balance the electrostatic trapping force (Le Quéau and Roux, 1987).

The inhomogeneity force can be determined from (1) and (3), while the local value of the electric field, which determines the trapping force, is linked to its value at the top by the WKB relations:

$$\mathbf{E}_\parallel/\mathbf{E}_\perp = k_\parallel/k_\perp\,, \qquad k_\perp \approx Cte\,,$$

and

$$v_g \mathbf{E}^2(z) = Cte\,.$$

One thus gets a relationship between the parallel electric field of the wave at the loop's top, x_d, the normalized distance at which electrons are suddenly detrapped, t, and the perpendicular energy W of the trapped electrons:

$$\mathbf{E}_{|\text{top}}/\mathbf{E}_{cr} = g(x_d; t; W/W_{cr}), \tag{4}$$

where $\mathbf{E}_{cr} = 4W_{cr}/L_0$, and $g(x; t; u)$ is given by a rather intricate expression.

Finally, the energy gained by the electrons can be related to the variation of the wave parallel phase velocity (or resonant velocity) between $x = 0$ and $x = x_d$

$$\Delta W = m/2[V_{\text{res}}^2(x_d) - v_{\text{res}}^2(0)] = W_{cr}h(x_d; t). \tag{5}$$

Using (2), (4), and (5), the gained energy can be related to the three input parameters of our model:

$$\Delta W/W_{cr} = \mathscr{F}(\mathbf{E}_{|\text{top}}/\mathbf{E}_{cr}; W_r/W_{cr}; W/W_{cr}). \tag{6}$$

3. Discussion

A numerical investigation of (6) has been performed (Figure 1), using the physical parameters characterising the above quoted loop model. One gets:

$$W_{cr} \approx 2.5 \text{ keV} \quad \text{and} \quad \mathbf{E}_{cr} \approx 1.5 \text{ mV m}^{-1}.$$

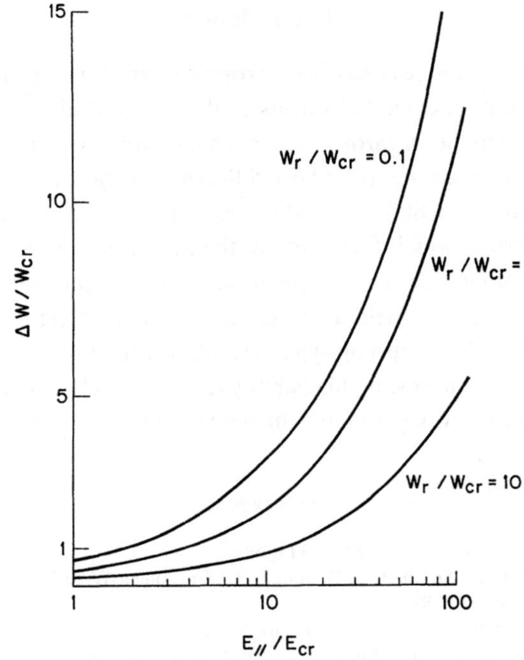

Fig. 1. Energy gain of the trapped electrons, versus the parallel electric field of the electrostatic whistler. Three values for the typical energies of the resonant electrons responsible for the generation of the electromagnetic whistlers have been chosen. The perpendicular energy of the accelerated electrons is there taken to be zero.

Thus, as shown on Figure 1, with a parallel electric field of the order of 100 times $E_{cr}(\approx 150 \text{ mV m}^{-1})$, the gain of energy is of the order of $\Delta W \approx 30$ keV, which is an order of magnitude higher than the estimated temperature.

On the other hand, an estimation of the maximal amplitude which can be reached by the electric field of the whistler wave can be done by using a trapping type saturation process (Le Quéau and Roux, 1987). Considering that the QMW is produced by a loss cone driven instability (Sharma and Vlahos, 1984), the linear growth rate can be estimated as:

$$\gamma_L \approx 10^{-3}\, \Omega_{ce} \,.$$

The inhomogeneity of the B-field being weak near the top, the saturation will occur whenever γ_L matches the electromagnetic trapping frequency Ω_T. This gives the following estimation for the maximal amplitude of the (perpendicular) electric field of the initial electromagnetic whistler wave:

$$\mathbf{E} \approx 800 \text{ mV m}^{-1} \,.$$

Even if the parallel electric field of the electrostatic whistler could be a little bit smaller than this value, one sees that it is likely to have an order of magnitude well corresponding to the above quoted evaluation.

4. Conclusion

An efficient acceleration process has been proposed which is able to produce a significant non-thermal tail of energetic electrons at the low-altitude part of a coronal loop.

This mechanism can be regarded as an energy transfer between two electronic populations initially characterized by very different energies, and being localized near the top of the loop. The first population exhibits a free energy source (thermal anisotropy, or loss-cone) which first leads to the generation of whistlers. These waves finally accelerate another electronic population along the coronal field lines.

The amplitude of the electric field which is needed for accelerating electrons up to tens of keV is consistent with a trapping-type saturation process.

All these facts give an interesting plausibility to this model, which has to be improved by further studies on the propagation of whistlers within the complex solar environment.

References

Le Quéau, D. and Roux, A.: 1987, *Solar Phys.* **111**, 59.

Norris, A. J., Johnson, J. F. E., Sotka, J. J., Wrenn, G. L., Cornilleau-Wehrlin, N., Perraut, S., and Roux, A.: 1983, *J. Geophys. Res.* **88**, 889.

Rauch, J. L. and Roux, A.: 1982, *J. Geophys. Res.* **87**, 8191.

Roux, A., Cornilleau-Wehrlin, N., and Rauch, J. L.: 1984, *J. Geophys. Res.* **89**, 2267.

Sharma, R. R. and Vlahos, L.: 1984, *Astrophys. J.* **280**, 405.

EVIDENCE FOR INTERACTING LOOP PROCESS IN A PHASE OF THE MAY 16, 1981 FLARE*

B. VRŠNAK[1], V. RUŽDJAK[1], M. MESSEROTTI[2], and P. ZLOBEC[3]

Abstract. The behaviour of the flare in the period of enhancement and maximum of hard X-ray, microwave and decimetric type IV continuum is analysed. The elongation of the Hα ribbons and microwave source disclose that the energy release site was shifting through a system of loops with a velocity less than 200 km s^{-1}, and that the energy was carried down the field lines with a velocity of about 1000 km s^{-1}, implying the thermal conduction front mechanism of energy transport. Several processes of energy release are considered and it is concluded that an explanation in terms of succeeding interactions of neighbouring loops, involving fast reconnection of their poloidal components is in best agreement with the observations.

1. Introduction

Although the large two-ribbon flare of May 16, 1981 has been studied in various aspects by many authors, interesting details still can be found in the rich accumulation of observational data. The flare was very complex, consisting of several distinct phases and exhibiting many interesting properties. The general aspects were described by Fárník *et al.* (1983) and Ishkov *et al.* (1985) while the aspect of the development of highly flare productive complex of activity was studied in Banin (1983) and Golovko (1983). Particular characteristics of this flare event were compared to other flares by Karlický (1984) and Ruždjak *et al.* (1984, 1986) and an analysis dealing exclusively with some special characteristics of the event was undertaken by Kaastra (1985).

In this paper we study the interesting phase lasting from 08 : 33 to 08 : 39 UT, during which a fast elongation of the Hα ribbons was observed and the maxima in hard X-rays, microwaves and decimetric type IV continuum flux were attained (Figure 1).

2. Observations

The period after 08 : 24 UT was characterized by a fast westward lengthening of the north-polarity ribbon, parallel with the neutral line. Due to the complex magnetic field configuration the process developed in several steps. Each step involved a distinct subsystem of loops as revealed by the observed Hα post-flare loops connecting north and south polarity regions in the late phases of the flare (Ruždjak *et al.*, 1984; Banin, 1983).

[1] Hvar Observatory, 58 450 Hvar, Yugoslavia.
[2] SISSA-ISAS, 34 014 Trieste, Italy.
[3] Trieste Astronomical Observatory, 34 131 Trieste, Italy.
* Proceedings of the Second CESRA Workshop on Particle Acceleration and Trapping in Solar Flares, held at Aubigny-sur-Nère (France), 23–26 June, 1986.

Solar Physics **111** (1987) 23–29.

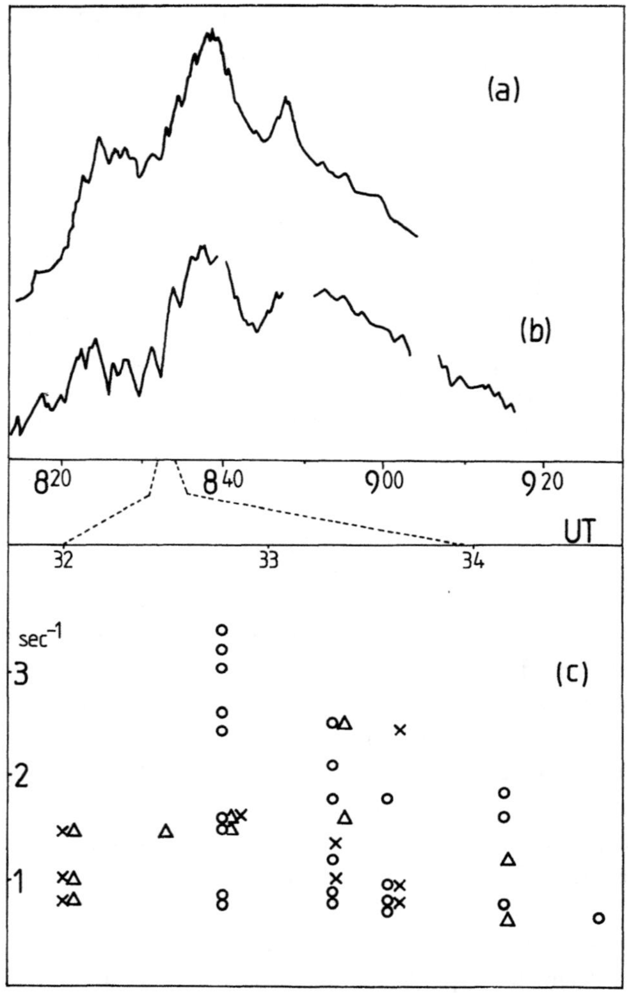

Fig. 1. The curves (a) and (b) represent the Westerbork microwave flux at 6 cm and Prognoz-8 X-ray flux between 20 and 40 keV (Fárník *et al.*, 1983), respectively. The peaks in power spectra of the flux at 237 MHz (triangles), 327 MHz (crosses), and 408 MHz (circles) are given in (c).

The phase which we analyse is the final step of elongation (Figure 2) which started at 08 : 34 UT and was related to the most powerful phase in the development of the flare. This phase began by the appearance of two bright Hα knots (marked C_1 and C_2 in Figure 2(a)), placed in the south and north polarity regions, respectively. Hα emission in the north polarity region elongated abruptly from the knot C_2 towards the leading spot (NL), with an average velocity of 240 km s^{-1} and protruded into its umbra. The emission in the south polarity region lengthened from knot C_1 in the north-west direction (Ishkov *et al.*, 1985). The elongation of the ribbons was preceded by a microwave impulsive burst at 08 : 03 UT which was located between the Hα knots (Fárník *et al.*,

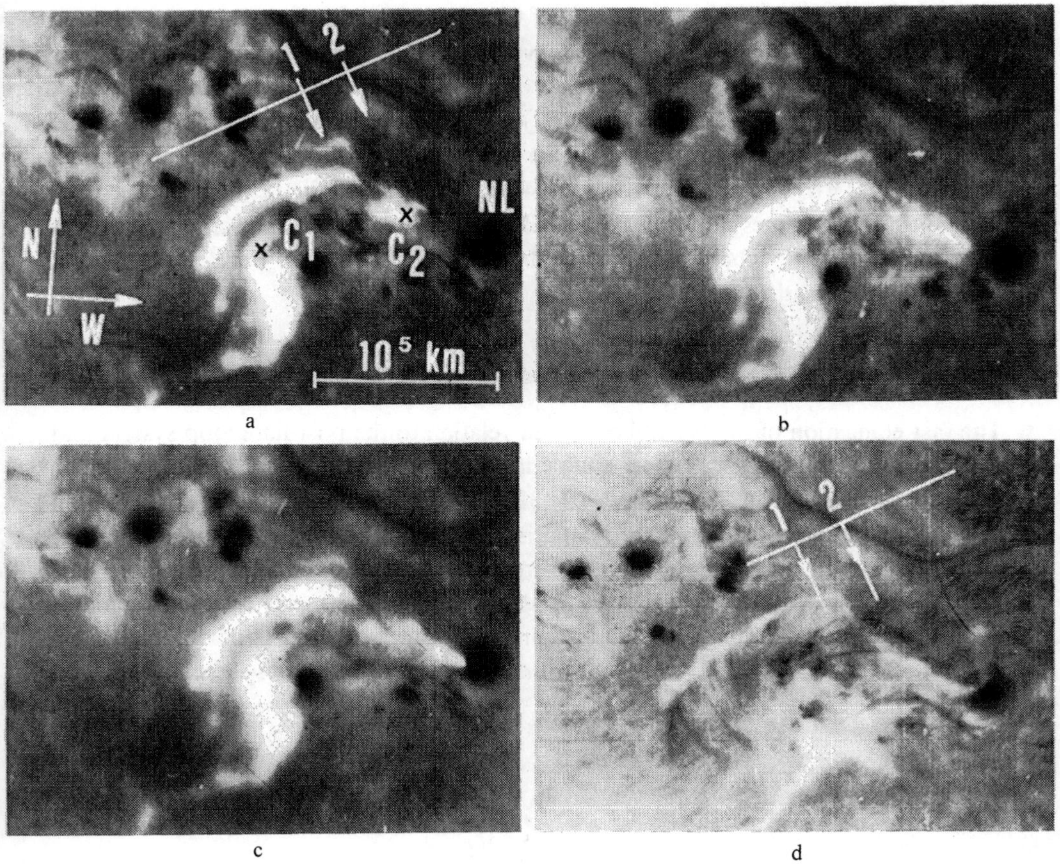

Fig. 2. The last phase of the Hα ribbon elongation process took place between 08 : 34 and 08 : 39 UT. Filtergram (a) was taken immediately after the bright knots appeared at positions marked by crosses C_1 and C_2. The direction of the Westerbork baseline is indicated together with the position of the microwave impulsive burst at 08 : 33 : 02 UT (arrow 1) which also corresponds to the starting position of the gradual source. The distance between positions 1 and 2 corresponds to the final length of the gradual source. Filtergrams (b) and (c) taken at 08 : 36 UT and 08 : 37 UT, respectively, illustrate the fast elongation of the Hα ribbon towards the leading spot NL. Filtergram (d) was taken at 10 : 45 UT when the post-flare Hα loop system connected with the process between 08 : 33 UT and 08 : 39 UT was partially visible in Hα line center. The Westerbork positions marked as (1) and (2) have the same meaning as in Figure 2a, and their intersections with the loops were used for the determination of the lengths L_2 and L_4.

1983). The burst was followed by appearance of a longlasting microwave source at roughly the same position (Figure 2(a)). Similar to the elongation of the north polarity ribbon, the microwave source was also elongating westwards, with a projected velocity to the Westerbork baseline of 180 km s^{-1}. The rise time of the 6 cm radiation at a certain position was somewhat less than one minute. The elongating ribbon reached the leading spots umbra at 08 : 37 UT and protruded above the umbra attaining maximum development at 08 : 38 : 30 UT. During this phase microwave and hard X-ray total fluxes were increasing and peaked at 08 : 39 UT.

In the decimetric range the beginning of the process was indicated by an increased number of bursts around 08 : 33 UT as revealed by the appearance of high frequency components in the power spectra of Trieste Observatory total flux measurements at 237, 327, and especially at 408 MHz (Figure 1). A still more pronounced increase in the number of bursts took place during the flare maximum. In the considered period, i.e., from 08 : 33 UT to 08 : 39 UT, pulsations (40% Rh polarized) in the range between 200 and 400 MHz were present, while type III like bursts were absent and spike activity was very low – just few of them appeared at about 08 : 35 UT (A. O. Benz, ETH Zürich, private communication).

3. Analysis and Interpretation

The fast elongation of the Hα ribbons, their relation to the post-flare loop system and the elongation of the microwave source indicate specific physical processes which occured between 08 : 33 UT and 08 : 39 UT. The magnetic field configuration documented by the post-flare loop system is sketched in Figure 3. We suggest that the energy release started at a point A somewhere on the field lines connecting the footpoints C_1 and C_2 where the Hα knots showed up at 08 : 34 UT. The energy released at point A was then led downwards along the field lines, producing the gross of observed 6 cm radiation at point B, and Hα emission when it reached the chromosphere. Meanwhile,

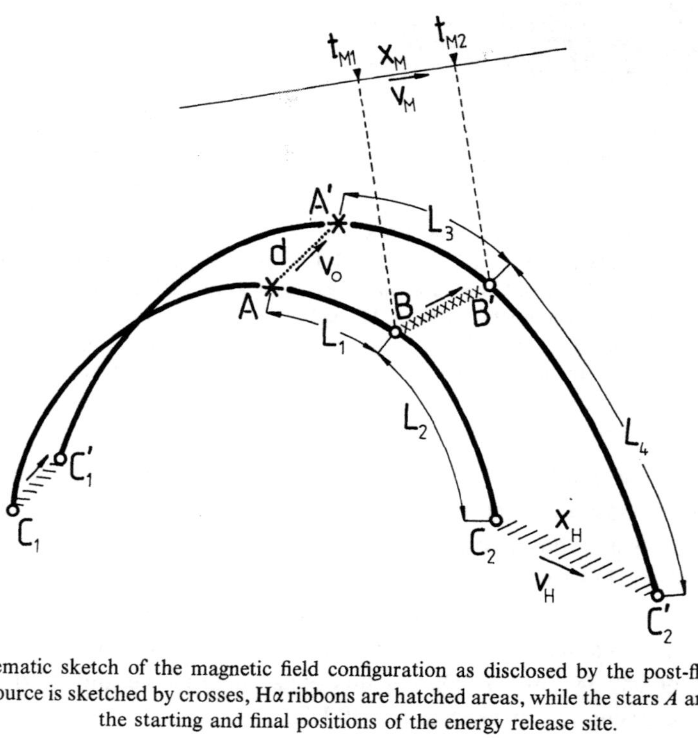

Fig. 3. Schematic sketch of the magnetic field configuration as disclosed by the post-flare loops. The microwave source is sketched by crosses, Hα ribbons are hatched areas, while the stars A and A' represent the starting and final positions of the energy release site.

the process of energy release protruded successively through the loop system, towards the point A' (Figure 3). The shift of the energy release site gave rise to the elongation of the microwave source from B to B' and of the north polarity ribbon from C_2 to C_2'. The projections of the distances between A and B; B and C_2; A' and B'; B' and C_2'; A and A' in the plane of the sky are marked as L_1, L_2, L_3, L_4, and d, respectively. The final projected length of the ribbon between C_2 and C_2' is marked as X_H.

The Westerbork positions of the elongating microwave 6 cm source were superposed on the Hvar Observatory filtergrams and the points B and B' were taken as the intersections of this positions and the post-flare loops connecting C_1 with C_2 and C_1' with C_2', respectively (Figure 2(d)). In this way we determined the values: $L_2 = 4 \times 10^4$ km, $L_4 = 5 \times 10^4$ km, $X_H = 3.6 \times 10^4$ km, and $X_M = 2.5 \times 10^4$ km, where X_M is the final length of the microwave source projected on the Westerbork baseline. The ratio of L_2 and the time difference between the appearance of the Hα kernels and the microwave source emission allow an estimate of the velocity V by which the energy was carried downwards. The Hα source appeared 30 to 40 s after the microwave source, which gives a velocity V between 900 and 1300 km s^{-1}.

To get a more reliable estimate of V, one has to eliminate the parameters whose determination is most inaccurate. We shall denote the time of the beginning of the process as $t_0 = 0$, the moment of the appearance of the Hα knot as t_{H1}, the time of appearance of the microwave source as t_{M1}, the times when the elongation of the Hα and microwave sources stopped as t_{H2} and t_{M2}, respectively, and the end of energy release as t_e. The following relations hold (v_0 is the energy release site velocity):

$$t_e = \frac{d}{v_0}\,; \qquad t_{M1} = \frac{L_1}{V}\,; \qquad t_{H1} = \frac{L_1 + L_2}{V}\,; \qquad t_{M2} = \frac{d}{v_0} + \frac{L_3}{V}\,; \qquad t_{H2} = \frac{d}{v_0} + \frac{L_3 + L_4}{V}\,. \tag{1}$$

The differences $\Delta t_M = t_{M2} - t_{M1}$ and $\Delta t_H = t_{H2} - t_{H1}$ are thus:

$$\Delta t_M = \frac{L_3 - L_1}{V} + \frac{d}{V_0} \quad \text{and} \quad \Delta t_H = \frac{L_3 + L_4 - L_1 - L_2}{V} + \frac{d}{v_0}\,. \tag{2}$$

Subtracting Δt_M and Δt_H and substituting $\Delta t_M = X_M/V_M$ and $\Delta t_H = X_H/V_H$ where V_M is the velocity of microwave source elongation projected on the Westerbork baseline, and V_H the velocity of the north polarity ribbon's elongation, one gets the velocity V as:

$$V = \frac{L_4 - L_2}{\dfrac{X_H}{V_H} - \dfrac{X_M}{V_M}}\,. \tag{3}$$

In this way the parameters t_{H1}, t_{H2}, t_{M1}, and t_{M2} are eliminated and replaced by X_H,

X_M, V_H, and V_M which can be measured more precisely. Also, for estimate (3) it is not necessary to know the position of the site of energy release, since the parameters L_1 and L_3 are eliminated as well. Using the measured values, one obtains the velocity $V = 900\ \text{km s}^{-1}$. Taking into account the uncertainty of measurement (especially of X_M) the estimated error amounts to some $300\ \text{km s}^{-1}$.

This estimate of the velocity V shows that the energy was not carried by electron beams, but probably by a thermal conduction wave (Rust and Somov, 1984). The microwave impulsive burst which appeared some ten seconds before the start of the gradual microwave component might have been the result of an electron beam emitted from the energy release site at the onset of the process. In that case the impulsive burst practically marks the onset of the process itself, since the fast electrons need less than one second to travel along the loop. This allows to estimate the length $L_1 = V \times t_M$, roughly as 10^4 km if the values $V = 10^3\ \text{km s}^{-1}$ and $t_{M1} = 10$ s are used, which locates the starting position of the energy release site close to the top of the loop between C_1 and C_2. Assuming that the site of energy release shifted across the summits of the field lines traced by the post-flare loop tops, the distance between A and A' is estimated as $d = 2 \times 10^4$ km, and the distance between A' and B' as $L_3 = 2 \times 10^4$ km. Then the second of relations (2) gives for the projected veloicty v_0 a value in the order of $100\ \text{km s}^{-1}$. Taking into account the geometry of the system, it can be concluded that the true velocity of the energy release site shift was less than $200\ \text{km s}^{-1}$.

4. Discussion

Summarizing the results of our analysis, we can say that the energy release site was located close to the summit of the field lines in a closed magnetic configuration, and that it has been shifting through the system with a velocity less than $200\ \text{km s}^{-1}$. The liberated energy was carried down along the field lines with a velocity of about $1000\ \text{km s}^{-1}$, which suggests a thermal front model of energy transport (Rust and Somov, 1984). The onset of the process was marked by an impulsive microwave burst and the increased number of bursts in the decimetric range. The gross of energy was liberated in the later, gradual phase, during which the site of energy release was shifting through the system.

There are several processes which could explain the observations, depending on the underlying magnetic field structure. If the currents were fragmented into current filaments to form a system of loops, then the observations can be described by the successive interactions of neighbouring loops. Such an interaction can proceed either mechanically (Emslie, 1981), or through inductive coupling of the currents flowing through the loops (de Jager, 1986). There are two important observational constraints: the energy release site shifted with a velocity less than $200\ \text{km s}^{-1}$, and the rise time of microwave emission at a certain location was about 1 min. This excludes the fast regime in the nonlinear development of coalescence instability (Lebœuf *et al.*, 1984; Tajima *et al.*, 1984) since it develops at the Alfvén time-scale. The same holds also for inductive mechanisms since the transmission of the signals of current and inductance changes is related to the Alfvén velocity (Spicer *et al.*, 1986) which is usually assumed to be considerably higher than 200 km s^{-1}.

Thus we propose that the process of driven reconnection of poloidal components of the magnetic field between neighbouring loops took place (Priest, 1985). The process was initiated by some fast time scale instability (Spicer and Brown, 1981) of the loop connecting the kernels C_1 and C_2 which resulted in the microwave impulsive burst and increased number of dm bursts. Impulsive heating of the loop led to its expansion due to labeled pressure gradient, thus initiating the driven reconnection process in the current sheet formed between the neighbouring loops. The maximal velocity of merging (u) in the Petschek regime is in the order of 10% of the Alfvén velocity based on inflow conditions, and thus is compatible with the observations. The reconnection led to heating of the outer layers of the second loop which then resulted in its expansion and initiation of reconnection of its poloidal component with the poloidal component of the next loop in the system. In this manner the observed velocities can be explained. On the other hand, the time needed for total reconnection of the poloidal components between two loops is approximately $T = R/u = 10\text{–}100$ s, R being the radius of the loop, and is in agreement with the recorded rise time of the microwave radiation at a certain position.

References

Banin, V. G.: 1983, *Issledovannija po geomagnetizmu, aeronomiji i fizike solnca* **65**, 129.

Emslie, A. G.: 1981, *Astrophys. Letters* **22**, 171.

Fárník, F., Kaastra, J., Kalman, B., Karlický, M., Slottje, C., and Valníček, B.: 1983, *Solar Phys.* **89**, 355.

Golovko, A. A.: 1983, *Issledovannija po geomagnetizmu, aeronomiji i fizike solnca* **65**, 121.

Ishkov, V. N., Markeev, A. K., Fomichev, V. V., Chernov, G. P., Chertok, I. M., Likin, O. B., Pisarenko, N. F., Karlický, M., Tlamicha, A., Fárník, F., Valníček, B., and Kalman, B.: 1985, *Bull. Astron. Inst. Czech.* **36**, 81.

de Jager, C.: 1986, *Space Sci. Rev.* **44**, 43.

Kaastra, J. S.: 1985, Thesis, Utrecht.

Karlický, M.: 1984, *Solar Phys.* **92**, 329.

Lebœuf, J. N., Brunel, F., Tajima, T., Sakai, J., Wu, C. C., and Dowson, J. M.: 1984, *Magnetic Reconnection in Space and Laboratory Plasmas*, Geophysical Monograph 30, American Geophysical Union, p. 282.

Priest, E. R.: 1985, *Reports on Progress in Phys.* **48**, 955.

Rust, D. M. and Somov, B. V.: 1984, *Solar Phys.* **93**, 95.

Ruždjak, V., Vršnak, B., Kotrč, P., Mercier, C., Schober, H. J., Schroll, A., Urpo, S., and Zlobec, P.: 1984, *Hvar Obs. Bull.* **8**, 25.

Ruždjak, V., Vršnak, B., Zlobec, P., and Schroll, A.: 1986, *Solar Phys.* **104**, 169.

Spicer, D. S. and Brown, J. C.: 1981, in S. Jordan (ed.), *The Sun as a Star*, NASA SP-450, Ch. 18.

Spicer, D. S., Mariska, J. T., and Boris, J. F.: 1986, in P. A. Sturrock, T. E. Holzer, D. M. Mihalas, and R. H. Ulrich (eds.), *Physics of the Sun II*, Geophysical and Astrophysical Monographs, D. Reidel, Dordrecht, Holland, Ch. 12.

Tajima, T., Brunel, F., Sakai, J. I., Vlahos, L., and Kundu, M. R.: 1984, in M. R. Kundu and G. D. Holman (eds.), 'Unstable Current Systems and Plasma Instabilities in Astrophysics', *IAU Symp.* **107**, 197.

EFFECTS OF TURBULENCE DEVELOPMENT IN SOLAR SURGES*

VINCENZO CARBONE[‡], GIORGIO EINAUDI[†], and PIERLUIGI VELTRI[‡]

[‡] *Dipartimento di Fisica-Universita' della Calabria, 87036 Arcavacata di Rende, Cosenza, Italy*

[†] *Scuola Normale Superiore, 56100 Pisa, Italy.*

Abstract. Several authors have claimed for correlations between surges (dark features) and various kinds of solar emissions (radio, microwave, X-ray). In this paper we propose a model to explain such correlations, in particular presenting the properties of the instabilities resulting from the coupling between material flow, connected to the appearance of a surge, and magnetic field topology. As a consequence of such instability a turbulent energy cascade to small characteristic lengths grows up. Depending of the relevant parameters of the surge (dark feature), different regimes can be found, producing different levels of electrons acceleration and mass motion deceleration. We try then to correlate the different developments of the instability with the behavior observed in type I and type III radio bursts related to surges.

1. Introduction

Material motions from the solar chromosphere up to the corona and then down again to the chromosphere, which occur in active regions with velocities ranging from 10 to 200 km s^{-1} are usually called surges, when observed at the limb, and dark features, when observed on the solar disk.

Several attempts to directly correlated such phenomenon with solar emission in the radio, microwave and X-ray ranges have been performed in the last years (Garczynska *et al.*, 1982; Verma, 1985; Mein and Avignon, 1983).

More recently the temporal association between dark features and production of radio Type III bursts has been extensively studied showing that radio emission occurs almost always during the ascending phase of the material motions. Moreover, a magnetic field parasitic polarity is observed between the dark feature and the sunspot toward which such dark feature is pointing (Chiuderi-Drago *et al.*, 1986).

From a theoretical point of view, a surge can be modeled as a jet of high density (chromospheric density) travelling in a sheared magnetic field. Such a jet is likely to be the site of instabilities which in turn trigger a nonlinear turbulent cascade with important effects on the behavior of the jet itself.

It is well known that in a sheared velocity field Kelvin–Helmholtz instabilities can arise whereas in a sheared magnetic field resistivity can drive reconnection.

In this paper we will show that when both a sheared velocity and magnetic fields are present, the instabilities which develop in the system can present, depending on the

* Proceedings of the Second CESRA Workshop on Particle Acceleration and Trapping in Solar Flares, held at Aubigny-sur-Nère (France), 23–26 June, 1986.

Solar Physics **111** (1987) 31–41.

small-scale features of the equilibrium configuration, a variety of different linear properties leading to considerably different nonlinear turbulent regimes.

In the following we will build up a simplified model for the surge (Section 2), we will analyze the dependence of the linear instability from the relevant physical parameters (Section 3) showing that in any case an electric field which can accelerate electrons is produced, we will set up a simulation of the nonlinear turbulent cascade (Section 4) which shows that at least in certain regimes an efficient deceleration of the jet is obtained, and finally we will discuss the consequence of our results from an observational point of view (Section 5).

2. A Simplified Model for the Surge

In order to discuss the possible role played by plasma instabilities in a surge, let us consider an idealized configuration in which the plasma is flowing in the z-direction along the magnetic field $\mathbf{B}$ with a velocity $\mathbf{v}$.

Both $\mathbf{B}$ and $\mathbf{v}$ vary in the x-direction. In order to take into account the shear of the magnetic field possibly induced by the presence of a parasitic polarity, we will consider a magnetic field of the form

$$B(x) = B_0 \,\mathrm{tgh}\,(x/a_B) . \tag{2.1}$$

For such magnetic field dissipative (resistive) effects can be important in a layer around $x = 0$, where the magnetic field is zero.

Since we have no direct information on the detailed structure of the velocity field, we will discuss two different sheared velocity equilibrium profiles with different stability properties:

$$v(x) = v_0(\mathrm{tgh}\,(x/a_v) + 1) , \tag{2.2}$$

which is stable against Kelvin–Helmholtz instabilities (equilibrium I), and

$$v(x) = v_0 \,\mathrm{sech}\,(x/a_v) , \tag{2.3}$$

which is unstable against Kelvin–Helmholtz instabilities (equilibrium II).

Since in this paper we are mainly interested to understand the effects produced by a mass motion in a sheared magnetic field we have not taken into account neither the density gradient due, to the propagation of chromospheric materials in the corona, nor the stratification, due to the solar gravity field.

3. Linear Stability Analysis

The linear stability analysis has been performed using a finite difference method described extensively elsewhere (Einaudi and Rubini, 1986).

Assuming the perturbations to behave like $\exp[i(k_{\parallel} z + \omega t)]$, the linear stability analysis of the above configurations shows that the relevant parameters are:

$$r = a_B/a_v , \quad M_A = V_0/V_A , \quad S = \tau_R/\tau_A , \quad R = \tau_v/\tau_A , \quad \alpha = ka_B ,$$

where M_A is the Alfvén Mach number, V_A is the Alfvén velocity associated to B_0, $\tau_R = 4\pi a_B^2 \sigma/c^2$ is the resistive time (σ being the electrical conductivity and c being the speed of light), τ_A is the Alfvén time and $\tau_v = \rho a_B^2/\eta_1$ is the viscous time (we have used the same notations as Braginskii (1965)). It can be shown that η_1, which is the smallest of the coefficients appearing in the stress tensor, is the only one important in this calculation (Mok, 1987).

The results can be summarized as follows:

(1) The frequency $\omega = \omega_R + i\gamma$ of the mode is complex and $\omega_R \simeq k_\parallel V(0)$, which means that the instability produces amplified oscillations in the perturbed quantities.

(2) There exists a transition from a behavior similar to the static tearing mode to a behavior similar to a Kelvin–Helmholtz mode. This transition occurs when $\varepsilon = rM_A \simeq M_A^{1/3}$ (here ε is a measure of the relative importance of the velocity gradient with respect to the magnetic gradient).

When $\varepsilon > M_A^{1/3}$, we find a stabilization (Figure 1) for equilibrium I and a rapid increase of the growth rate for equilibrium II. This difference is due to the different stability properties against ideal Kelvin–Helmholtz modes of the two configurations.

(3) It is well known that the viscosity stabilizes the static tearing mode. In presence of flows this is still true only wen $\varepsilon \ll 1$ or $S/R < 1$, otherwise we have to distinguish between equilibrium I and II.

For equilibrium I, with $\varepsilon \simeq M_A^{1/3}$ and $S/R > 1$, the growth rate of the mode increases with the viscosity and presents a maximum value for $R \simeq 20$ which is independent of the resistivity. For $R > 20$ the viscosity again stabilizes the mode.

For equilibrium II, with $\varepsilon > M_A^{1/3}$, the viscosity is important and represents a stabilization factor only when $R \simeq 1$.

(4) In all cases there is a non-zero x-component of the magnetic field perturbation at $x = 0$ and, therefore, all these modes are reconnecting modes. The level of reconnection varies as a function of all the parameters along with the efficiency in accelerating electrons by the parallel (to **B**) electric field created in the resistive layer.

(5) When $\varepsilon > M_A^{1/3}$ in all the perturbations are not localized close to the resistive layer as in the static tearing mode and, therefore, the typical scale length of the mode is a_B and not the width δ of the resistive layer as in the static case.

We can easily identify several different regimes, characterized by different velocity profiles and different values of the relevant parameters.

Two of these regimes occur when $\varepsilon \simeq 1$, respectively, in configuration I and II. The main differences of the two regimes are that:

(i) $\gamma\tau_A \simeq S^{-1/2}$ in the first regime while $\gamma\tau_A \simeq S^0$ in the second regime.

(ii) $(\gamma\tau_A)_{max}$ occurs at $\alpha \lesssim 0.1$ in the first regime while $(\gamma\tau_A)_{max}$ occurs at $\alpha \gtrsim 1$ in the second regime.

(iii) The velocity and magnetic field perturbations are much more correlated in the first than in the second regime.

We have studied two specific cases one for each linear regime mentioned above. The values of the relevant parameters, and the results of the linear analysis for the two cases are summarized in the Table I (α_m is the value of α at which the maximum growth rate occurs).

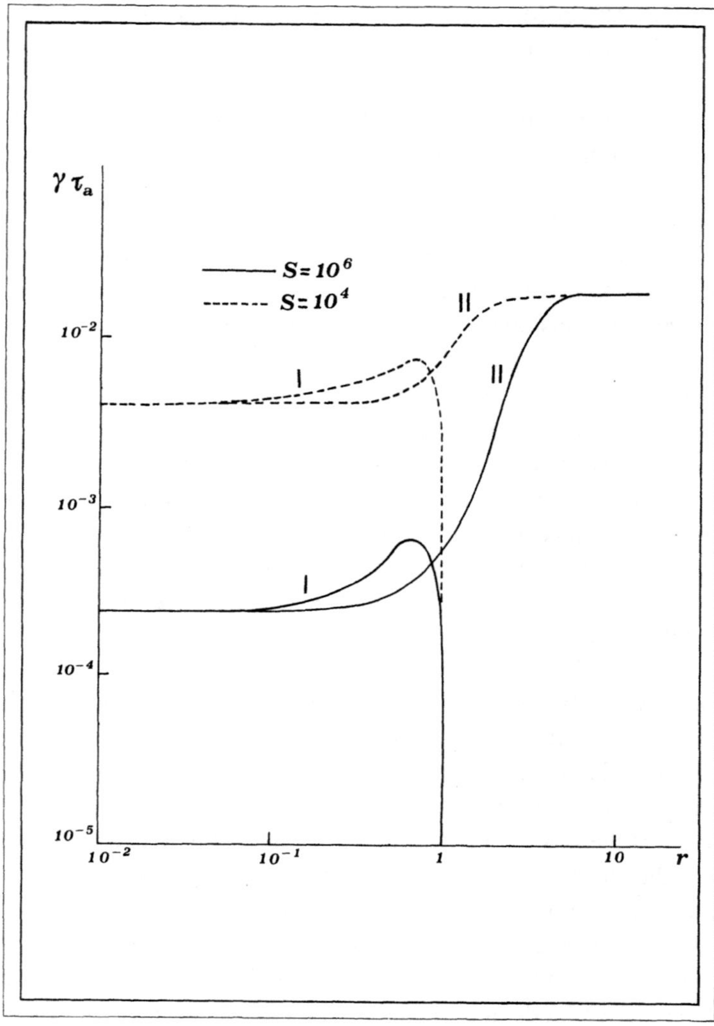

Fig. 1. Normalized growth rate vs r for different values of S and the two equilibrium configurations; $M_A = 1$, $\alpha = 0.5$, and $R = 0$.

TABLE I

Parameters and results of the linear analysis for the two cases considered

Equilibrium	I	II
M_A	1	1
r	0.9	6
S	4×10^7	10^9
R	5×10^4	3×10^7
$\gamma \tau_A$	7.25×10^{-4}	1.5×10^{-2}
$\omega_R \tau_A$	0.2	1
α_m	0.2	1

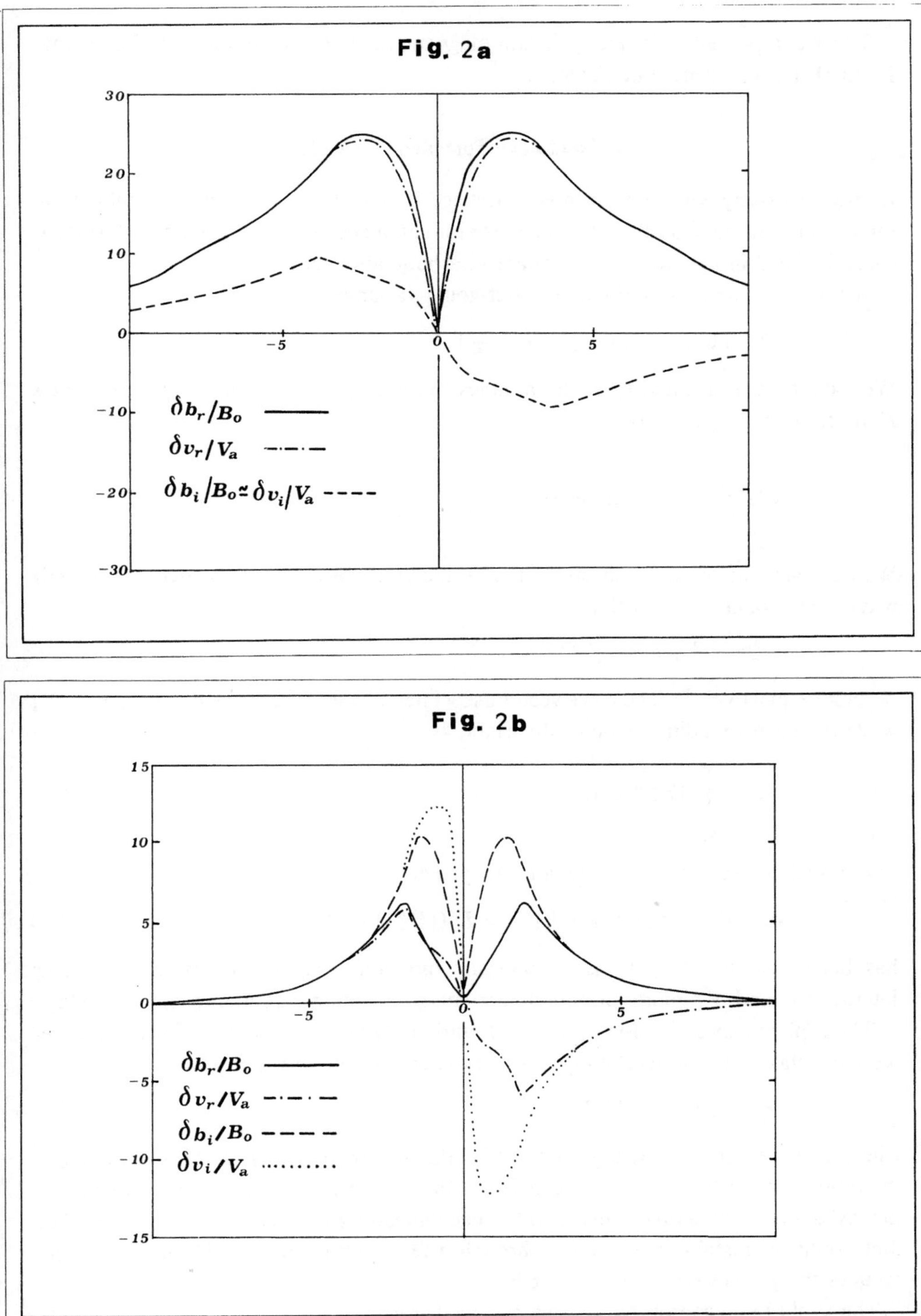

Fig. 2. Behavior of the real and imaginary part of the x-component of magnetic and velocity perturbations, normalized respectively to B_0 and V_A, for: (a) Equilibrium I; (b) Equilibrium II.

The x-components of velocity δv and magnetic field δb perturbations in the regimes I and II are represented in Figure 2.

4. Nonlinear Turbulent Cascade

In order to study what are the effects induced by the start of the linear instability on the equilibrium configuration we have used a simplified model of nonlinear turbulent cascade, local in the wave vector space (Carbone and Veltri, 1987).

We have defined the in-going and out-going variables:

$$z^\sigma = \delta\mathbf{b}/B_0 + \sigma\delta\mathbf{v}/V_{\mathrm{A}}\,, \quad \sigma = \pm 1\,.$$

We have then associated to such variables the energy density spectra per unit mass $E^\sigma(k, t)$ defined as follows:

$$\langle (z^\sigma)^2 \rangle /2 = \int_0^\infty \mathrm{d}k\, E^\sigma(k, t)\,.$$

We have also introduced a discretization in the wave vector space by defining the nth wave vector band Δ_n such that:

$$k_0 2^{n-1/2} \le k \le k_0 2^{n+1/2}$$

(k_0 being equal to a_B^{-1}). The wave vector associated to such band is evidently $k_n = k_0 2^n$, while the corresponding dynamical variable is:

$$U_n^\sigma = \int_{\Delta_n} \mathrm{d}k\, E^\sigma(k, t)\,.$$

For such set of variables a dynamical system

$$\mathrm{d}U_n^\sigma/\mathrm{d}t + k_n^2(1/S + 1/R)U_n^\sigma = \Gamma_n^\sigma(\{U_m^\sigma\}) + \mathscr{L}_n^\sigma \tag{4.1}$$

has been written where Γ_n^σ is a quadratic functional which models the nonlinear interactions and $\mathscr{L}_n^\sigma$ represents the linear energy injection rate due to the instability.

The explicit expression for Γ_n^σ can be found in Carbone and Veltri (1987), moreover we have chosen to model the linear energy injection rate in the form:

$$\mathscr{L}_n^\sigma = 2\gamma_0(E^f/E_0^f)U_n^\sigma\,, \tag{4.2}$$

where γ_0 represents the linear growth rate of the instability and (E^f/E_0^f) is nothing but the ratio between the free energy existing in the equilibrium structure and its value at the instability starting time. This ratio has been introduced to take into account the fact that nonlinear instability evolution, through a feedback on the equilibrium structure, reduces the growth rate of the instability.

The evolution equation for E^f has assumed to be:

$$\mathrm{d}E^f/\mathrm{d}t = a(t)\,(E^f/E_0^f)\,, \tag{4.3}$$

with $a(t)$ chosen in such a way to ensure energy conservation:

$$a(t) = -2\gamma_0 \sum_n U_n^\sigma .$$

Finally in our simulation the initial values for dynamical variables U_n^σ has been chosen in order to reproduce the spectral distribution of the instability eigenmodes.

In Figure 3 we have represented the time evolution of the spectra for the dynamical variable

$$W_n^\sigma = 2^{n/2} U_n^\sigma$$

which is more useful to display the eventual occurrence of a Kraichnan's spectrum.

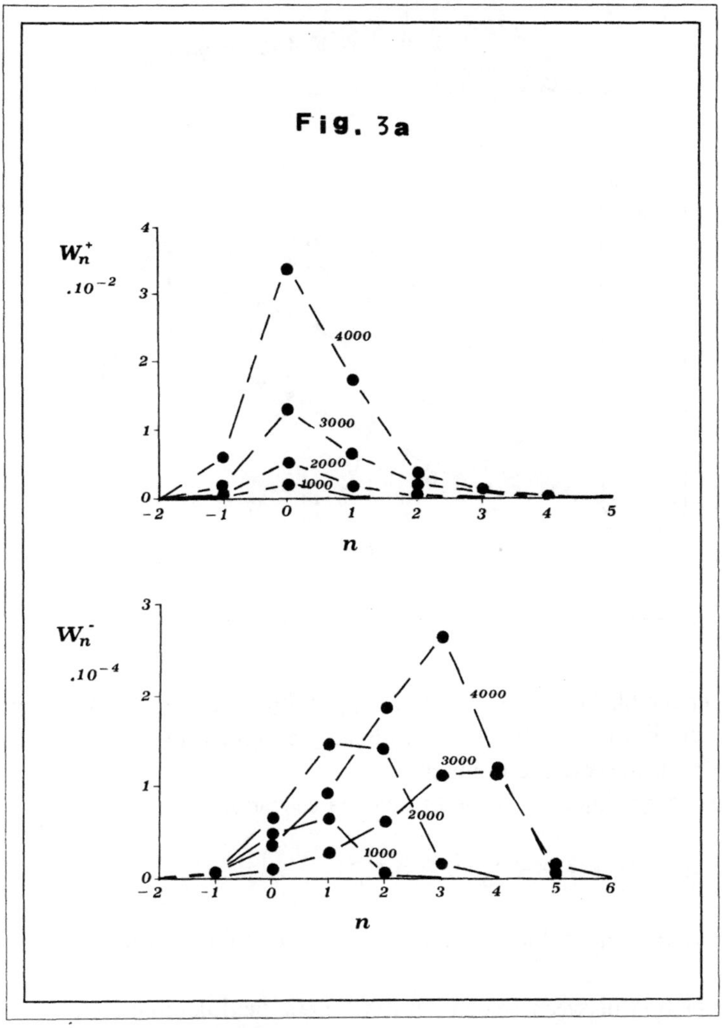

Fig. 3. Time evolution of the energy spectra W_n^σ, with $\sigma = \pm 1$, for: (a) Equilibrium I; (b) Equilibrium II. Time values are marked on the figure.

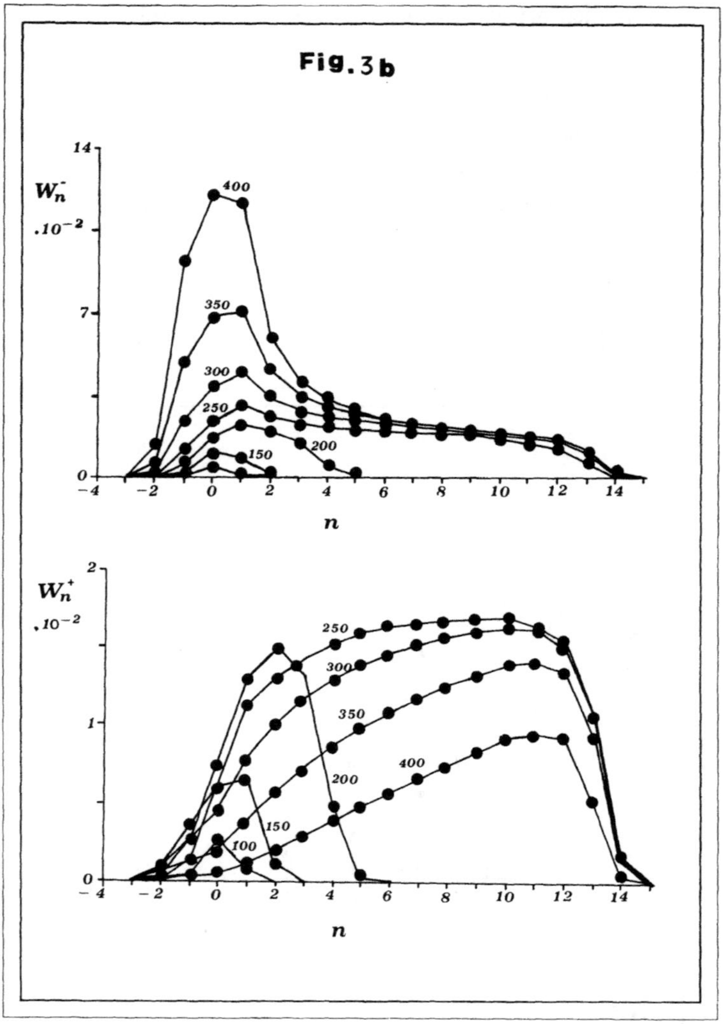

After a linear phase during which energy is injected without modifying the initial spectrum, nonlinear effects start producing a spectrum of the Kraichnan type $E(k) \sim k^{-3/2}$ in the intertial domain.

The total energy injected in the turbulence cascade,

$$E = E^+ + E^- = \sum_n (U_n^+ + U_n^-),$$

is not totally dissipated but partially (about $\frac{2}{3}$) remains in form of turbulent energy (Figure 4).

The correlation between velocity and magnetic field fluctuations,

$$h = (E^+ - E^-)/(E^+ + E^-),$$

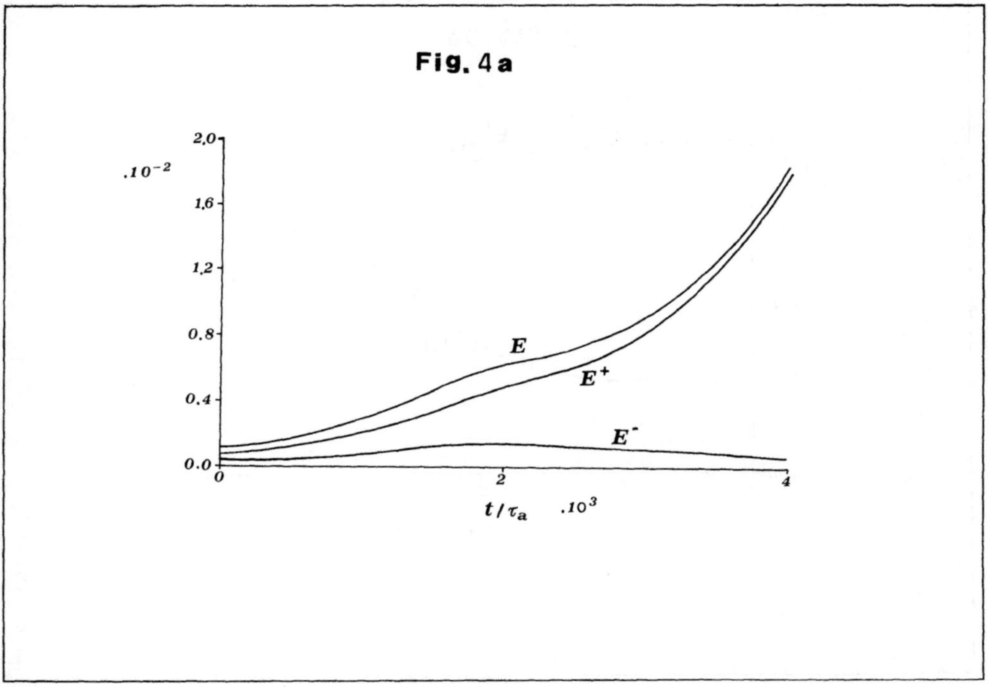

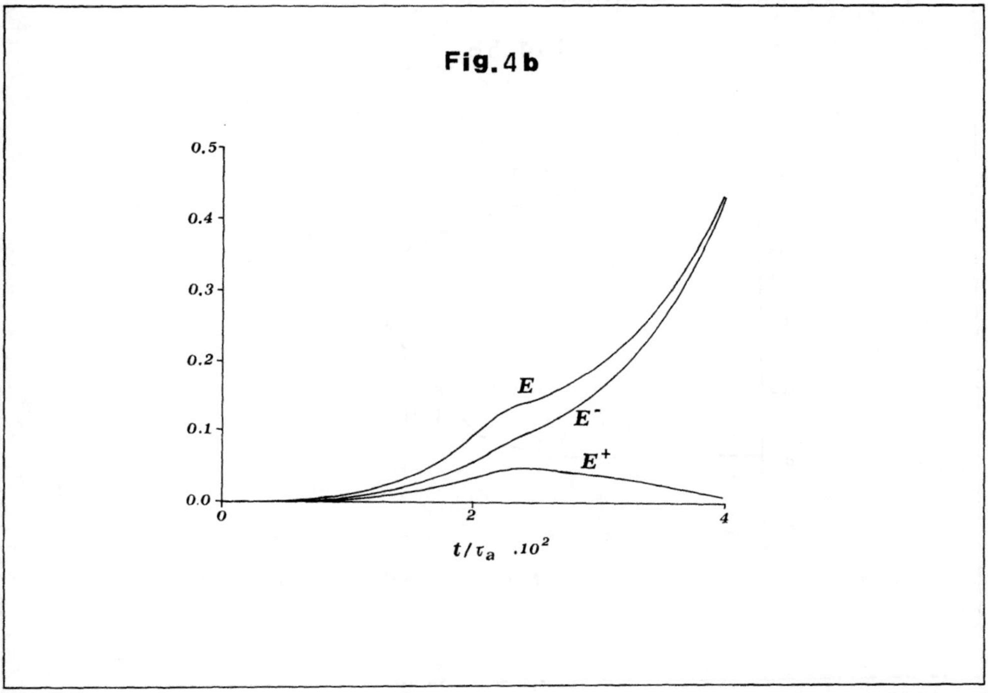

Fig. 4. Time evolution of the turbulent energy for: (a) Equilibrium I; (b) Equilibrium II.

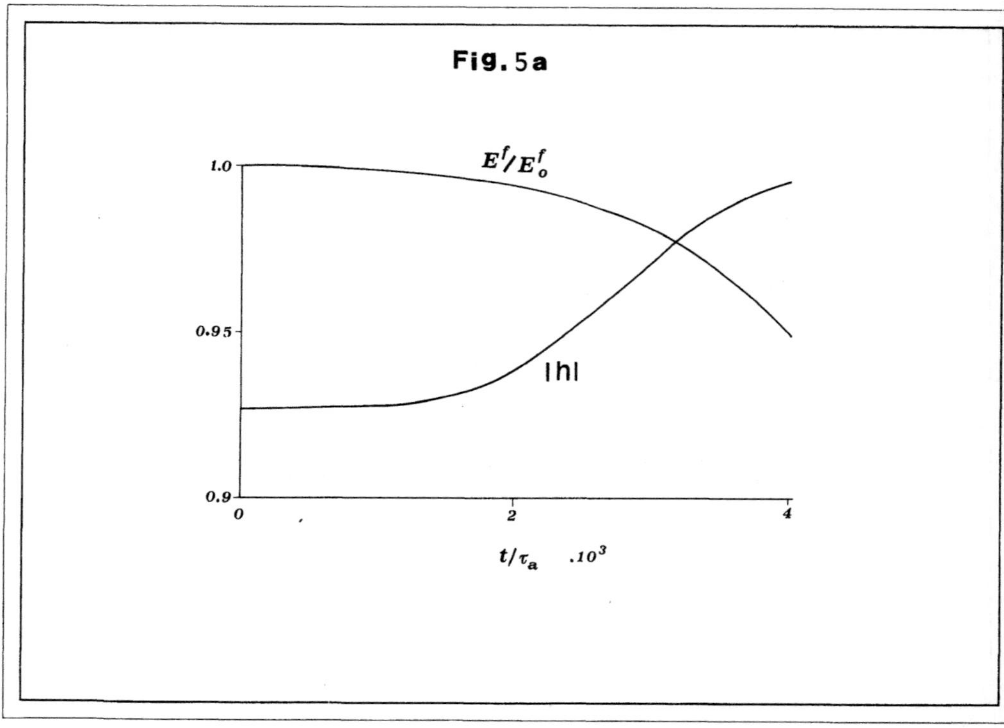

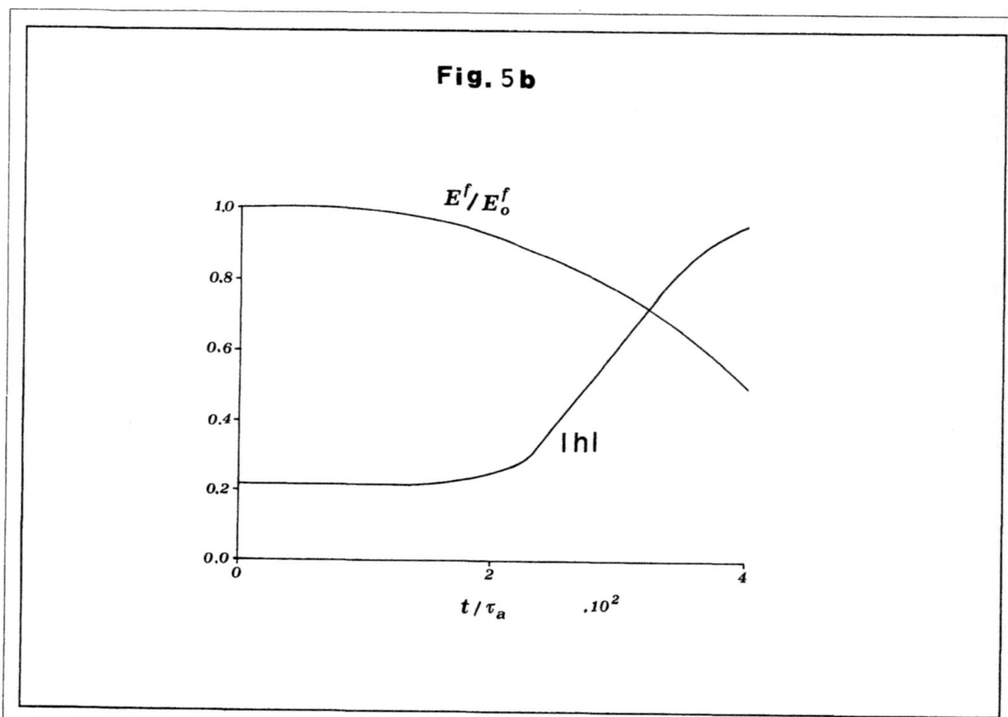

Fig. 5. Time evolution of the correlation between velocity and magnetic field and of the free energy of the structure for: (a) Equilibrium I; (b) Equilibrium II.

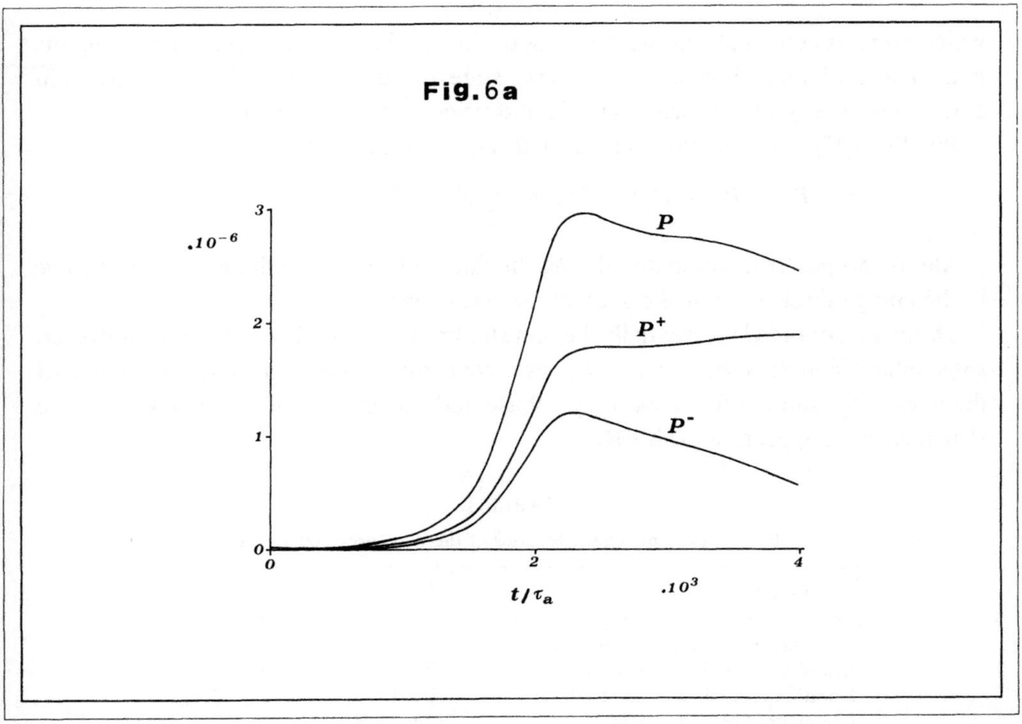

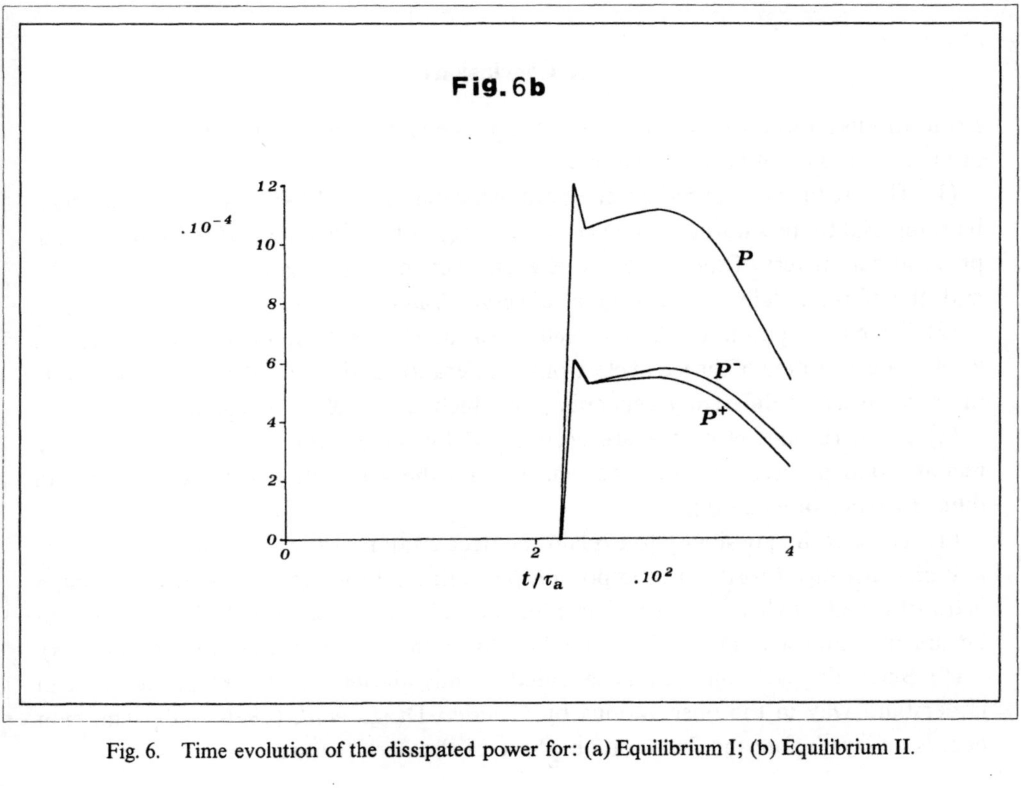

Fig. 6. Time evolution of the dissipated power for: (a) Equilibrium I; (b) Equilibrium II.

which remains essentially constant during the linear phase, grows up to one during the nonlinear evolution (Figure 5). The final state of turbulence is one of maximum correlation ($h \simeq \pm 1$) between magnetic and velocity field fluctuations.

Finally in Figure 6 we have represented the dissipated power:

$$P = P^+ + P^- = (1/S + M_A/R) \sum_n 2^{2n}(U_n^+ + U_n^-) .$$

Almost no power is dissipated during the linear phase, while there is a sudden rise in the energy dissipation at the start of nonlinear phase.

The picture outlined is essentially the same for both cases studied. There are, however, substantial differences between the two cases concerning the characteristic times of both the linear (τ_L) and the nonlinear (τ_{NL}) phase and the amount of energy taken-off the structure, as it appears in Table II.

TABLE II

Energy and time characteristics of the two cases considered

Equilibrium	I	II
Free energy taken-off the structure	3%	55%
Duration of the linear phase (τ_L)	$2 \times 10^3 \tau_A$	$2 \times 10^2 \tau_A$
Duration of nonlinear phase (τ_{NL})	$2 \times 10^3 \tau_A$	$2 \times 10^2 \tau_A$
Energy level in fluctuations	2×10^{-2}	0.45

5. Conclusions

Let us discuss now the consequences of the previous theoretical analysis from the point of view of mass motion observations:

(1) The life time of a surge (or of a dark feature) is about 10–30 min. This time must be compared to the saturation time ($\tau_S = \tau_L + \tau_{NL}$) of the instability. Depending on the physical parameters of the surge, the relevant Alfven time ranges between 0.2 and 2 s and, therefore, τ_S falls in the range of observed values.

(2) We have found that different regimes of the instability can occur, with different levels of deceleration of the jet, of electrons acceleration and of turbulence. Consequently the surge behaves differently depending on which is the relevant regime.

(3) In all regimes electrons are accelerated by the electric field produced in the reconnection process, and this fact can explain the association between surges and different types of emission.

(4) There is the possibility to explain the deceleration of the surge since, when $\gamma \tau_A$ is sufficiently high ($\simeq 10^{-2}$) an important percentage of the kinetic energy of the surge is transformed into heat and turbulence (in particular in the example II the deceleration occurs on a time-scale (τ_{NL}) of 3 min if the Alfvén time is estimated of the order of 1 s).

(5) Since the electrons are accelerated mainly during the linear phase we can understand why in the observations by Chiuderi-Drago *et al.* (1986) radio emission occurs always before the deceleration phase of the dark feature.

(6) Garczynska *et al.* (1982) have claimed for a correlation between chains of type I bursts and surges; in such case it is perhaps possible to correlate the repetivity of the bursts to the oscillation frequency of the instability, whereas the frequency drift of the chains (De Groot *et al.*, 1976) can be related to the fact that, on a time-scale of the growth of the instability (γ^{-1}), the electrons are more ad more efficiently accelerated (Mangeney and Veltri, 1976). In particular in the example I, taking $\tau_A = 0.2$ s, the frequency drift $f/(\partial f/\partial t) \simeq 5$ min and the repetivity time is $\simeq 3$ s, while in the example II, taking $\tau_A = 1$ s, we have, respectively, 1 min and 3 s, in good agreement with the observations.

Which kind of phenomena is associated to the surge and the efficiency of its deceleration depends on: velocity profile, velocity and magnetic field scale length, local value of the magnetic field when the surge is propagating.

It is very difficult to have direct informations on these quantities since the relevant scales are very small, of the order of 10–50 km. In order to achieve a better understanding of the phenomenology associated to a surge, it would be necessary to establish deeper links between the theory and observable quantities and, therefore, to obtain new observational data and more defined theoretical results. The present work indicates that:

(a) From an observational point of view it would be interesting to study:

(i) the correlation between deceleration time of the surges and the turbulence level;

(ii) the correlation between frequency drift times and deceleration times when chaines of Type I bursts are present;

(iii) the correlation between repetivity and drift times in a chain.

(b) From a theoretical point of view it would be important:

(i) to improve the model taking into account the density gradients and the anisotropy in the turbulent cascade;

(ii) to better understand the electron acceleration mechanism.

Acknowledgements

It is a pleasure to thank F. Chiuderi-Drago, N. Mein, M. Pick, and A. Mangeney for fruitful discussions. One of us (G.E.) would like to thank the DASOP Observatoire de Paris-Meudon for the hospitality offered to him when this work was started.

This work was supported in part by Ministero della Pubblica Istruzione on National Projects Foundation.

References

Braginskii, S. I.: 165, *Rev. Plasma Phys.* **1**, 205.

Carbone, V. and Veltri, P.: 1987, *Astron. Astrophys.* (in press).

Chiuderi-Drago, F., Mein, N., and Pick, M.: 1986, *Solar Phys.* **103**, 235.

De Groot, T., Loonen, J., and Slottje, C.: 1976, *Solar Phys.* **48**, 321.

Einaudi, G. and Rubini, F.: 1986, *Phys. Fluids* **29**, 2563.

Garczynska, I. N., Rompolt, B., Benz, A. O., Slottje, C., Tlamicha, A., and Zanelli, C.: 1982, *Solar Phys.* **77**, 277.
Mangeney, A. and Veltri, P.: 1976, *Astron. Astrophys.* **47**, 181.
Mein, N. and Avignon, Y.: 1985, *Solar Phys.* **95**, 331.
Mok, Y.: 1987, *Astron. Astrophys.* **172**, 327.
Verma, V. K.: 1985, *Solar Phys.* **97**, 381.

TYPE II SOLAR RADIO EMISSION – A SELF-CONSISTENT APPROACH*

G. THEJAPPA

Indian Institute of Astrophysics, Bangalore 560034, India

Abstract. An attempt is made to construct a self-consistent model for type II radio bursts. It is proposed that a majority of the type II shocks are super-critical and the reflected ions from such type II shock fronts are described by the drifted Maxwellian in the upstream and by the Dory–Guest–Harris distribution in the downstream. The low-frequency waves excited by these ions accelerate electrons resonantly along the field lines both in the upstream as well as in the downstream, which are responsible for the lower-frequency and upper-frequency bands in the dynamic spectrum of a type II radio burst. The functional behaviour of the distribution functions of the accelerated electrons is the same in both the cases whereas the number densities of the accelerated electrons in the downstream is smaller than that in the upstream.

1. Introduction

Type II bursts, which are characterized by the slow drift in their spectral features from high to low frequencies are generally accompanied by a large flare. On the plasma hypothesis the frequency drift is found to correspond to a velocity of the order of 10^3 km s^{-1} and the moving source had been identified with a flare-associated collisionless MHD shock wave (Pikel'ner and Gintzburg, 1963). Later, direct evidence for the generation of type II radiation by shock waves was given by the space observations of interplanetary type II bursts and interplanetary shock waves (Malitson *et al.*, 1973; Cane *et al.*, 1982). The estimates of type II shock speeds are subject to considerable error because of the uncertainty in the coronal density models. Velocities derived from heliograph observations of source positions are even more uncertain because of the imperfectly known effects of coronal scattering and refraction and of ionospheric refraction (see Nelson and Melrose, 1985).

In this paper, we propose, contrary to the usually accepted picture that the majority of the type II shocks are supercritical since the estimated Mach numbers of the most of the type II shocks exceed the recently revised critical Mach number M_c by Edmiston and Kennel (1984), which lies between 1 and 2 for typical solar wind parameters. This fact is also supported by the studies of the forward interplanetary shocks observed by ISEE-3 (Bavassano-Cattaneo *et al.*, 1986); where it has been shown that a majority of the interplanetary shocks are supercritical. We also propose that many of the shocks responsible for type II radiation have an overshoot, foot and ramp in the magnetic field structure in consistence with their supercritical character. The reflected ions which behave like a beam in the foot and ramp of the shock front, and like a ring in velocity

* Proceedings of the Second CESRA Workshop on Particle Acceleration and Trapping in Solar Flares, held at Aubigny-sur-Nère (France), 23–26 June, 1986.

Solar Physics **111** (1987) 45–51.

space just behind the overshoot (Leroy *et al.*, 1981, 1982) accelerate electrons to ultra-relativistic energies both in the upstream as well as in the downstream through resonantly excited low-frequency waves.

Regarding the shock propagation angle relative to the magnetic field, Dulk *et al.* (1971) suggested that propagation is likely to be more parallel than perpendicular. These authors, however, could not arrive at any firm conclusions. Later Stewart and Magun (1980) analyzed a case where a transverse shock seemed to be required in order to explain the herringbone structure of a type II burst. Therefore, the observational evidence is not consistent with either strictly perpendicular or parallel shock propagation for all type II events. The range of shock angles over which type II bursts can be produced seems to be quite large. The advantage of the proposed acceleration mechanism over the fast Fermi process or the shock drift acceleration (Wu, 1984; Leroy and Mangeney, 1984; Holman and Pesses, 1983) is that it is applicable for a wide range of shock angles.

It is well known that in some cases the fundamental and harmonic bands of type II bursts are split in two. The present model is consistent with the qualitative model suggested by Smerd *et al.* (1975), who attributed the split-band structure of type II bursts to simultaneous plasma-frequency emission from plasma ahead of and behind a type II shock front. Nelson and Robinson (1975) reported that at a given frequency the U and L sources were circular, essentially the same size and in the same position where U and L represent the upper frequency band and lower frequency band, respectively. At the fixed observing frequency the L band is observed earlier in time than the U band and so it is inferred that at a given time the L source is further from the Sun than the U source confirming the model of Smerd *et al.* (1975) for band splitting.

2. Electron Acceleration

Uchida (1974) proposed that the type II radiation comes both at fundamental and harmonic from the low-V_A regions, where the shock has a relatively high Mach number. Gary *et al.* (1984) reported that the Mach number of type II shock lies in the range 1.3 to 3. Kennel *et al.* (1982) first reported that among 10 interplanetary shock, the fast wave Mach number M_F ranged from 1.3 to 4.7, where the critical Mach number $M_F \approx 2.5$. However, Bavassano-Cattaneo *et al.* (1986) reported that among 34 forward interplanetary shock waves observed by ISEE-3 during 1978 and 1979, 19 were supercritical, seven had a Mach number close to the critical Mach number, four were subcritical and the remaining four shocks were ambiguous. However, the average critical Mach number $\langle M_c \rangle$ was taken as 1.5 due to its strong dependence on shock parameters.

Kennel *et al.* (1982) as well as Bavassano-Cattaneo *et al.* (1986) have detected large-amplitude low-frequency electrostatic noise, whistler turbulence and a high-frequency ($\gtrsim f_p$) continuum near each shock and for up to several hours downstream. In the cases observed by Kennel *et al.* (1982) no type II bursts were observed at 1 AU, although intense impulsive Langmuir waves were observed an hour upstream from one shock. Impulsive Langmuir waves were present for a few minutes on either side for other shocks.

It is well known that the structure of a shock wave propagating perpendicular to the ambient magnetic field in a collisionless plasma undergoes a distinct change of shape when the Alfvén-mach number is increased above a critical value M_c. In addition to the ramp, the spacecraft observations (Russel and Greenstadt, 1979) show that there is a precursor structure of length equal to a few c/ω_{pi} which appears closely associated with the presence of ions reflected off the ramp, and is usually called the 'foot' where c is the velocity of light and ω_{pi} is the ion plasma frequency. Also it is seen (Russel and Greenstadt, 1979) that in the immediate post-ramp region the magnetic field exceeds its downstream value (magnetic field overshoot) and develops further downstream a somewhat oscillatory behaviour with a scale length of an ion gyroradius. Leroy *et al.* (1981, 1982) from numerical computations showed that the reflected ions behave essentially like a beam in the shock front (in the foot and ramp region), whereas they tend to form a gyrating stream in the downstream region behind the overshoot. Krasnosels'kikh *et al.* (1985) approximated the distribution function of the reflected ions in the foot and ramp as:

$$f_b^u = n_b \left(\frac{1}{2\pi \Delta V_b}\right)^{3/2} \exp\left(-\frac{(\mathbf{v} - \mathbf{v}_b)^2}{2\Delta V_b^2}\right), \tag{1}$$

where n_b is the reflected ion beam density, V_b and ΔV_b are the beam velocity and velocity spread, respectively. More realistic representation of the reflected ion distribution function in the downstream, which includes thermal effects, is the Dory–Guest–Harris distribution (Dory *et al.*, 1965):

$$f_b^q(V_\perp, V_\parallel) = \frac{1}{\pi^{3/2}(N+1)!}\,\frac{1}{\Delta V_{b_\parallel} \Delta V_{b_\perp}^2}\left(\frac{V_\perp}{\Delta V_b \perp}\right)^{2N} \times$$
$$\times \exp -\left(\frac{V_\parallel^2}{\Delta V_{b_\parallel}^2} + \frac{V_\perp^2}{\Delta V_{b_\perp}^2}\right), \tag{2}$$

where N is the anisotropy index and $\Delta V_{b_\perp}$ ($\Delta V_{b_\parallel}$) is the thermal speed $\perp$ ($\parallel$) to the ambient magnetic field (Thejappa, 1986; Akimoto *et al.*, 1985). These ion beams whose distribution functions are given by (1) and (2) in upstream and downstream, respectively, excite low-frequency waves described by the following dispersion relation (Galeev, 1984a; Krasnosels'kikh *et al.*, 1985):

$$\omega^2 = \frac{\omega_{LH}^2}{(1 + \omega_{pe}^2/k^2c^2)}\left[1 + \frac{m_i}{m_e}\,\frac{k_\parallel^2}{k^2}\,\frac{1}{(1 + \omega_{pi}^2/k^2c^2)}\right], \tag{3}$$

where $\omega_{LH} \approx \Omega_e \Omega_i$ is the lower hybrid frequency; Ω_j is the cyclotron frequency of jth kind particles; ω_{pe}, electron plasma frequency; m_j, mass of the jth plasma species; $K_\parallel$, the component of the wave vector parallel to magnetic field. The above dispersion relation is derived under the conditions $\Omega_e \gg \omega \gg \Omega_i$, $\omega_{pe} \gg \Omega_e$, and $K_\parallel \ll K_\perp$. Krasnosels'kikh *et al.* (1985) estimated the growth rate of these waves in the upstream as

$$\gamma_b^u = \frac{\pi}{2}\,\omega\,\frac{m_e}{m_i}\,\frac{\Omega_e^2}{(K\Delta V_b)^2}\,\frac{n_b/n_0}{(1+\omega_{pe}^2/k^2c^2)}\,. \tag{4}$$

The growth rate of the same waves due to ring ions in the downstream is given by:

$$\frac{\gamma_b^d}{\omega} = -\frac{2}{(N+1)!}\,\frac{n_b}{n_0}\,\frac{m_e}{m_i}\,\frac{\Omega_e^2}{\omega^2}\,\frac{t^3\,e^{-t^2}}{(1+\omega_{pe}^2/k^2c^2)}\times$$
$$\times\int_0^\infty \mathrm{d}x\,e^{-x^2}(t^2+x^2)\left(1-\frac{N}{t^2+x^2}\right), \tag{5}$$

where $t = \omega/K_\perp V_\perp$ (see Barbosa *et al.*, 1985; Thejappa, 1986). The growth rate due to the ion beam at $N = 1$ is given by:

$$\left(\frac{\gamma_b^d}{\omega}\right)_{\max} = \frac{\pi}{2}\,\frac{n_b}{n_0}\,\frac{m_e}{m_i}\,\frac{\Omega_\alpha^2}{(K\Delta V_b)^2}\,\alpha\,. \tag{6}$$

Here the maximum growth occurs when $t = t_0 \approx 0.52$ (Barbosa *et al.*, 1985) and $\alpha = 2/\sqrt{\pi}$ to $e^{-t_0^2}(0.5 - t_0^2) \approx 0.095$.

Papadopoulos (1981) proposed that these waves can accelerate electrons in the ambient plasma to very high energies due to the resonance $\omega = K_\parallel V_\parallel$, since the phase velocities along the magnetic field is much greater than that of perpendicular. Vaisberg *et al.* (1983) and Galeev (1984a) later successfully applied such mechanism for Earth's bow shock and galactic jets, respectively. Krasnosels'kikh *et al.* (1985) and Thejappa (1986) could explain many features of type II radiation adopting the above mechanism. If we assume that the energy taken by the waves from the reflected ions is completely given to the electrons by accelerating them, the solution of the wave balance equation gives us the distribution function of the accelerated electrons, i.e., by solving

$$\gamma_b^{u,d} + \gamma_e = 0\,, \tag{7}$$

where γ_e is the damping of waves due to electron acceleration and is given by:

$$\frac{\gamma_e}{\omega} = \frac{\pi}{2}\,\frac{\Omega_e^2}{k^2}\left(1+\frac{\omega_{pe}^2}{k^2c^2}\right)^{-3}\frac{1}{n_0}\,\frac{\partial F}{\partial V_\parallel}\bigg|_{V_\parallel=\omega/K_\parallel}\,. \tag{8}$$

By assuming $k^2c^2 \gg \omega_{pe}^2$, i.e., short wave limit, we get for the distribution functions in the upstream F_u as well as in the downstream F_d, after solving Equation (7):

$$F_{u(d)} = n_{s(d)}(V_h - |V_\parallel|)\,, \quad V_\parallel \le V_h\,, \tag{9}$$

where

$$n_{s_u} = n_b \frac{m_e}{m_i} \quad \text{and} \quad n_{s_d} = n_{s_u} \alpha .$$

It is clear that $\alpha \ll 1$ for the values of $N > 1$. One should notice that the functional behaviour of the distribution functions remains the same both in the upstream and in the downstream. By substituting these distribution functions again into the quasilinear equation the spectral density of the excited waves can be easily derived and a general analysis shows that the distribution of the energy spectrum of the excited waves in the k-space is of streamer type (see Breizman *et al.*, 1973). The value of V_h calculated by conserving the energy fluxes of reflected ions and electrons (Galeev, 1984b),

$$V_h^2 = \left[\frac{(m_i/m_e)^3}{10 \tan \theta}\right]^{2/5} V_b^2 , \tag{10}$$

where θ is the angle between the magnetic filed and the shock front plane, is used in Equation (9) above.

3. Discussion

The frequency splitting, which is one of the main characteristics of the type II bursts is due to the character of the reflected ions, which excite low frequency waves both in the upstream as well as in the downstream. Due to anisotropy in the phase velocities the electrons are accelerated by these waves to very high energies along the field lines. A similar acceleration mechanism was applied by Lampe and Papadopoulos (1977) for type II bursts; these authors considered the same low-frequency waves excited by the modified two-stream instability excited in the shock front, which can accelerate electrons to very large energies in the downstream. The detailed observations of the electron and ion distribution functions in the downstream of the Earth's bow shock as well as interplanetary shocks will be a crucial test to the present model. The energy density of the Langmuir waves excited by these electron beams can be approximately written as

$$W_{L_{u(d)}} \approx 0.1 n_{s_{u(d)}} m_e \frac{V_h^2}{2} \approx 10^{-4} n_0 T_e .$$

If one assumes that all the energy is converted into transverse waves and the source size is $\approx 1\, R_\odot$, the brightness temperature can be estimated which lies in the range 10^9 to 10^{11} K. Nelson and Robinson (1975) reported that $T_B^{(L)}/T_B^{(u)} \approx 2.3$. The difference in the number density of the electron beams in the upstream and downstream will account for this difference in the brightness temperature in the two bands. If the anisotropy index N is large, the value of α is decreased which will in turn decrease the electron number density in the beam. This beam density will not be sufficient to excite observable radiation leading to disappearance of the bandsplitting.

4. Conclusions

(1) The majority of the shocks responsible for type II radiation are supercritical, and they are characterized by the ion reflection.

(2) The reflected ions behave like a beam in the foot and the ramp whereas they behave like a ring in the downstream. They resonantly excite low frequency waves whose frequency is near the lower-hybrid frequency both in the upstream as well as in the downstream.

(3) The electrons are accelerated by these low-frequency waves to ultra-relativistic energies due to the anisotropy in the phase velocities.

(4) The present model fully agrees with the suggestion of Smerd *et al.* (1975) regarding the generation of the *L* and *U* bands in the upstream and downstream respectively.

(5) The brightness temperature in the *L* and *U* bands depends on the number density in the accelerated beams. Since the number density of the electron beam in the downstream is less than that of upstream, the *U* band is fainter than *L* band as experimentally observed (Nelson and Robinson, 1975).

Acknowledgement

I am grateful to the organizers of the Aubigny Workshop for providing me the financial assistance to attend and encouraging me to prepare this article. I also gratefully acknowledge the valuable comments by the referee, which helped me to improve the paper. I thank Prof. Ch. V. Sastry for constant encouragement.

References

Akimoto, A., Papadopoulos, K., and Winske, D.: 1985, *J. Plasma Phys.* **34**, 445.

Barbosa, D. D., Coroniti, F. V., Kurth, W. S., and Scarf, F. L.: 1985, *Astrophys. J.* **289**, 392.

Bavassano-Cattaneo, M. B., Tsurutani, B. T., Smith, E. J., and Lin, R. P.: 1986, *J. Geophys. Res.* **91**, 11929.

Breizman, B. N., Musher, S. L., and Zhakarov, V. E.: 1973, *Zh. Experim. Teor. Fiz.* **64**, 1297.

Cane, H. V., Stone, R. G., Fainberg, G., Steinberg, J. L., and Hoang, S.: 1982, *Solar Phys.* **78**, 187.

Dory, R. A., Guest, G. Z., and Harris, E. G.: 1965, *Phys. Rev. Letters* **14**, 131.

Dulk, G. A., Altschuler, M. D., and Smerd, S. F.: 1971, *Astrophys. Letters* **8**, 235.

Edmiston, J. P. and Kennel, C. F.: 1984, *J. Plasma Phys.* **32**, 429.

Galeev, A. A.: 1984a, *Zh. Experim. Teor. Fiz.* **86**, 1655.

Galeev, A. A.: 1984b, *Adv. Space Res.* **4**, 255.

Gary, D. E., Dulk, G. A., House, L., Illing, R., Sawyer, C., Wagner, W. J., McLean, D. J., and Hildner, E.: 1984, *Astron. Astrophys.* **134**, 222.

Holman, G. D. and Pesses, M. E.: 1983, *Astrophys. J.* **267**, 837.

Kennel, C. F., Scarf, F. L., Coroniti, F. V., Smith, E. J., and Gurnett, D. A.: 1982, *J. Geophys. Res.* **87**, 17.

Krasnosels'kikh, V. V., Kruchina, E. N., Thejappa, G., and Volokitin, A. S.: 1985, *Astron. Astrophys.* **149**, 232.

Lampe, M. and Papadopoulos, K.: 1977, *Astrophys. J.* **212**, 886.

Leroy, M. M., Goodrich, C. C., Winske, D., Wu, C. S., and Papadopoulos, K.: 1981, *Geophys. Res. Letters* **8**, 1268.

Leroy, M. M., Goodrich, C. C., Winske, D., Wu, C. S., and Papadopoulos, K.: 1982, *J. Geophys. Res.* **87**, 5081.

Leroy, M. M. and Mangeney, A.: 1984, *Ann. Geophys.* **2**, 449.
Malitson, H. H., Fainberg, J., and Stone, R. G.: 1973, *Astrophys. Letters* **14**, 111.
Nelson, G. J. and Melrose, D. B.: 1985, in D. J. McLean and N. R. Labrum (eds.), *Solar Radio Physics*, Cambridge University Press, Cambridge, p. 333.
Nelson, G. J. and Robinson, R. D.: 1975, *Proc. Astron. Soc. Australia* **2**, 370.
Papadopoulos, K.: 1981, in T. D. Guyenne and G. Levy (eds.), *Plasma Astrophys.*, ESA Sp. 161, p. 313.
Pikel'ner, S. B. and Ginstzburg, M. A.: 1963, *Astron. Zh.* **40**, 842.
Russel, C. T. and Greenstadt, E. W.: 1979, *Space Sci. Rev.* **23**, 3.
Smerd, S. F., Sheridan, K. V. and Stewart, R. T.: 1975, *Astrophys. Letters* **16**, 23.
Stewart, R. T. and Magun, A.: 1980, *Proc. Astron. Soc. Australia* **4**, 53.
Thejappa, G.: 1986, *Kodaikanal Obs. Bull.* **6**, 97.
Uchida, Y.: 1974, *Solar Phys.* **39**, 431.
Vaisberg, O. L., Galeev, A. A., Zastenker, G. N., Klimov, S. I., Norzdrachev, M. N., Sagdeev, R. Z., Sokolov, A., and Shapiro, A.: 1983, *Zh. Experim. Teor. Fiz.* **85**, 1232.
Wu, C. S.: 1984, *J. Geophys. Res.* **89**, 8847.

A TYPE IV BURST ASSOCIATED WITH A CORONAL STREAMER DISRUPTION EVENT*

M. R. KUNDU

Astronomy Program, University of Maryland, MD 20742, U.S.A.

Abstract. A type IV burst was observed on February 17, 1985 with the Clark Lake Radio Observatory multifrequency radioheliograph operating in the frequency range 20–125 MHz. This burst was associated with a coronal streamer disruption event. From two-dimensional images produced at 50 MHz, we show evidence of a type II burst and a slow moving type IV burst. The observations of the moving type IV burst suggests that a plasmoid containing energetic electrons can result from the disruption of a coronal streamer.

1. Introduction

Type IV bursts are observed over the entire wavelength range, centimeter to meter-decameter wavelengths (Boischot, 1957; Kundu, 1961; Pick, 1961). They can be both moving and stationary (e.g., Kundu, 1965), and are usually associated with flares. The association of type II bursts with disappearing filaments has been reported; however, no type IV burst has ever been reported in association with filament activity alone. Similarly, there has been no report of a type II or type IV burst activity in association with a coronal streamer disruption. Indeed, type IV bursts are not known to be entirely of coronal origin. They are usually associated with and preceded by microwave type IV which occurs in the upper chromosphere or lower corona. In this paper we describe a type IV burst at decameter wavelengths which is associated with only a coronal streamer disruption event.

The observations were made with the Clark Lake Radio Observatory multi-frequency radioheliograph in the frequency range 20–125 MHz. The instrument is in the form of a 'Tee' 3.0 km (E–W) and 1.8 km (N–S), and is composed of 720 conical spiral antennas. Its use as a radioheliograph has been described by Kundu *et al.* (1983).

The radioheliograph produces 64 × 64 pixel images (of 0.5 HPBW × 0.5 HPBW per pixel) within the frequency range 20–125 MHz. The field of view and the angular resolution of the telescope are both frequency-dependent. The field of view is approximately $2\overset{\circ}{.}3 \times 1\overset{\circ}{.}9$ at 80 MHz, when observing at the zenith. It scales inversely with frequency (in both dimensions), and is larger because of foreshortening when observing away from the zenith. The angular resolution ranges from 2.7 arc min at 125 MHz to 17 arc min at 20 MHz. The telescope is electronically steered for pointing in different sky directions, and is continuously tunable across the entire frequency range. In practice, one is restricted to observe within the interference-free bands. A number of such bands are available, those most commonly used are centered on 25.6, 30.9, 38.5, 57.5, 73.8,

* Proceedings of the Second CESRA Workshop on Particle Acceleration and Trapping in Solar Flares, held at Aubigny-sur-Nère (France), 23–26 June, 1986.

and 110.6 MHz. Several instantaneous bandwidths are available to the observer, ranging from 0.15 to 3.0 MHz. The sensitivity of the system is about 1 Jy (10^{-4} solar flux units) at all frequencies. The images obtained at present are calibrated by observing sidereal sources along with the Sun in close succession.

2. Results

At 20 : 06 UT, 17 February, 1985, the GOES satellite recorded a very small X-ray flare (B1.2). Since no Hα flaring was reported (which is not surprising, in view of the weakness of the X-rays) in association with this event, the location of the X-ray source is unknown. However, it should be noted that two days later, an active region appeared on the east limb, so it is not possible to rule out the possibility that a flare or sub-flare occurred behind the limb. The Sagamore Hill Observatory reported two type II bursts associated with this event, the first starting at 20 : 07 : 18 UT and lasting until

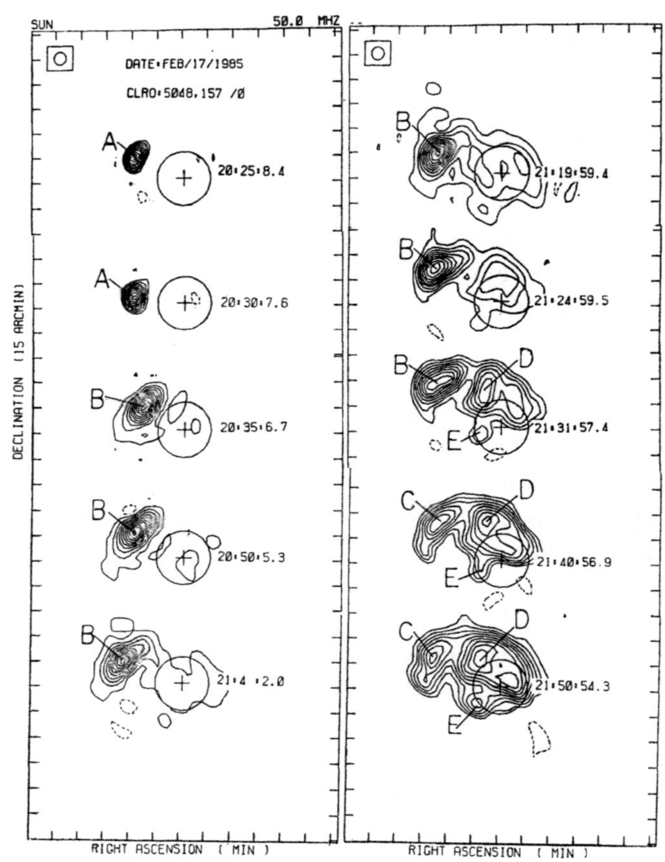

Fig. 1. Snapshot (0.6 s) maps of the Sun at 50 MHz, showing the evolution of the type II–type IV event associated with a coronal streamer disruption event.

20 : 15 : 06 UT and the second one starting at 20 : 24 and ending at 20 : 35 : 36 UT (*Solar Geophysical Data*, 1985).

The CLRO observations at 50 MHz started at 20 : 225 UT and at this time, a source (type II) was seen 16′ arc above the NE limb. This source is denoted by ‘A’ in Figure 1. From 20 : 25 to 20 : 33 UT the brightness temperature of this source declined from about 2×10^9 to 3×10^8 K (Figure 2). At 20 : 35, motion appeared in another source close below it, as shown in the height-time diagram of Figure 3. In this and the other figures, the moving source is denoted by the letter ‘B’. Both Figures 1 and 3 show that the source ‘B’ started at a height 2–3′ arc below the source ‘A’.

As the moving source rose (at a rate of some 200 km s^{-1}), its brightness temperature decreased from a few times 10^7 K to several times 10^5 K (Figure 2). After about an hour of motion, at approximately 21 : 30 UT, the source finally stopped at a height of 31′ arc above the limb (‘C’ in Figures 1, 2, and 3). Its brightness temperature declined less rapidly after its motion halted.

During the outward progress of the moving source, two stationary sources became visible (‘D’ and ‘E’ in Figure 1). It is difficult to assign a start time to these sources because the moving source was so much brighter (near the upper limit of the dynamic range of the radioheliograph images). The source ‘D’ was seen as early as 21 : 16 UT, and the source ‘E’ may also have been present at that time. (We cannot rule out the possibility that these sources existed much earlier.) The stationary sources have locations relative to the moving source such that they could be the footpoints of a large loop with the moving source at its apex.

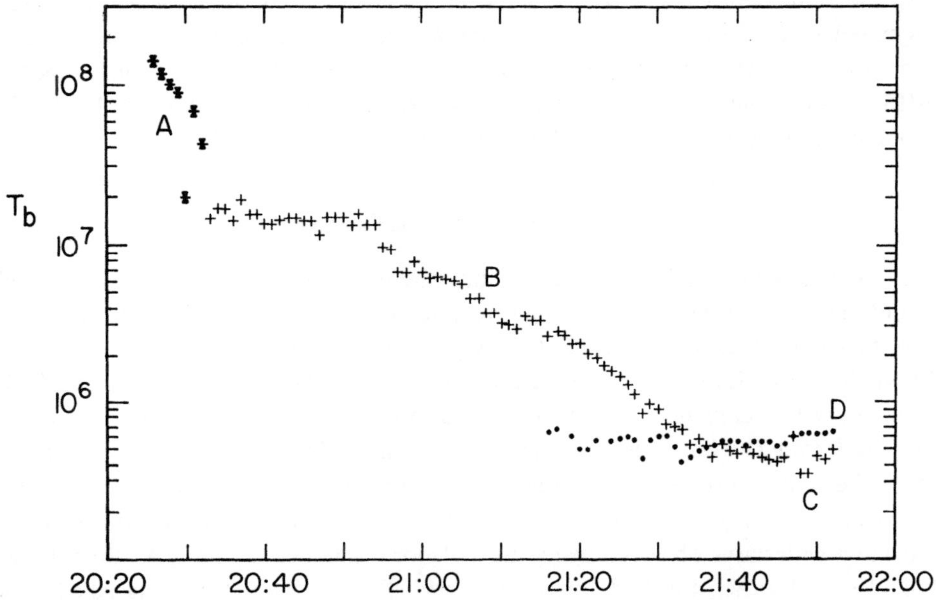

Fig. 2. Brightness temperature plotted as a function of time for the type II and various phases of the type IV event.

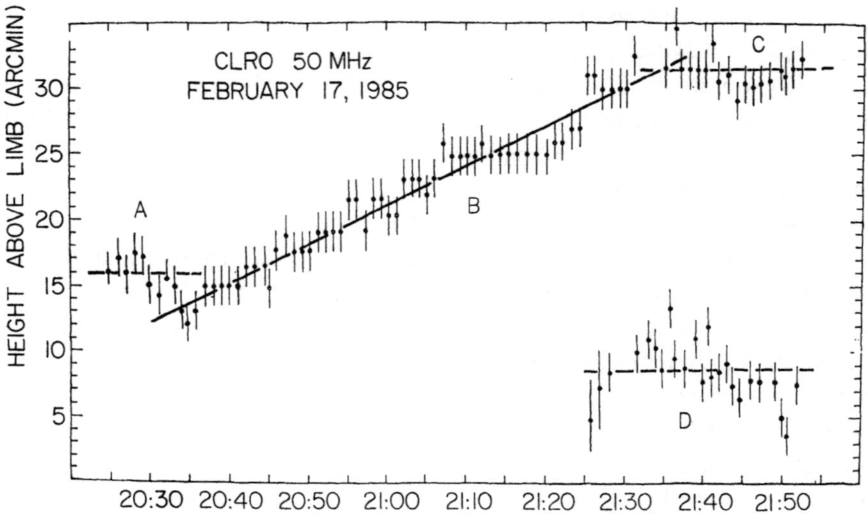

Fig. 3. Height-time plot of the 50 MHz burst sources in the type II–type IV event.

While the moving source rose into the corona, a filament disappeared (21 : 03 UT) on the disk at the location S 22 W 35 (D. Speich, private communication, SMM forecast center). However, the optical observers reported that the filament had been fading for the last day or two, so perhaps it is not wise to assign too much credence to the exact disappearance time.

As mentioned earlier, the type II–type IV burst observed by us at 50 MHz was associated with a coronal streamer disruption event. It was also associated with a coronal mass ejection (CME) event observed by the HAO C/P experiment aboard SMM-II. A full joint analysis of the Clark Lake radio data along with the HAO-SMM coronograph data will be published elsewhere (Kundu *et al.*, 1987).

3. Discussion and Conclusions

Examining the solar activity near the NE limb where the type II–type IV burst occurred we find that on Feburary 15, 1986 there were no active regions near E-limb, but just a small prominence above it. On February 19 there appeared a small active region (AR 4629) at N 02 E 77 with several flares reported. This region was 14° behind the limb on February 17. As mentioned earlier, it is not possible to exclude the possibility that our type II–type IV event could be associated with a chromospheric event in this region behind the limb. However, we note that the moving type IV flux in this event was more than an order of magnitude weaker (and the speed 2–5 times slower) than typical bursts associated with flares observed using the Culgoora radioheliograph (e.g., Smerd and Dulk, 1971). Further, the CME associated with the radio burst was classified as a streamer disruption, a type of event which is associated with filament eruptions without flares (Illing and Hundhausen, 1986).

A coronal streamer disruption results from a slow temporal evolution of the streamer over a period of 1 or 2 days prior to the disruption. The structural forms displayed by a helmet-like streamer during this slow evolution suggest that the magnetic field is distorted away from a potential configuration by currents associated with the addition of material to the helmet (see the description by Illing and Hundhausen, 1986 for the August 18, 1980 coronal streamer disruption event). It has been suggested that this 'pre-event' evolution leads the corona to a state which is ripe for disruption and the CME may be viewed as a result of this slow evolution rather than an abrupt or local event (Illing and Hundhausen, 1986). Our radio results provide the evidence that following this disruption, both a shock wave and a plasmoid containing energetic electrons can be formed in the corona without any Hα flare which are respectively responsible for the type II and type IV bursts observed with the Clark Lake radioheliograph. In all respects the type IV observed in this event behaves as an 'isolated' type IV usually associated with a visible flare. We estimate, following Ramaty and Lingenfelter (1968) that in a magnetic field of ~ 1 G, $\sim 10^{31}$–10^{32} electrons of energy ~ 300 kV were involved in producing gyrosynchrotron emission responsible for the type IV burst. This number is more than an order of magnitude smaller than inferred for flare-associated moving type IV bursts observed at 80 MHz using the Culgoora radioheliograph.

Acknowledgements

This research was supported by NASA contract NAG5–572 and NSF grant ATM 86–15304. I wish to thank Dr E. J. Schmahl for many useful discussions.

References

Boischot, A.: 1957, *Compt. Rend.* **244**, 1326.

Illing, R. J. M. and Hundhausen, A. J.: 1986, *J. Geophys. Res.* **91**, 10951.

Kundu, M. R.: 1961, *Astrophys. J.* **134**, 96.

Kundu, M. R.: 1965, *Solar Radio Astronomy*, John Wiley-Interscience, New York, 660 pp.

Kundu, M. R., Erickson, W. C., Gergely, T. E., Mahoney, M. J., and Turner, P. J.: 1983, *Solar Phys.* **83**, 385.

Kundu, M. R., Schmahl, E. J., and Hildner, E.: 1987, in preparation.

Pick, M., 1961, *Ann. Astrophys.* **24**, 183.

Ramaty, R. and Lingenfelter, R. E.: 1968, *Solar Phys.* **5**, 531.

Smerd, S. F. and Dulk, G. A.: 1971, in R. Howard (ed.), 'Solar Magnetic Fields', *IAU Symp.* **43**, 618.

QUASI-MONOCHROMATIC WAVE-PARTICLE INTERACTIONS IN MAGNETOSPHERIC PLASMAS*

(Invited Review)

D. LE QUÉAU and A. ROUX

CRPE/CNET/CNRS, Issy-lès-Moulineaux, France

Abstract. The natural waves observed within magnetospheres are often emitted over a very narrow spectral range. They thus can interact coherently with the charged particles forming the weakly collisional magnetospheric plasmas. In this review some aspects of these quasi-coherent Wave Particle Interactions, usually known as trapping processes, are underlined.

A particular emphasis is here put on the situation effectively encountered within natural plasmas. Hence, the effects of plasma inhomogeneities on the resonant WPI are thoroughly studied. We focus on three effects:

(i) The inhomogeneity of the medium changes the wave number and correspondingly the relative phase between the wave and the particle's motion. Thus, resonant conditions cannot be fulfilled over a large wave path, which tends to decrease the efficiency of the WPI. On the other hand, this effect leads a monochromatic wave to interact with a larger set of charged particles, which tends to restore a quasi linear-like diffusion process.

(ii) Even when trapping occurs, the inhomogeneity changes the saturation level of the kinetic instabilities.

(iii) Trapping can then be an efficient acceleration process.

1. Introduction

Beyond their dense, collision-dominated, internal boundaries (ionospheres, chromosphere, or inner solar corona), the outer ionized regions of magnetospheric systems are usually collision-free. Their dynamical characteristics are, therefore, mainly dominated by collective effects, which self-consistently link the various ionized species and the electromagnetic fields and play a leading role in determining the dynamics of the plasma: most of the thermalization, diffusion, energetization and acceleration phenomena can only be achieved via 'anomalous' processes involving wave-particle interactions.

The natural waves which are recorded within the planetary magnetospheres have often a relatively narrow spectral bandwidth: for instance, the chorus, which propagates in the whistler mode waves, at frequencies lying between f_{LH}, the lower hybrid frequency, and f_{ce}, the electron gyrofrequency (Helliwell, 1969). In the ULF range, pc-1 emissions, in the protonic gyrofrequency range, are also nearly monochromatic (Wehrlin *et al.*, 1973). Narrow-banded electrostatic plasma waves, mainly confined to the Earth's equatorial region are observed in the VLF range, close to f_{UH} (upper hybrid frequency), or to the odd harmonics of the electronic gyrofrequency (Ronmark *et al.*, 1978). Finally, the Earth's non-thermal radio emissions, either at kilometric wavelength (Auroral Kilometric Radiation or A.K.R. hereafter), or at myriametric wavelength (the so-called

* Proceedings of the Second CESRA Workshop on Particle Acceleration and Trapping in Solar Flares, held at Aubigny-sur-Nère (France), 23–26 June, 1986.

Solar Physics **111** (1987) 59–80.

'Continuum Radiation') exhibit fine spectral structures (Gurnett *et al.*, 1979, Etcheto *et al.*, 1982).

The plasma waves phenomena, which have been then observed within the Jovian and Chronian magnetospheres, exhibit remarkable similarities with the terrestrial emissions (Kurth, 1986; Genova, 1986). Chorus emissions, which also appear as narrow-banded risers, have been recorded in the inner Jovian magnetosphere, near the Io plasma torus. Electron cyclotron electrostatic emissions at $(n + \frac{1}{2}f_{ce}$, and at f_{UH} have been observed in the Io torus and in the outer Jovian magnetosphere, as well as near the Chronian plasma ring, in the vicinity of the equatorial plane. They are also present within the Titan's wake. Jovian and Chronian non-thermal 'Continua' are very reminiscent of the terrestrial myriametric emission. Finally, it is known for a long time that the bursty decametric emissions of Jupiter have an extremely complex dynamical spectrum, exhibiting very monochromatic emissions with a very large intensity. This is also the case of the Saturnian kilometric radiation which presents strong similarities with AKR.

In Section 2 one will study, in the case where the medium is homogeneous, the essential differences between the coherent WPI's due to quasi-monochromatic waves (trapping processes) and the stochastic WPI's linked to broad-banded emission (quasi-linear interaction). An emphasis will be put on the way such processes saturate linear kinetic instabilities. In Section 3 the role of the plasma inhomogeneities will be reviewed. The inhomogeneity broadens the resonance which can lead coherent WPI's to become stochastic. On the other hand, trapping processes can then lead to particle acceleration. Section 4 is devoted to conclusive remarks concerning all these processes.

2. WPI in Homogeneous Plasmas

Within the collisionless magnetospheric plasmas, the Vlasov equation must be used for describing the collective behavior of the plasma. Various instabilities that can give rise to small-scale turbulence are indeed produced by free energy features linked to the velocity space of the charged species forming the plasmas such as loss cone, 'gentle' bump-in-the-tail or thermal anisotropies. Resonant particles playing an important role, a fluid description is generally inadequate.

For the sake of simplicity, we will first restrict our analysis to the study of one-dimensional electrostatic plasma waves. It should be kept in mind that a similar analysis can be made for waves having a more complex polarisation (Roux and Solomon, 1970). Moreover, since we wish to focus on the differences between quasi-linear and trapping-type processes, we will finally also neglect the wave-wave interactions. Such effects have been included in Adam *et al.* (1980), where it is shown that they are, in fact, necessary to fully understand the quasi-linear theory. On the other hand, when a strong level of turbulence can develop throughout the plasma, the modulational instability (Thornhill and Ter Haar, 1978) can create collapsing self-sustained structures known as cavitons. Since the evidence of their existence within the magnetospheres are still controversial, we will not discuss these phenomena further here. These simplifications allow us to work with well-defined spatial Fourier component for each quantities studied in this paper.

2.1. Treatment of the Vlasov Equation

Let us consider a one-dimensional, homogeneous, one component plasma, with a motionless background of ions, embodied in a box of length L, in the direction Oz. The electrostatic plasma waves which one wants to study are characterized by an electric field $\mathbf{E}(z, t)$, directed along the same axis Oz. The Vlasov equation for the one--dimensional electronic distribution function $\mathbf{f}(v, z, t)$ then writes:

$$\partial \mathbf{f}/\partial t + v \cdot \partial \mathbf{f}/\partial z - (e\tilde{\mathbf{E}}/m) \cdot \partial \mathbf{f}/\partial v = 0 . \tag{1}$$

After performing a spatial Fourier transform ($k = 2n\pi/L$ with $n = (0, \pm 1, \ldots)$:

$$\partial \mathbf{f}_k/\partial t - ikv\mathbf{f}_k = (e/m)\left[\tilde{\mathbf{E}}_k \cdot \partial \mathbf{f}_0/\partial v + \sum_{k' \neq 0} \tilde{\mathbf{E}}_{k-k'} \cdot \partial \mathbf{f}_{k'}/\partial v\right], \tag{2a}$$

$$\partial \mathbf{f}_0/\partial t = (e/m) \sum_{k \neq 0} \tilde{\mathbf{E}}_k^* \cdot \partial \mathbf{f}_k/\partial v , \tag{2b}$$

$\tilde{\mathbf{E}}_k(t)$ and $\mathbf{f}_k(v, t)$ are the spatial Fourier components of the electric field and of the distribution function; $\mathbf{f}_0(v, t)$ is the homogeneous part of the distribution function ($k = 0$) and $\tilde{\mathbf{E}}_0(t) = 0$, due to the global neutrality inside the box. The field being a real vector: $\tilde{\mathbf{E}}_k = \tilde{\mathbf{E}}^*_{-k}$. Neglecting, in Equation (2a), the 2nd mode-coupling term, which means that the nonlinearity is sufficiently weak so that no energy transfer occurs between different scales, one finally gets:

$$\partial \mathbf{f}_0/\partial t = (e/m)^2\, \partial/\partial v \left\{ \int_0^t \mathrm{d}u \sum_{k \neq 0} \left[e^{i(\Omega_k - kv)u} \mathbf{E}_k^*(u) \mathbf{E}_k(t-u)\, \partial \mathbf{f}_0(t-u, v)/\partial v \right] \right\} . \tag{3}$$

Here $\mathbf{E}_k(t)$ is the slowly-varying amplitude of the Fourier mode:

$$\tilde{\mathbf{E}}_k(t) = \mathbf{E}_k(t) \exp(i\Omega_k t) , \tag{4}$$

where $\Omega_k = \Omega_{pe}[1 + 3/2(k\lambda_d)^2]$ is the solution of the electrostatic dispersion equation, λ_d being the Debye length.

The dominant contributions to the integrals in (3) come from the resonant particles, for which $\Omega_k - kv = 0$. Contributions from other electrons rapidly phase-mix to zero, due to the rapid oscillations of the phazors. Due to the summation over k in (3), the temporal width of the resonant function is given by:

$$\tau_{ac} = \{\Delta k[\Omega_k/k - \partial \Omega_k/\partial k]\}^{-1} , \tag{5}$$

where Δk is the spectral width of the electrostatic waves. This auto-correlation time corresponds to the time over which any given resonant electron, with a velocity $v_z = \Omega_k/k$, can feel the wave field as a quasi-monochromatic one.

Two limiting cases can be considered, according to the relative size of τ_{ac}, at one hand, and of the characteristic times of evolution of both $\mathbf{f}_0$ and $\mathbf{E}_k(t)$: τ_f, τ_E.

2.2. Quasi-linear regime

When $\tau_{ac} \ll \{\tau_f, \tau_E\}$, the slowly varying functions $\mathbf{E}_k^*$, $\mathbf{E}_k$ and $\partial \mathbf{f}_0/\partial v$ can be removed from the integral in (3). Then one gets (see Laval and Pellat, 1972) for a more detailed discussion):

$$\partial \mathbf{f}_0/\partial t = \partial/\partial v[\mathbf{D}(v, t)\, \partial \mathbf{f}_0/\partial v]\,, \tag{6a}$$

$$\mathbf{D}(v, t) = \pi(e/m)^2 \sum_{k \neq 0} \mathbf{E}_k^2(t)\delta(\Omega_k - kv) \approx \delta v^2/\tau_{ac}\,, \tag{6b}$$

where $\delta v^2 \approx (e\mathbf{E}\tau_{ac}/m)^2$ is the mean-squared velocity increment suffered by an electron during the short autocorrelation time. Equation (6a) is a diffusion equation which describes the electronic random walk in velocity space due to the stochastic interaction with the phase wandering electric fields.

An important application of the quasi-linear theory is the ability of the diffusive evolution of $\mathbf{f}_0$ to quench the instability which gives rise to the electrostatic waves. Let us briefly comment the usual example of the 'gentle' bump-in-the-tail instability (Figure 1(a)).

To close up our system of equations, one needs an equation for the evolution of the slowly varying electric field amplitude:

$$\partial \mathbf{E}_k^2(t)/\partial t = 2\gamma_k \mathbf{E}_k^2\,, \tag{7a}$$

$$\gamma_k(t) \approx (\Omega_k^3/k^2)\, \partial \mathbf{f}_0(v, t)/\partial v|_{v = \Omega_k/k}\,. \tag{7b}$$

In (7b), the instantaneous values of $\mathbf{f}_0$ are used. Over the phase velocity range Δv where $\partial \mathbf{f}_0/\partial v > 0$, the electrostatic waves grow up, which leads to an enhancement of the diffusion coefficient $\mathbf{D}(v, t)$ (Figure 1(b)). Due to the diffusion process, the gradients of f_0 are diminished over Δv (Figure 1(c)). Finally a plateau occurs, the instability is quenched, and the wave amplitude saturates. The saturation level can be determined by balancing the loss of kinetic energy suffered by the electronic population (both resonant and non-resonant particles then intervene) and the gain of electrostatic energy embodied in the wave bath. We will here restrict ourselves to an order of magnitude estimate.

From (6) and (7), one can get:

$$\gamma_k(0) = -(\Omega_k^3/k^2)\, \partial^2/\partial v^2\{(k^2/\Omega_k^3)\mathbf{D}(v = \Omega_k/k, \infty)\}\,, \tag{8}$$

which writes approximately, thus giving an estimation of the saturation level:

$$\mathbf{D}(\infty) \approx \gamma_k(0)\Delta v^2(\infty)\,, \tag{9}$$

where $\Delta v(\infty)$ is the final extension of the velocity plateau.

Thus the time for the diffusion process to smooth out the bump whose typical velocity width is $\Delta v(\infty)$ with an electric field having the saturation amplitude, which is also, roughly, the saturation time of the whole process ($\tau_{\text{sat}} \approx \Delta^2 v/\mathbf{D}$), must be of the order of the initial growth time of the instability ($\tau_{\text{growth}} \approx \gamma_L^{-1}$). It is worth noticing that the saturation time here scales like $\mathbf{E}^{-2}$, where $\mathbf{E}$ is the saturation level.

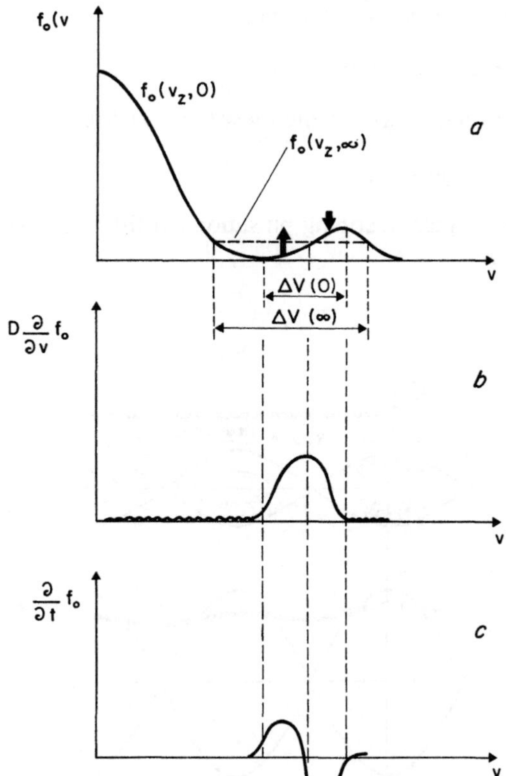

Fig. 1. Initial (full line) and final (dashed line) 1-D distribution function for the gentle bump-in-the-tail instability (Figure 1(a)). Unstable waves have initially phase velocities in the range $\Delta v(0)$. This leads to a finite diffusion coefficient on the same velocity range (Figure 1(b)), over which $\partial f/\partial v > 0$. The time derivative of the distribution function (Figure 1(c)) then tends to smooth the bump (arrows in 1(a)), until the instability saturates, due to the formation of a plateau over the velocity range $\Delta v(\infty)$.

2.3. Trapping regime

Conversely, when $\tau_{ac} \gg \{\tau_f, \tau_E\}$, resonant electrons essentially 'see' a unique Fourier component of the wave electric field. The relative phase between the field and the particle motion is now slowly varying, thus leading to a coherent WPI: the electrons oscillate back and forth within the potentials wells of the wave. The global energy exchange between the wave and the population of resonant charged particles (otherwise characterized by the Landau growth or damping rate) tends to be quenched, simply because the velocity gradients of the distribution function are rapidly smoothed by these particles oscillations.

τ_E is then likely to be larger than τ_f which is of the order of the period of the particles oscillations, which allows to assume that the electric field amplitude remains quasi constant.

Let us consider the equation for the one-dimensional motion of an individual electron within a monochromatic electrostatic wave:

$$m\ddot{z} = -e\mathbf{E}\sin(kz - \Omega t), \tag{10}$$

which gives, in the resonant wave frame ($y = k[z - (\Omega/k)t]$):

$$\dot{y}^2 - 2\Omega_T^2 \cos(y) = J; \tag{11}$$

$\Omega_T = (ke\mathbf{E}/m)^{1/2}$ is the typical trapping pulsation. In the phase space (y, $\dot{y}$), electrons follow lines of equal J.

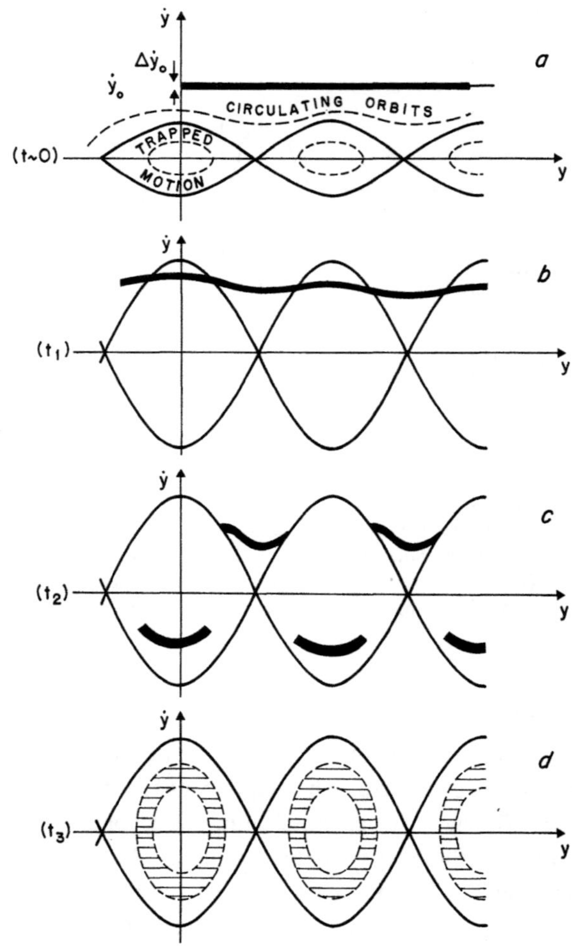

Fig. 2. A quasi-monokinetic electron beam (mean velocity $\dot{y}_0$, velocity dispersion $\Delta\dot{y}_0$, in the resonant frame) destabilizes a quasi monochromatic wave. This leads to the possibility of trapping motion in the electronic phase space (Figure 2(a)), here sketched in the resonant frame. When the wave grows further, the size of the trapping cells (delimited by the full-lined separatrix) increases until the destabilizing beam become trapped (at time t_1, Figure 2(b)). The electron of the beam then rotate in phase space (time $t_2 > t_1$, Figure 2(c)), until the distribution is smoothed such that equally number of electrons give energy to the wave and take energy from it, then leading to the saturation of the instability (Figure 2(d)).

For $J > 2\Omega_T^2$ the particle is freely circulating and its velocity oscillates around a mean value, when, for $J < 2\Omega_T^2$, it is trapped: the range of y over which it can move is limited to a fraction of a wavelength (Figure 2(a)). The period $T(J)$ of the phase space motion of a trapped electron depends on its energy J:

$$T(J) = 4\Omega_T^{-1}\mathbf{K}[(1 + J/2\Omega_T^2)], \tag{12}$$

where $\mathbf{K}[x]$ is the complete elliptic integral of the first kind. $T(J)$ has a logarithmic divergence for $J \approx 2\Omega_T^2$.

Even if the distribution function is initially well ordered (Figure 2(a–b)), after a few trapping periods, it will be ergodically smoothed over a velocity range corresponding to the width of the trapping cell $\Delta v \approx \Omega_T/k$.

In particular, when the bump-in-the-tail is not so 'gentle', e.g., if it is characterized by a narrow velocity range, the auto-correlation time of the growing wave is very large (Drummond *et al.*, 1970). Thus it can trap the driving electrons, when its amplitude reaches a sufficient level (Figure 2(c)). The electrons of the beam then begin to rotate in the phase space. Due to their different rotation periods within the trapping cells, the initial distribution function is smeared out, which leads to the smoothing of its velocity gradients and to the quenching of the instability (Figure 2(d)).

The nonlinear time over which the trapping begins to occur must be of the order of γ_L^{-1} because the linear exponential growth is such a steep function of time that not much saturation is expected to occur until the last e-folding. The maximal wave amplitude will be roughly given by the condition:

$$\Omega_T \approx \gamma_L\,, \tag{13}$$

which means, at least intuitively, that, when the trapping effect begins to saturate the wave, the electric field amplitude is sufficient to 'give' energy to the electrons at the same rate than those 'give' energy to the wave (Nielson, 1969). After that time, the amplitude begins to oscillate about this level, due to the periodic energy exchange between electrons and wave. Figure 2(c) thus corresponds to a maximum of the wave amplitude. These

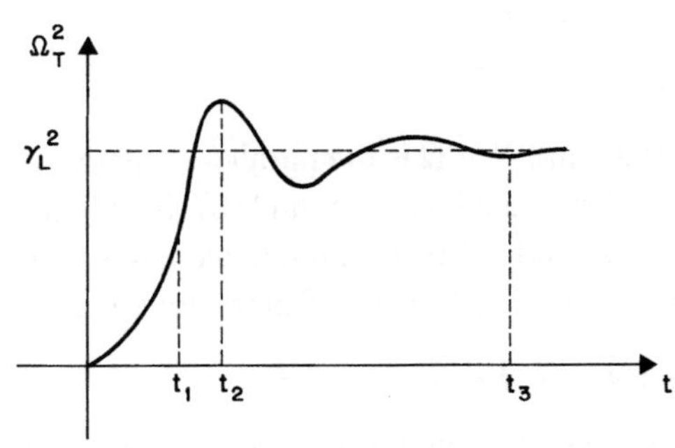

Fig. 3. Schematic view of the evolution of the unstable wave amplitude (or trapping pulsation), during the trapping type saturation process. Times t_1, t_2, and t_3 schematically correspond to those of Figure 2. Oscillations at the trapping frequency finally damp away, due to the mixing phenomena described in the text.

oscillations then further damp, due to the above-quoted mixing effect (O'Neil, 1968) (Figure 3). Condition (13) can be obtained in a more quantitative way by using energy conservation arguments (Galeev and Sagdeev, 1975; Al'Tshul and Karpman, 1966).

According to the preceding section, the condition for the quasi-linear saturation process to occur writes:

$$\tau_f \approx \tau_{\text{sat}} \approx (k\Delta v)^2 (m/keE)^2 \tau_{ac}^{-3} \Omega_T^{-4} \gg \tau_{ac}\,, \tag{14}$$

or, simply: $1 \gg (\Omega_T \tau_{ac})^4$, while the validity of the trapping theory is obviously linked to the converse inegality.

From (13), the trapping saturation time is $\tau_{\text{sat}} \approx O(\mathbf{E}^{-1/2})$ where $\mathbf{E}$ is the saturation amplitude. Thus, when trapping can occur, it will lead to a given saturation level much faster than quasi linear effects would have. Such a situation will occur, in natural plasmas, when the growth rate of any instability occurs over a narrow spectral range and when simultaneously the spontaneous noise, which provides initial conditions for the wave growth, is very weak. The competition between the different unstable modes then greatly endeavours the most unstable one and the autocorrelation time of the turbulence is large.

2.4. Resonant WPI's with electromagnetic waves

Resonant quasi-linear diffusion processes, as well as trapping type processes can also occur when charged particles interact with electromagnetic waves. To illustrate these phenomena we here restrict our calculations to the interaction of electrons with circularly polarized waves (right- or left-handed) of frequency Ω, propagating along the direction of the magnetic field $B_0 e_z$ (Palmadesso and Schmidt, 1971).

The resonant electrons must satisfy the Doppler-shifted cyclotron conditions:

$$\Gamma(\Omega - k_\parallel v_\parallel) = \pm \Omega_{ce} \tag{15a}$$

(+ for right-handed wave, and − for left-handed wave), which means that the electromagnetic fields of the wave must be stationary in the particle rest frame. Here $\Gamma = (1 - v^2/c^2)^{-1/2}$ is the Lorentz factor of the electronic motion.

When $\beta = v/c \ll 1$ (15a) rewrites:

$$\Delta\Omega - \beta_\parallel N_\parallel + (\beta_\parallel^2 + \beta_\perp^2)/2 = 0\,, \tag{15b}$$

where $\Delta\Omega = (\Omega - \Omega_{ce})/\Omega_{ce}$ and $N_\parallel c/\Omega$ is the parallel refractive index. It is clear that the inclusion of relativistic effects (the last term in (15b)) is only necessary when the wave dispersion allows both $\Delta\Omega$ and $N_\parallel$ to be simultaneously small.

In the opposite case, in which $N_\parallel \gg 1$ (infraluminous wave) (15a) simply writes:

$$v_\parallel = v_{\text{res}} = (\Omega \pm \Omega_{ce})/k_\parallel\,, \tag{15c}$$

which defines a resonant parallel velocity v_{res}: resonance curves in the $(v_\parallel, v_\perp)$-plane are straight lines parallel to the $v_\perp$-axis (Figure 4(b)).

By using a Lorentz transform it is simple to see that in a frame moving at the parallel phase velocity $v_{\phi\parallel} = \Omega/k_\parallel$ both the wave electric field and the wave frequency vanish.

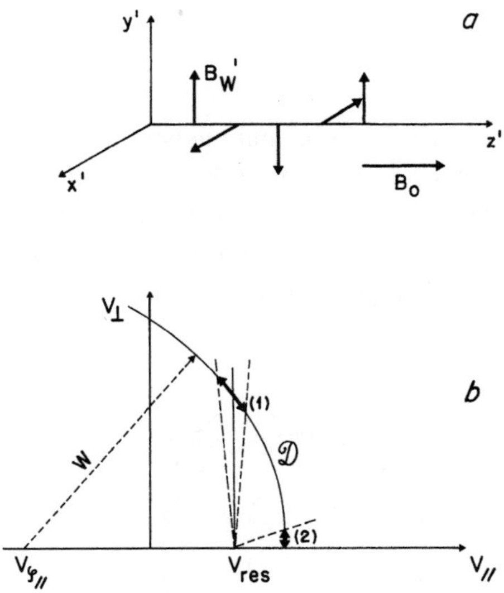

Fig. 4. Resonant interaction with an infraluminous RH wave. In a frame moving at the parallel phase velocity, the wave electric field vanishes, and the wave becomes a stationary, spatially varying magnetic field (Figure 4(a)). The energy W of the particle is thus conserved, in this frame, and in the velocity plane (Figure 4(b)) the electrons are constrained to follow the diffusion curve. The expression of their trapping pulsation depends on their initial pitch angle (region (1) or (2) over) (see text).

In this frame, the infraluminous wave is felt as a stationary, spatially varying, magnetic field B_w (Figure 4(a)), which implies the constancy of the electronic energy:

$$v_\perp^2 + (v_\parallel - \Omega/k_\parallel)^2 = W. \tag{16}$$

In the $(v_\parallel, v_\perp)$-plane, the interacting electrons are constrained to follow the 'diffusion curves' defined by Equation (16). This fact is also true in a statistical sense and, in particular, the gradients of the distribution function which lead to the linear damping or growth of the monochromatic wave must necessarily be taken along these curves.

As in the electrostatic case, another constant of the electronic motion can be found:

$$\dot{y}^2 - 2k_\parallel v_\perp \Omega_c \sin(y + \Phi) = J, \tag{17}$$

where $y = k_\parallel(z - v_{res}t)$, $\Omega_t = eB_w/m$ is the gyrofrequency linked to the wave magnetic field and $\Phi = tg^{-1}(v_y/v_x)$. From Equations (16) and (17) and the equation of motion along z, the equation for $y(t)$ can be solved. This leads to the possibility of axial, bounded-y, trapped motions.

Two trapping regimes are available (Roux and Solomon, 1970) (Figure 4(b)):

(i) For large initial pitch angle ($\dot{y}_0 \ll k_\parallel v_{\perp 0}$), the perpendicular velocity remains quasi-constant, and the trapping frequency is approximately given by:

$$\Omega_T \approx (k_\parallel v_{\perp 0} \Omega_c)^{1/2}. \tag{18a}$$

(ii) For small pitch angle ($\dot{y}_0 = k_\parallel v_{\perp 0}$), $v_\perp$ is altered and the trapping frequency becomes:

$$\Omega_T \approx 2(\Omega_{ce}\,\Omega_c^2)^{1/3}\,. \tag{18b}$$

Conversely, when the wave is supraluminous and has a frequency very close to the gyrofrequency, the resonance curves can no longer be approximated by straight lines: they become circles, when $\beta^2 \ll 1$, or ellipses, when β is not very small.

In the $(\beta_\parallel, \beta_\perp)$ plane the radius of the resonance circle is $(N_\parallel^2 - 2\Delta\Omega)^{1/2}$ while it is centered at $\beta_\parallel = N_\parallel, \beta_\perp = 0$ (Figure 5(a)).

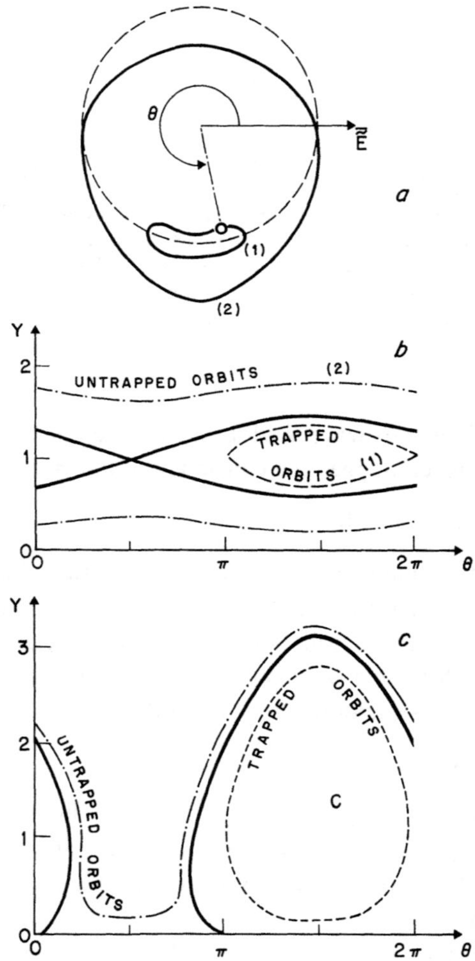

Fig. 5. Resonant interaction of electrons with a supraluminous RH wave. In the frame moving at velocity $N_\parallel c$, along the magnetic field line, and rotating at the wave pulsation, the wave becomes a stationary electric field (Figure 5(a)). In this frame, the electronic motion, which follows, at zero order, a Larmor circle, can be trapped (1) or circulating (2). Depending of the size of the parameter r (see text), the trapping cells in phase space (Θ, Y) can have two different aspects, and the trapping pulsations then behave differently (Figure 5(b) and 5(c)) (see text).

For such a wave, a right handed X-mode for instance, there is no reference frame in which the wave electric field vanishes. Conversely, in the frame which moves at the velocity $N_\parallel c$ along the B-field, both the wave magnetic field and the parallel wave vector vanish. The wave becomes a spatially homogeneous rotating electric field (Figure 5(b)). As a consequence the relativistic parallel momentum of the electron is conserved:

$$\Gamma(1 - \beta_\parallel/N_\parallel) = \pi_\parallel = Cte\,. \tag{19}$$

The diffusion curves are now ellipses tangent to the light circle which degenerate into straight lines $v_\parallel = Cte$, for weakly relativistic particles. In the above quoted frame, the resonance condition (15a) becomes simply: $W/mc^2 = \Omega_{ce}/\Omega_w$, where Ω_w is the transformed wave frequency, and W the relativistic electron energy.

Therefore, an X-mode wave, propagating in a low density plasma $[(\Omega_{pe}/\Omega_{ce})^2 \ll 1]$, can be destabilized at frequencies close to the X-mode cut-off frequency $(\Omega_x \approx \Omega_{ce}[1 + (\Omega_{pe}/\Omega_{ce})^2])$, by an electronic population presenting an inversion of population in perpendicular energy $(\partial \mathbf{f}/\partial v_\perp > 0)$, over some range of $v_\parallel$. The emitted wave must have $k_\parallel$ very small, but still sufficient to allow the unstable frequency band to be doppler shifted above Ω_x. One thus can understand why the observational results (Benson and Calvert, 1980), that demonstrated that Auroral Kilometric Radiation (AKR) is generated, on the X-mode, inside cold plasma depleted region $(\Omega_{pe}/\Omega_{ce} < 0.2)$ have triggered a series of theoretical papers (Wu and Lee, 1979; Dusenbery and Lyons, 1982; Omidi and Gurnett, 1982; Le Quéau *et al.*, 1984), on the cyclotron maser instability, which is directly linked to the above quoted resonant process.

The study of the coherent electronic motion in the above mentioned frame leads to another constant of the motion (Le Quéau *et al.*, 1984), which writes, in the weakly relativistic limit:

$$(Y - 1)^2 + rY^{1/2}\sin(\theta) = C\,, \tag{20}$$

where $Y = \pi_\perp/\pi_{\perp 0}$, $\pi_{\perp 0} = [\pi_\parallel^2 - 2\Delta\Omega]^{1/2}$ is the resonant perpendicular momentum and $\theta = \Omega_w t - \Phi$ is the relative phase between the rotating electric field vector and the electronic zero-order gyromotion (Figure 5(a)). Here $r = (8e\mathbf{E}_w/mc\Omega_w)\pi_{\perp 0}^{-3}$ measures the wave amplitude.

The motion of the electron can be drawn in the phase space (θ, Y) (Figure 5(b–c)), and corresponds to an azimuthal bunching motion. Again, two trapping regimes are possible, depending of the size of r.

(i) For $r \ll 1$ (large value of $\pi_{\perp 0}$, Figure 5(b))

$$\Omega_T \approx r^{1/2}\pi_{\perp 0}^2\Omega_w\,. \tag{21a}$$

(ii) For $r \geq 1$ Figure 5(c))

$$\Omega_T \approx \Omega_w^{1/3}(eE/mc)^{2/3}\,. \tag{21b}$$

This trapping effect has been invoked for determining the maximal intensity which could be reached by the natural planetary radio emissions (Le Quéau *et al.*, 1984). This is linked to the fact that the above quoted maser cyclotron instability likely produces a very narrow banded X-mode wave, which is conforted by the observations. In the AKR case, the B-field inhomogeneity plays an important role in the determination of the emitted spectral intensity. The effect due to such plasma inhomogeneities will be emphasized in the following sections.

When the waves exhibit a wide band spectrum, the above conservation laws are only locally valid. In the case of infraluminous wave, for example, (16) leads to the differential relation:

$$v_\perp \, dv_\perp + (v_\| - \Omega/k_\| \, dv_\| = 0 . \tag{16b}$$

In other words, the change of the parallel velocity which is locally associated to the diffusion process leads a resonant particle to interact successively with different Fourier components of the wave. Then the diffusion curves are obtained by solving (16b) together with the resonance condition (15c) and the wave dispersion equation. By simply plotting diffusion curves and resonance curves, one can also easily determine whether a given distribution function is stable or unstable (Gendrin, 1983).

The quasi-linear WPI has been studied, for instance in (Kennel and Petschek, 1966). The diffusing waves were electromagnetic whistlers, when the characteristic free energy which drives the waves was the anisotropy of the electronic distribution function. The consequences of this WPI upon the maximum flux of the charged particles populations trapped within the geomagnetic field are discussed in Kennel and Petschek (1966) and Gendrin (1968). These problems have been thoroughly reexamined in Etcheto *et al.* (1973), where the spectrum of the whistler mode at the equilibrium is calculated self-consistently. A general formulation of quasi-linear equations for any kind of resonant wave can be found in Kennel and Engelmann (1966).

3. Resonant WPI in Inhomogeneous Plasma

When the plasma is inhomogeneous it is no more possible to consider even a quasi-monochromatic wave to be simply characterized by a unique spatial Fourier component: any propagating wave encounters various plasma conditions which modify its wavelength. Furthermore, the charged particle motion is also modified by the various forces resulting, for instance, from the change of the magnetic field intensity. Both effects result in a detuning of the wave particle resonance and, thus, to a natural limitation of the time over which the resonant WPI can take place.

3.1. Space-time limtations and broadening of the resonance in inhomogeneous media

Let us consider a monochromatic wave, with frequency Ω, which propagates along the principal B-field of a weakly inhomogeneous magnetized plasma. The relative phase between the wave fields and the position of a given charged particle then writes:

$$\Psi(z) = \int^{z} k_\|(z') \, [1 - v_{\rm res}(z')/v_\|(z')] \, dz' , \tag{23}$$

where z is the space coordinate along $\boldsymbol{B}$-field, and v_{res} the resonant velocity ($\Omega/k_{\parallel}(z)$ for an electrostatic wave, and $(\Omega - \Omega_{ce}(z))/k_{\parallel}(z)$, for an infraluminous RH polarized wave). A WKB approach is here used for describing the wave propagation, and $k_{\parallel}(z)$ is the local value of the parallel wave number.

Wave and particle will be resonant at the point z_r where the relative phase is stationary:

$$\partial\Psi/\partial z|_{z=zr} = k_{\parallel}(z_r)\,[1 - v_{\text{res}}(z_r)/v_{\parallel}(z_r)] = 0\,. \tag{24}$$

Due to the plasma inhomogeneities the resonant interaction only occurs over the coordinate range Δz over which Ψ changes by a significant amount, say ≈ 1 radian. Developing Ψ about the resonance point $z = z_r$ this writes:

$$\begin{aligned}\Delta z = L_i &\approx [(1/2)\,\partial^2\Psi/\partial z^2]^{-1/2}\\ &= \{[k_{\parallel}(z_r)/2v_{\text{res}}]\,[\partial(v_{\parallel} - v_{\text{res}})/\partial z]\}^{-1/2}\,.\end{aligned} \tag{25}$$

The detuning of the resonance is linked to the variation of the resonant velocity and to the variation of the parallel velocity of the particle, which is due to the various forces linked to the plasma inhomogeneities. On the other hand, at a given place, corresponding to a given value of the resonant velocity, the wave can resonate with a larger set of particles, having a finite velocity-range:

$$\Delta v_{\parallel} \approx L_i\,\partial(v_{\parallel} - v_{\text{res}})/\partial z = 2v_{\text{res}}/(k_{\parallel}L_i)\,. \tag{26}$$

The resonance function which leads to the determination of the linear convective growth or damping rate is no more an infinitely thin function, as in the homogeneous case. It has a width $\Delta v_{\parallel}$, and the convective rate is no more given by the usual Landau's expression (Eidman, 1966). It has recently been shown that such effects play an important role for determining the linear growth rate of the instability responsible for the Auroral Kilometric Radiation (Le Quéau *et al.*, 1985): the $\boldsymbol{B}$-field inhomogeneity then naturally limits the altitude range over which the amplification can take place, which allows to determine directly the AKR spectral intensity (Zarka *et al.*, 1986).

The finite interaction length, L_i, provides a natural autocorrelation time for the WPI:

$$\tau_{ac}^{inh} \approx L_i/v_{\text{res}}\,, \tag{27}$$

which corresponds to the time over which the particle can coherently interact with the wave. As in the homogeneous situation this autocorrelation time has to be compared with the trapping time $\tau_T \approx \Omega_T^{-1}$. If the wave amplitude is sufficiently weak ($\tau_{ac}^{inh} \ll \Omega_T^{-1}$), the coherent trapping process has no time (more exactly no place...) to occur during the WPI, and the motion of the particle under the wave action can be taken as an incoherent random motion. Then a quasi-linear like process can occur, despite the fact that the wave is a quasi-monochromatic one. An example of such a phenomenon will be studied in Section 3.2. Conversely, if $\tau_{ac}^{inh} > \Omega_T^{-1}$ the charged particle has time to suffer a coherent interaction and trapping phenomena eventually occur. This will be emphasized in Section 3.3.

3.2. An example of quasi-linear like diffusion process due to a quasi-monochromatic wave

We here study the interaction between protons bouncing within the geomagnetic field and quasi monochromatic ion cyclotron waves which propagate more or less along the field lines. The wave frequency belongs to the ULF range (pc1 pulsations), and the amplitude of these ICW is insufficient to trap the protons ($\Omega_T^{-1} \ll \tau_{ac}^{inh}$).

Near the geomagnetic equator, the static magnetic field intensity varies quadratically ($\partial B/\partial z|_{z=0} = 0$). Thus the relative phase between the monochromatic ICW and a proton which resonates with it in the equatorial vicinity can be developed as:

$$\Psi(z) = \Psi(z_r) + (z - z_r)q - (p/6)(z^3 - z_r^3), \tag{28}$$

where

$$q = \partial\Psi/\partial z|_{z=0} = k_{\parallel 0}(1 - v_{\mathrm{res}|0}/v_{\parallel 0}),$$

and

$$\begin{aligned} p &= -\partial^3 \Psi/\partial z^3|_{z=0} \\ &= (k_{\parallel 0}/2\Omega_{\mathrm{H}^+})[3/1 - \Omega/\Omega_{\mathrm{H}^+}) + (v_{\perp 0}/v_{\parallel 0})^2]\partial^2 \Omega_{\mathrm{H}^+}/\partial z^2|_{z=0}, \end{aligned}$$

z_r is the point at which resonance occurs, the subscripts 0 standing for quantities taken at the equator. For determining p, the dispersion equation for the ICW has been taken into account.

At each time the protons reach the point of coordinate $\pm z_r$, they can resonnate with the ICW. As they have a trapped bouncing motion along the field lines, such an interaction can occur many times. But there is no reason for conserving a fixed phase relation between the two waves with which a given proton successively interacts. Thus after many elementary resonant interactions, the proton motion can be seen as a random walk in the velocity space, which can be statistically treated as a quasi-linear diffusion process. From (28) one can evaluate a pitch angle diffusion coefficient $D_{\alpha\alpha} \approx \langle \Delta\alpha^2 \rangle / 2\Delta t$, where $\Delta\alpha$ is the characteristic pitch angle variation due to a 'collision' between the ICW having a B-field $\mathbf{b}$ and the proton,

$$\Delta\alpha \approx \Delta v_{\parallel}/v_{\parallel} \approx (e/mv_{\parallel}) \int^{z} \mathbf{b}\cos(\Psi)\,\mathrm{d}z',$$

and Δt the proton bounce time. The proton distribution function $\mathbf{f}_0$ then follows a Fokker–Planck equation (29):

$$\begin{aligned} &\partial\mathbf{f}_0/\partial t = (\sin\alpha)^{-1}\,\partial/\partial\alpha\{\sin(\alpha) D_{\alpha\alpha}\,\partial\mathbf{f}_0/\partial\alpha\}), \\ &D_{\alpha\alpha} = [(2^{2/3}\pi^2\mathbf{b}^2)/(v_{\parallel}^2 p^{2/3})]\,\{\mathbf{Ai}[-2^{1/3}k_{\parallel 0}q/p^{1/3}]\}^2, \end{aligned} \tag{29}$$

where $\mathbf{Ai}[x]$ is the Airy function.

This result, here obtained for protons and ICW (Welti *et al.*, 1973), is very similar to what can be found for electrons and whistlers waves (Ashour-Abdalla, 1972).

The diffusion coefficient is large only in the near proximity of the equatorial plane. The inhomogeneity is indeed very weak there, thus allowing a quite large interaction length $L_i \approx (\partial^3 \Psi/\partial z^3)^{1/3}$.

In such a weakly inhomogeneous plasma, the resonant WPI between a finite amplitude monochromatic wave and resonant protons, both trapped on a given flux tube, can be described by a quasi-linear equation. This remains true as long as the wave amplitude does not exceed the threshold at which the wave begins to trap the protons:

$$\tau_T \approx 2\pi/\Omega_T \gg L_i/v_{\text{res}} \,. \tag{30}$$

Comparing with (14) one can say that the inhomogeneity provides a 'natural' spectral width $\Delta k \approx L_i^{-1}$, even for a monochromatic wave. Of course, for a wideband spectrum characterized by $\Delta k > L_i^{-1}$, the above discussion becomes useless since the standard quasi-linear theory works, as in an homogeneous medium.

From the expression of the diffusion coefficient, it can also be said that the effect of the inhomogeneity upon the resonance results in a resonance broadening, since now a finite range of parallel velocities is concerned in the resonance with a monochromatic wave.

3.3. Electrostatic trapping in an inhomogeneous magnetized plasma

For the sake of simplicity we shall again restrict our detailed analysis to the case of an electrostatic wave, polarized along the geomagnetic field $B_0(z)e_z$. If the plasma inhomogeneities are sufficiently weak, one can use a WKB description for the wave propagation, and determine the parallel electronic motion by the equation:

$$\ddot{z} = -(e\mathbf{E}(z)/m)\sin\left[\Omega t - \int^{z} k_{\parallel}(z')\,\mathrm{d}z'\right] - (\mu/m)\partial B_0/\partial z \,, \tag{31}$$

$k_{\parallel}(z)$ is the local value of the parallel wave number, and $\mathbf{E}(z)$ a slowly varying amplitude. The acceleration due to the variation of the B-field intensity has been also included, μ being the electronic magnetic moment. Let Z_r be the instantaneous position of the local resonance frame. It is defined by the equation:

$$\dot{Z}_r = \Omega/k_{\parallel}(Z_r) \,. \tag{32}$$

Let $y = k_{\parallel}(Z_r)(z - Z_r)$ be the normalized position of the electrons, measured in this frame. To the first order in $y \ll k_{\parallel} Z_r$, the equation for y writes:

$$\ddot{y} = k_{\parallel}\Gamma_{\text{inh}} + [k_{\parallel} eE(Z_r)/m]\sin(y) \,, \tag{33}$$

where $\Gamma_{\text{inh}} = (\Omega^2/k^3)\,\partial k_{\parallel}/\partial z|_{z=Z_r} - (\mu/m)\partial B_0/\partial z|_{z=Z_r}$ is an acceleration linked to both the non-Galilean character of the resonant frame (firt term) and the adiabatic acceleration due to the convergence of the field lines (second term). Integrated once, (33) provides a local constant of the motion very similar to (11):

$$\dot{y}^2 - V(Z_r, y) = J(Z_r) \tag{34a}$$

with

$$V(Z_r, y) = 2k_\parallel \Gamma_{\text{inh}}[\,y - q\cos(y)]\,, \tag{34b}$$

$q = \Omega_T^2/k_\parallel \Gamma_{\text{inh}}$, where $\Omega_T = (k_\parallel eE/m)^{1/2}$ is the local value of the trapping pulsation. For $q > 1$ a trapped motion is possible, when for $q < 1\,V(Z_r, y)$ becomes monotonic which prevents any trapping to occur (Figure 6(a–b)).

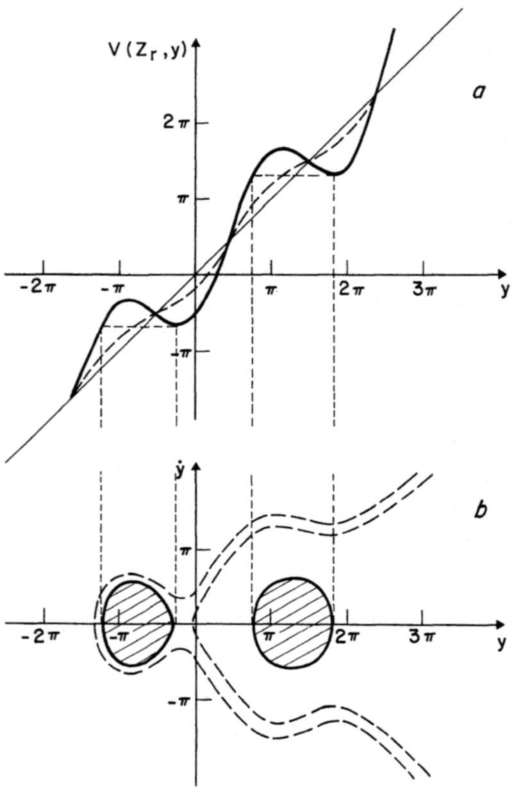

Fig. 6. A sketch of the effective potential locally felt by an electron influenced by an electrostatic wave propagating along the magnetic field of an inhomogeneous magnetospheric plasma (Figure 6(a)). For $q < 1$ (see text), the potential is monotonic, and the trapping is impossible (dashed line $q = 0.5$). For $q > 1$ trapping cells do exist, in the phase space $(y, \dot{y})$, as shown in Figure 6(b)), which corresponds to $q = 2$ (full line in Figure 6(a)), and hatched areas in 6(b)).

As Z_r varies along the wave path, the above quoted equation is equivalent to that of a pendulum with varying parameters. From (34(a–b)) an adiabatic invariant can be found (Laval and Pellat, 1970; Asseo *et al.*, 1972):

$$\Lambda(Z_r) = \int [J(Z_r) - V(Z_r, y)]^{1/2}\,\mathrm{d}y\,, \tag{35}$$

where the integration is to be done over a trapping period. This leads to conclude that, as long as Z_r varies adiabatically $[(Z_r)^{-1}\,\partial Z_r/\partial t \ll \Omega_T]$, only a few particles are detrapped. Conversely, a rapid detrapping will occur whenever $q \approx 1$, or:

$$\Omega_T^2 \approx k_\parallel \Gamma_{\text{inh}} . \tag{36}$$

3.4. Modification of the Trapping Type Saturation Process, in Inhomogeneous Plasmas

When the inhomogeneity forces are strong ($q < 1$) the trapping is quenched and the saturation of instability due to the trapping does not take place: the saturation level given by (15) can be exceeded. Roughly speaking, then, the wave can grow up to the level at which the trapping force just balances the inhomogeneity forces and the saturation level must simply be given by (36).

This effect has been widely studied in Asseo *et al.* (1972) and Roux (1974). We here briefly present the main conclusions of these works, in the simple situation where both the linear growth rate γ_L and the inhomogeneity parameter $k_\parallel \Gamma_{\text{inh}}$ are supposed to be constant all over the wave path. Let us consider a convectively growing quasi-monochromatic wave having an initial amplitude such that $\Omega_T^2 \ll \inf\{k_\parallel \Gamma_{\text{inh}}, \gamma_L^2\}$.

If $k_\parallel \Gamma_{\text{inh}} > \gamma_L^2$ the wave amplitude, and, correspondingly, the trapping frequency,

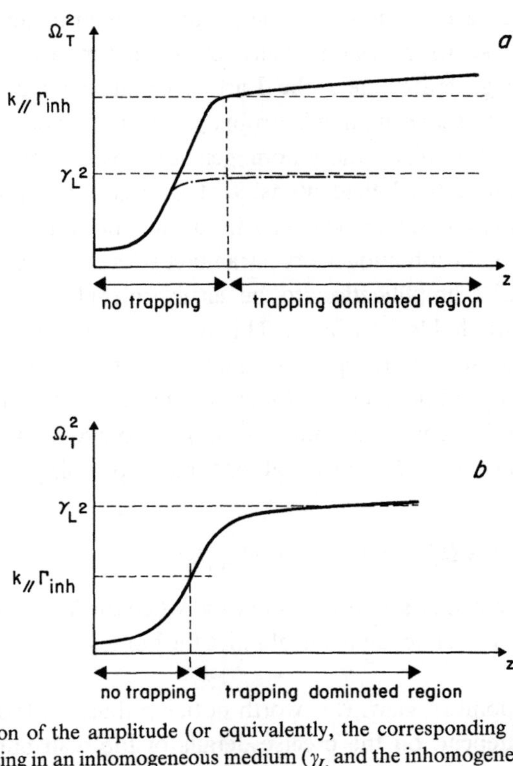

Fig. 7. Spatial evolution of the amplitude (or equivalently, the corresponding trapping pulsation) of an unstable wave propagating in an inhomogeneous medium (γ_L and the inhomogeneity forces being constant). (a) $k_\parallel \Gamma_{\text{inh}} > \gamma_L^2$. (b) $k_\parallel \Gamma_{\text{inh}} < \gamma_L^2$.

grows exponentially, with a convective spatial growth rate $\approx \gamma_L/v_g$, where v_g is the wave group velocity. When the wave amplitude reaches the theoretical saturation level (13), the inhomogeneity force still overcomes the trapping force, the trapping remains quenched and the exponential growth further proceeds until the wave electric fields reaches the critical amplitude given by (36). The trapping then begins to work and reduces the wave growth. After a few trapping periods, two regimes successively occur:

(i) At short distances, when the variation of the resonant velocity due to the inhomogeneity is small as compared to the velocity trapping extension $\Delta v \approx \Omega_T/k$ the wave grows linearly with the distance.

(ii) At large distances, when the converse becomes true, the wave grows like the squared root of the distance.

These nonlinear growth or damping remaining very weak, the level effectively reached by the wave is roughly given by (36).

Conversely, when $k_\parallel \Gamma_{\text{inh}} < \gamma_L^2$ (Figure 7(b)), trapping begins at a time, and a position at which the trapping frequency exceeds $(k_\parallel \Gamma_{\text{inh}})^{1/2}$. Then, the inhomogeneity forces being 'small', the saturation proceeds like in the homogeneous situation and the wave reaches the level determined by (13): when the threshold is reached for the trapping motion to begin, a sufficient amount of free energy still remains in the amplifying medium to allow the further growth of the wave.

In natural media such as the Earth's magnetosphere the inhomogeneity parameter $k_\parallel \Gamma_{\text{inh}}$ is often an increasing function of the distance, taken along the wave path (e.g., instability occurs close to a region where B_0 is minimum). This is the case of ion-cyclotron waves generated near the Earth's equatorial plane, by a cyclotronic instability, due to the presence of an anisotropic proton population (Roux, 1974). In the immediate vicinity of this plane the inhomogeneity is weak, due to the symmetrical variations of, for instance, the B-field intensity, at each sides of it (34). Then, at equator, trapping tends always to dominate, whereas, far from it, no trapping can occur, because of the rapid increase of the inhomogeneity parameter with the geomagnetic latitude. One thus expects the ICW amplification to be easily quenched at the equator, while exponential growth still holds far from it. This quasi-saturation will only be efficient if particles suffer at least one trapping oscillation, before being detrapped by the latitudinally increasing inhomogeneity force. In this case, this leads to the further condition that the ICW amplitude must be large enough for the trapping force to dominates the inhomogeneity force over at least half a trapping length on each side of the equatorial plane:

$$k_\parallel \Gamma_{\text{inh}}(z_T) \approx \Omega_T^2 \quad \text{with} \quad z_T \approx \pi V_{\text{res}}/\Omega_T \,. \tag{37}$$

The above-quoted considerations are only valid when the instability under consideration is a convective one. For absolute instabilities the homogeneous saturation condition (13) still holds.

From a practical point of view, it is worth noticing that, contrary to the estimation (13), (36) does not depend on the precise details of the distribution function of the particles which generate the wave: both $k_\parallel \Gamma_{\text{inh}}$ and Ω_T indeed only depend of the wave dispersion and of gross features of the plasma, which are, by far, easier to determine. Thus, (36) generally gives a very useful estimate for the maximal amplitude of the wave, which has the further advantage of being less model-dependent than (13).

3.4. Particle acceleration by monochromatic waves

As in the preceding section, let us consider a finite amplitude monochromatic wave propagating through an inhomogeneous plasma. Let us further suppose, for the sake of simplicity, that its amplitude remains constant all over its ray path, but that the inhomogeneity parameter $k_{\|}\Gamma_{\mathrm{inh}}$ increases along the ray path. Then, particles which are trapped within the wave potential throughs in the regions where the trapping is possible, will suddenly be detrapped when the inhomogeneity parameter reaches the square of the wave trapping frequency Ω_T^2 (Figure 8(a)).

As long as they remain trapped, the charged particles have a mean velocity which is equal, at each instant, to the wave-particle resonant frequency v_{res}. Yet this resonant velocity can experience a gradual increase. Thus, when they are finally detrapped, these

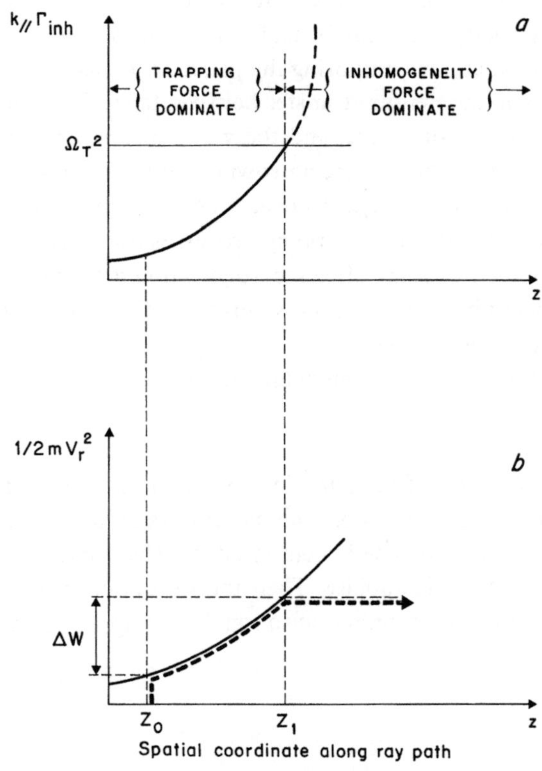

Fig. 8. Acceleration of charged particles trapped by a wave which propagates in an inhomogeneous plasma (its amplitude being considered as constant, and the inhomogeneity force increasing along the wave path). Particles are trapped at z_0. For $z_0 < z < z_1$, they remain trapped (Figure 8(a)), and their mean energy increases, as the resonant velocity v_r (dashed curve in Figure 8(b)). For $z > 2_1$, they are detrapped, and their energy remains constant.

particles have gained an energy

$$\Delta W \approx (m/2)\,[v_{\mathrm{res}}^2(z_1) - v_{\mathrm{res}}^2(z_0)]\,, \tag{38}$$

where z_0 is the point at which they have been initially trapped, and z_1 the point at which they are finally detrapped (Figure 8(b)).

This very simple process provides a very efficient way for increasing the particle energy within the inhomogeneous magnetospheric plasmas. Such a process has already been invoked by Roux *et al.* (1984) for explaining an observed acceleration of electrons to energies $\approx 20, 30$ eV in the Earth's magnetosphere and by Le Quéau and Roux (1987) to explain electronic acceleration up to tens of keV in solar coronal loops.

4. Conclusions

The outer part of the various magnetospheres is formed by weakly collisional plasmas in which wave-particle interactions play the role which are otherwise assigned to the collisional dissipation mechanisms.

Even when the electromagnetic level is relatively low, resonant interactions can lead to very efficient energy and momentum transfer between the electromagnetic field and the various ionized populations composing the plasmas. *In fine*, these phenomena also could lead to energy and momentum transfer between these different populations, one of the species giving rise to the wave, and the wave energetizing another species via resonant interactions. Depending on the narrowness of the wave spectrum, two type of phenomena are known to occur: quasi-linear stochastic diffusion in velocity space, when the wave is broad-banded and trapping processes for narrow-banded emissions. As quasi-monochromatic waves are often recorded within magnetospheric plasmas, the role played by trapping phenomena in the saturation of the kinetic micro-instabilities has been emphasized in this review.

The inhomogeneity of the plasma medium can be considered as low when

$$k_\parallel \Gamma_{\mathrm{inh}} \ll \gamma_L^2\,,$$

where Γ_{inh} is the acceleration of the particles, as seen in the resonant frame, which is linked to the gradients of the various plasma parameters influencing the wave propagation and the particle motion (magnetic field intensity, electronic density...), and γ_L is the linear growth rate of the instability. Then the saturation proceeds, such that the saturation level (linked to the trapping pulsation Ω_T) is approximately determined by the relation

$$\Omega_T \approx \gamma_L\,.$$

Conversely, when

$$\gamma_L^2 \ll k_\parallel \Gamma_{\mathrm{inh}}\,,$$

the inhomogeneity of the medium is strong enough to quench the trapping motion (and, correspondingly, the saturation process) until the wave level reaches an higher value, determined by the condition

$$\Omega_T^2 \approx k_{\|} \Gamma_{\text{inh}} .$$

The above quoted formulas are very useful, because they allow to determine very simply the maximal levels which could be nonlinearly reached by the quasi-monochromatic wave, by using only the gross parameters of the plasmas.

When the wave level is not strong enough for allowing trapping motions, the wave-particle interaction can behave in a quasi-linear way, even if the wave is a quasi-monochromatic one. This appears when both the wave and the particles are confined for a while within a localized magnetospheric trap and can interact many times together. The inhomogeneity of the medium then provides the necessary finite interaction time which links to a broadening of the resonance, and to quasi-linear diffusion process.

Conversely, when the wave level is strong, trapping processes in nonhomogeneous media can lead to an efficient acceleration mechanism for charged particles. The mean velocity of the trapped particles indeed closely follows the resonant frame velocity which can increase as the wave propagates throughout the magnetospheric medium.

References

Adam, J. C., Laval, G., and Pesme, D.: 1980, *J. Physique Suppl.* **41**, C3–383.
Al'tshul', L. M., and Karpman, V. I.: 1966, *Soviet Phys. JETP* **22**, 361.
Ashour-Abdalla, M.: 1972, *Planetary Space Sci.* **20**, 639.
Asseo, E., Laval, G., Pellat, R., Welti, R., and Roux, A.: 1971, *J. Plasma Phys.* **8**, 341.
Benson, R. F. and Calvert, W.: 1980, *Geophys. Res. Letters* **6**, 479.
Drummond, W. E., Malmberg, J. H., O'Neil, T. M., and Thompson, J. R.: 1970, *Phys. Fluids* **13**, 2422.
Dusenbery, P. B. and Lyons, L. R.: 1982, *J. Geophys. Res.* **87**, 7467.
Eidman, V. Ya.: 1966, *Soviet Phys. JETP* **22**, 370.
Etcheto, J., Gendrin, R., Solomon, J., and Roux, A.: 1973, *J. Geophys. Res.* **78**, 8150.
Etcheto, J., Christiansen, P. J., Gough, M. P., and Trottignon, J. G.: 1982, *Geophys. Res. Letters* **9**, 1239.
Galeev, A. A. Sagdeev, R. Z.: 19**00**, in M. A. Leontovich (ed.), *Reviews of Plasma Physics*, Vol. 7, Consultants Bureau, New York.
Gendrin, R.: 1968, *J. of Atm. Terr. Phys.* **30**, 1313.
Gendrin, R.: 1983, *Space Sci. Rev.* **34**, 271.
Genova, F.: 1986, in D. Le Quéau, B. M. Pedersen, A. Roux, and A. Saint Marc (eds.), *Comparative Study of Magnetospheric System*, CEPADUES-éditions, Toulouse.
Gurnett, D. A., Anderson, R. R., Scarf, F. L., Fredricks, R. W., and Smith, A. G.: 1979, *Space Sci. Rev.* **23**, 763.
Helliwell, R. A.: *Rev. Geophys.* **7**, 281.
Kennel, C. F. and Petschek, H. E.: 1966, *J. Geophys. Res.* **71**, 1.
Kennel, C. F. and Engelmann, F.: 1966, *Phys. Fluids* **9**, 2377.
Kurth, W. S.: 1986, in D. Le Quéau, B. M. Pedersen, A. Roux, and A. Saint Marc (eds.), *Comparative Study of Magnetospheric Systems*, CEPADUES-éditions, Toulouse.
Laval, G. and Pellat, R.: 1970, *J. Geophys. Res.* **75**, 3255.
Laval, G. and Pellat, R.: 1972, in De Witt and Peyroud (eds.), *Quasi-Linear Theory*, in Proceedings of the Summer School at Les Ouches.
Le Quéau, D., Pellat, R., and Roux, A.: 1984, *Phys. Fluids* **27**, 247.
Le Quéau, D., Pellat, R., and Roux, A.: 1985, *Ann. Geophys.* **3**, 273.
Melrose, D. B., Ronmark, K. G., and Hewitt, H. G., 1982, *J. Geophys. Res.* **87**, 5140.
Nielson, C. W.: 1969, *Phys. Fluids* **12**, 22418.

O'Neil, T.: 1968, *Phys. Fluids* **8**, 2255.
Omidi, N. and Gurnett, D. A.: 1982, *J. Geophys. Res.* **87**, 2397.
Palmadesso, P. and Schmidt, G.: 1971, *Phys. Fluids* **14**, 1411.
Ronmark, K., Borg, H., Christiansen, P. J., Gough, M. P., and Jones, D.: *Space Sci. Rev.* **22**, 401.
Roux, A. and Solomon, J.: 1970, *Ann. Geophys.* **26**, 279.
Roux, A.: 1974, in B. M. McCornac (ed.), *Magnetospheric Physics*, D. Reidel Publ. Co., Dordrecht, Holland.
Roux, A., Cornilleau-Wehrlin, N., and Rauch, J. L.: 1984, *J. Geophys. Res.* **89**, 2267.
Thornhill, S. G. and ter Haar, D.: 1978, *Phys. Reports* **43**, 43.
Wehrlin, N., Gendrin, R., and Roux, A.: 1973, *J. Geophys. Res.* **78**, 763.
Wu, C. S. and Lee, L. C.: 1979, *J. Astrophys.* **230**, 621.
Welti, R., Pellat, R., and Roux, A.: 1973, *Ann. Geophys.* **29**, 263.
Zarka, P., Le Quéau, D., and Genova, F.: 1986, in D. Le Quéau, B. M. Pedersen, A. Roux, A. Saint Marc (eds.), *Comparative Study of Magnetospheric Systems*, CAPADUES-éditions, Toulouse.

MICROWAVE SOURCES OF SOLAR FLARES: LOOP TOP OR FOOT POINTS?*

KEIZO KAI

Tokyo Astronomical Observatory, University of Tokyo, Mitaka, Tokyo 181, Japan

Abstract. The microwave images of solar flares obtained with the VLA are reviewed from a standpoint that the microwave source is near the top or foot point(s) of a flaring loop. The review is focused on whether extended structure is missed due to the lack of short baseline components leading to an incorrect interpretation of the processed images. We conclude that at short cm (< 2 cm) there is no conclusive evidence for the source near the loop top whereas at longer cm (6, 20 cm) the source tends to occupy a significant portion of a loop. The observed bipolar structure could be unambiguously interpreted as evidence for the source at the loop top, only when it is confirmed that a more extended structure has not been missed.

Introduction

For better understanding of the acceleration and confinement of energetic electrons in solar flares it is crucial to know whether hard X-ray and microwave sources are either near the top or the foot points of a magnetic loop. From X-ray observations made with the HXIS on the SMM satellite and some initial microwave observations with the VLA and the WSRT, an idea that hard X-ray sources are at the foot points and microwave sources at the top of a loop has attained popularity.

However, hard X-ray imaging observations with the SXT on the Hinotori satellite and subsequent VLA observations are not always compatible with this simplified view. On the other hand the close similarity of time profile as between hard X-rays and microwaves strongly suggests that the two kinds of emission are given rise by a single population of electrons and further that the sources of these emissions are co-spatial or close to each other. Both a detailed comparison of time variation between microwaves (17 GHz) and hard X-rays at various energies (Nitta and Kosugi, 1986), and a comparison of flux between microwaves (17 GHz) and hard X-rays (Kosugi *et al.*, 1986) lead us to estimate the effective energy of microwave emitting electrons to be ~ 100 keV, and thence to require strong magnetic field of ~ 1000 G in the microwave source. These results may not be compatible with the loop top hypothesis of the microwave source at least at short cm wavelengths.

In the present paper I will review microwave imaging observations of flares made with the VLA which have suggested the loop top or the foot points in turn, as well as indirect observation relevant to this subject and theoretical models of microwave emission. A review of simultaneous microwave and hard X-ray imaging observations with regard to the question of relative positions of microwave and hard X-ray sources as well as the loop top versus foot point location of these sources has already been published by

* Proceedings of the Second CESRA Workshop on Particle Acceleration and Trapping in Solar Flares, held at Aubigny-sur-Nère (France), 23–26 June, 1986.

Solar Physics **111** (1987) 81–87.

Kundu (1984). Further Alissandrakis (1986) reviewed the current status of our knowledge on the theory of gyro-synchrotron emission from solar flares as well as recent high-resolution observations of microwave bursts and theoretical modelling of microwave sources. Here I present a re-examination of similar problems to stimulate further discussion on these important topics.

2. Evaluation and Interpretation of Microwave Imaging Observations

Observations which are regarded to lend support to the loop top model of microwave sources have been presented in a number of papers: Alissandrakis and Kundu (1978), Marsh and Hurford (1980), Marsh *et al.* (1981), Velusamy and Kundu (1982), Hoyng *et al.* (1983), etc. On the contrary, observations which favour the footpoint model of microwave sources have been presented in several papers: Willson (1984), Willson and Lang (1984), Shevgaonkar and Kundu (1985), Dulk *et al.* (1986), etc.

I will attempt in the following to evaluate these observations in some detail and then show possible interpretation of microwave images. I restrict myself to published VLA observations because of the ambiguity in one-dimensional observations. When we evaluate images obtained with the VLA, we consider the following points:

(1) *Aliasing effect.* The area of the field of view F is determined by the lowest Fourier components $u_{\min}$ and $v_{\min}$ on the $u-v$ plane (the shortest baseline components): $F \sim 1/u_{\min} \cdot 1/v_{\min}$. The ratio of F to the true area of the source S should be greater than 1: $F/S > 1$. If this condition is not satisfied, serious missing flux would result (the flux measured with a full-Sun radiometer minus that measured with an antenna pair of the shortest baseline). Structures extended more widely than F would be missed so that what we observe is only smaller scale structures (irregularities) superimposed on the undetected extended structures.

(2) *Image quality.* A crude evaluation of the quality of obtained images can be made by a parameter $\eta = N_{\text{obs}}/(F/B)$, where N_{obs} is the number of observed Fourier components and $B = 1/u_{\max} \cdot 1/v_{\max}$ is the area of the synthesized beam ($u_{\max}$ and $v_{\max}$ are the highest Fourier components). When $\eta \ll 1$, the uniqueness of synthesized images would not be assured. We need a priori extra information of source structure to restore an image from observed Fourier components (FCs). A more precise evaluation of the image quality would not be possible unless the distribution of FCs on the $u - v$ plane is given.

2.1. Observations suggesting microwave sources at the loop top

2.1.1. *Microwave Sources between* Hα *Kernels*

Marsh and Hurford (1980) is the first paper presenting two-dimensional VLA maps for three weak (0.4–5 s.f.u.) solar bursts. From the comparison with Hα pictures, they interpret microwave sources to be near the top of magnetic loops (see, Figures 3–5 in their paper). The authors say, 'In the impulsive phase of all three flares, the microwave emission was dominated by a compact ($\sim 2''$) source located between the Hα kernels', and interpret these results to imply that the energetic electrons are confined to the top of the loop. However, at least in the first example (7 September) the source is located close to one of the Hα kernels, and in the second example (8 September) the source would not be between the kernels but closer to one of the kernels if we assume, as the authors did, the source of the post-burst increase to be in between the kernels. The source is exactly midway between the two kernels in the third extremely weak burst (9 September). But it seems to me that the source is unusually small ($\sim 2''$) compared with the extension of the presumed loop which connects the Hα bright regions ($\sim 20''$). These observations were made at 15 and 23 GHz with only 10 antennas, and $\eta \sim \frac{1}{25}$. Because observed FCs on the $u - v$ plane are not indicated, we cannot quantitatively judge the reliability of the obtained images. But there is the possibility that extended structure has been missed with this antenna configuration. We need a more detailed examination.

Willson (1983) presented a 6 cm source located between two Hα bright regions (Figure 1 of his paper). The source has an extension of $> 30''$ and occupies a significant portion of the Hα bright regions with the intensity maximum in between. The observation was made with the C-configuration with 22 antennas, and, therefore, images of good quality could be obtained. A comparison with the photospheric magnetogram or V-map (not shown in his paper) is necessary to confirm that the source is near the top of a magnetic loop connecting the two bright regions. On the other hand a 2 cm compact ($\sim 5''$) source located between well-separated ($\sim 3'$) sunspots, was demonstrated as evidence for a loop top source (Figure 2 of Willson's paper). However, the interpretation does not seem to me likely for the same reason as that given for the third example of Marsh and Hurford (note that in the latter case the observation was made with only 10 antennas). In addition the high circular polarization degree ($\sim 80\%$) may not be compatible with this interpretation.

2.1.2. *Bipolar Sources Spanning Neutral Lines*

Marsh *et al.* (1981) presented a 15 GHz source of bipolar structure: the RH and LH circularly polarized components overlapping each other, span the neutral line of the underlying photospheric magnetic field (Figure 3 of their paper). The authors suggest that the result is consistent with emission near the top of a magnetic loop. Whether the interpretation is correct depends on what scale of a magnetic loop they assume: it would be equally possible to interpret from the contour plot that the footpoints are bright, too. Also we have to check whether more extended structure has been missed, because the observation was made presumably with the A-configuration with a smaller number of antennas (21).

Hoyng *et al.* (1983) presented a similar source structure at 15 GHz from simultaneous observations of a flare with both the VLA and the HXIS. The microwave source is compact ($\sim 3''$) located between two X-ray sources which they presume the foot points of a loop ($\sim 20''$ in extension, see, Figures 2 and 5 of their paper). From the source configuration they conclude that the microwave source is near the top and the X-ray sources near the foot points. The conclusion may be correct, but I would like to point out that any structure larger than $\sim 3''$ would disappear because the observation was made with the A-configuration at 15 GHz. Indeed the missing flux is very large: the polarimeter at Nobeyama recorded 214 s.f.u. at 17 GHz, ~ 5 times more than the flux measured with the shortest baseline.

Further examples of bipolar structure observed at 6 cm were presented by Kundu *et al.* (1981, 1982). A multiple spike burst of moderate intensity (~ 300 s.f.u.) consists of eight sources in a complex active region, the strongest source of which is bipolar: the RH and LH sources span the radio neutral line ($V \sim 0$) with the intensity maximum located above the neutral line. A loop is defined here referring to the radio neutral line regardless of Hα images and magnetograms. Thence Kundu *et al.* (1982) suggest that 'the microwave burst source is located near the top and occupies a significant portion of the sides of the flaring loop'. The antenna configuration used for this observation is not indicated, but it is to be noted that the flux measured with the patrol instrument at Sagamore Hills is ~ 4 times larger than that measured with the antenna pair of the shortest baseline. Similar structures were found in 6 cm images for one of 4 multiple spike bursts (Kundu *et al.*, 1981). In the remaining three cases the source is either unipolar or more complex. Since only 8 antennas were used for the observations, we need to know the distribution of FCs on the $u - v$ plane to judge the quality of the images ($\eta \sim \frac{1}{20}$).

In conclusion no convincing evidence supporting the loop top model has been obtained, at least from observations at < 2 cm. We really need more observations with the C-configuration. In order to consider the bipolar structure found for 6 cm bursts as evidence supporting the loop top model, we have to clarify that more extended structure has not been missed with the antenna configurations used for the observations. In this regard we must be careful about what scale of a magnetic loop we are considering for the interpretation of the images.

2.2. Observations suggesting microwave sources at foot points

Recently, Dulk *et al.* (1986) presented simultaneous observations at 5 and 15 GHz of an impulsive burst which occurred near the limb. The 15 GHz source is compact ($\sim 5''$), located near one of Hα bright regions while the 5 GHz source is broader ($\sim 15''$), located between two Hα bright regions (Figures 2 and 3 of their paper). The separation of the centroid between the 5 and 15 GHz sources is more than $10''$. They estimate the height of the 15 GHz source to be less than 3000 km. The result strongly suggests that the 15 GHz source is near one of the foot points of a loop whereas the 5 GHz source occupies a much larger portion of the loop. The observation was made with the C-configuration although 27 antennas were divided to two sub-arrays for the two frequency observation, so that the restoration of large structure would not be deteriorated (the missing flux is negligible even at 2 cm). Using the VLA in the C-configuration simultaneously at 2 and 6 cm, Shevgaonkar and Kundu (1985) provided an example in which the source was located at two foot points of a loop at 2 cm whereas at 6 cm the source occupies almost the entire loop including the top. However, the missing flux is so large ($\sim \frac{1}{60}$) at 2 cm that extended structure as large as the 6 cm source would have been missed.

Willson (1984) presented a good example suggesting 20 cm sources near the foot point of a loop. He observed two spike bursts superimposed on more gradual enhancements. The second spike burst shows two separated ($\sim 30''$) sources of opposite polarity (Figure 4 of his paper). The author suggests that the two sources represent the legs of a loop. The observation was made with the *B*-configuration but extended structures would not be missed because of the long wavelength.

The first example presented by Willson and Lang (1984), a spike burst superposed on gradual enhancement also shows two separated sources of opposite polarity at 20 cm (Figure 2 of their paper).

In conclusion there are some observations which strongly suggest microwave source(s) near the foot point(s) of a loop at 2 cm and even at 20 cm. These observations are reliable, but it is hard to say that the majority of impulsive flares shows such source structure at 2 cm (generally it is not true at 20 cm).

3. Indirect Evidence Relevant to Loop Top vs Foot Point Argument

Recently the Nobeyama group has made an effort to estimate the effective energy of electrons responsible for microwave emission and thence the magnetic field intensity in microwave emitting regions, using simultaneous observations of microwaves (17 GHz) and hard X-rays measured at various energies (Kai *et al.*, 1985; Kai, 1986; Nitta and Kosugi, 1986; Kosugi *et al.*, 1986). I summarize briefly the quantitative estimate developed in the last two papers.

Nitta and Kosugi compare time profiles between 17 GHz and X-rays at various energies recorded with the hard X-ray spectrometer on the Hinotori satellite. On the assumption that the electrons giving rise to hard X-rays whose time profile is most similar to that of microwave emission also give rise to 17 GHz emission, they estimate for impulsive flares the effective energy of these electrons to be < 200 keV and, thence, require relatively strong magnetic field of 500–1000 G.

Based on about 400 flares recorded with the HXRBS on the SMM Kosugi *et al.* examined the correlation between 17 GHz flux and hard X-ray flux measured at various energies, and found that the correlation between the two types of emission is most excellent for X-rays below 80 keV for impulsive flares and above 360 keV for extended flares. Thence, they derive the following values for the electron energy and magnetic field intensity in the microwave (17 GHz) emitting region: < 200 keV and 900 G for impulsive flares and ~ 1 MeV and < 100 G for extended flares.

In the two papers lower energy of microwave emitting electrons has been derived for impulsive flares than that previously thought, and consequently stronger magnetic field (500–1000 G) is required in microwave (17 GHz) emitting regions. These estimates favour the low altitude of the microwave (17 GHz) source: near the leg or legs of a loop. On the contrary for extended flares electrons of ~ 1 MeV trapped in a weak magnetic loop have been suggested.

4. Concluding Remarks

Microwave sources are expected to be brightest near the top or the foot points of a loop depending on the structure of flaring loop (either small scale or large scale), the distribution of accelerated electrons (isotropic or beam-like), and the direction of the observer. The source configuration may change also with observing frequency: the part of a loop where the magnetic field is strong is seen at high frequencies while almost the entire loop is seen at low frequencies.

Petrosian (1982) has made a model calculation of microwave emission to simulate VLA observations suggesting the microwave source at the loop top, and imposed a constraint to the magnemtic field model. Subsequently Alissandrakis and Preka-Papadema (1984) and Klein and Trottet (1984) have made similar calculations of gyro-synchrotron emission assuming particular magnetic field models. The result and discussion were well described in a review paper presented at the last CESRA Workshop (Alissandrakis, 1986) and, therefore, I will not discuss here further. Only what I would like to point out is that gyro emission strongly depends on magnetic field intensity (B), the viewing angle (θ), and also the spatial, energy and pitch angle distributions of radiating electrons. For instance, the mirowave flux decreases by 1.5–2 orders of magnitude with the decrease of B by only a factor of 2 (cf. Figure 3 of Kai *et al.*, 1985), and 1–1.5 orders of magnitude with the decrease of θ from ~ 85 to 15 deg. Consequently either the loop top or the foot points could be bright depending on what configuration of a magnetic loop we assume. In addition the electron distribution in space makes the situation more complicated.

On the other hand from the VLA observations presently available, we cannot tell whether the observed source configuration clearly indicates the top or the foot points of a loop. However, at frequencies higher than the turn-over frequency there has been no conclusive evidence which favours the loop top model whereas at frequencies lower than the turn-over frequency there have been some observations suggesting that the entire or a significant portion of a loop brightens.

Acknowledgements

The author thanks Drs M. R. Kundu and T. Kosugi for their useful comments and the referee for critical comments.

References

Alissandrakis, C. E.: 1986, *Solar Phys.* **104**, 207.
Alissandrakis, C. E. and Kundu, M. R.: 1978, *Astrophys. J.* **222**, 342.
Alissandrakis, C. E. and Preka-Papadema, P.: 1984, *Astron. Astrophys.* **139**, 507.
Dulk, G. A., Bastian, T. S., and Kane, S. R.: 1986, *Astrophys. J.* **300**, 438.
Hoyng, P., Marsh, K. A., Zirin, H., and Dennis, B. R.: 1983, *Astrophys. J.* **268**, 865.
Kai, K.: 1986, *Solar Phys.* **104**, 235.
Kai, K., Kosugi, T., and Nitta, N.: 1985, *Publ. Astron. Soc. Japan* **37**, 155.
Klein, K.-L. and Trottet, G.: 1984, *Astron. Astrophys.* **141**, 67.

Kosugi, T., Dennis, B. R., and Kai, K.: 1986, *Astrophys. J.* (submitted).
Kundu, M. R.: 1984, *Adv. Space Res.* **4**, 157.
Kundu, M. R., Bobrowsky, M., and Velusamy, T.: 1981, *Astrophys. J.* **251**, 342.
Kundu, M. R., Schmahl, E. J., and Velusamy, T.: 1982, *Astrophys. J.* **253**, 963.
Marsh, K. A. and Hurford, G. J.: 1980, *Astrophys. J.* **240**, L111.
Marsh, K. A., Hurford, G. J., Zirin, H., Dulk, G. A., Dennis, B. R., and Frost, K. J.: 1981, *Astrophys. J.* **251**, 797.
Nitta, N. and Kosugi, T.: 1986, *Solar Phys.* **105**, 73.
Petrosian, V.: 1982, *Astrophys. J.* **255**, L85.
Shevgaonkar, R. K. and Kundu, M. R.: 1985, *Astrophys. J.* **292**, 733.
Velusamy, T. and Kundu, M. R.: 1982, *Astrophys. J.* **258**, 388.
Willson, R. F.: 1983, *Solar Phys.* **83**, 285.
Willson, R. F.: 1984, *Solar Phys.* **92**, 189.
Willson, R. F. and Lang, K. F.: 1984, *Astrophys. J.* **279**, 427.

PLASMA EMISSION: A REVIEW*

(*Invited Review* presented by G. Dulk)

D. B. MELROSE

School of Physics, University of Sydney, Sydney NSW 2006, Australia

Abstract. The theory of plasma emission is reviewed emphasizing general concepts rather than the details of the analysis. The generation of the Langmuir turbulence, its evolution due to nonlinear processes, and the plasma emission processes are described. Several outstanding difficulties in the applications to solar radio bursts are discussed, concentrating on those with implications on local density inhomogeneities in the sources.

1. Introduction

Plasma emission is the generation of escaping radiation near the local plasma frequency or its harmonics. Most meter-wave solar radio bursts are interpreted in terms of plasma emission at the fundamental (F) or second harmonic (H) (e.g., McLean and Labrum, 1985). Plasma emission is also observed ahead of the Earth's bow shock and of other interplanetary shocks. Plasma emission is necessarily a multistage process, unlike the direct emission processes such as cyclotron and synchrotron radiation. These stages include the generation of Langmuir turbulence, its nonlinear evolution and its partial conversion into escaping radiation. There are differences in detail in various treatments depending, for example, on whether or not phase-coherent or narrow-band effects are important, and whether or not strong turbulence effects are important. The simplest viewpoint is that these effects are not important, and this is the viewpoint adopted here. The generation of the Langmuir waves is then attributed to an instability involving negative absorption and saturation due to quasi-linear effects, and the evolution of the Langmuir turbulence is attributed to three-wave processes whose saturation may be discussed within the framework of weak-turbulence theory. The main emphasis in this review is in presenting the general ideas underlying this form of plasma emission. The phase-coherent and strong-turbulence effects are discussed only briefly.

Plasma emission remains at best a semi-quantitative theory. There have been some notable successes for the theory, e.g., in explaining the harmonic structure and the predominantly o-mode polarization, but there remain many difficulties. In discussing the applications here, it is argued that several difficulties can be overcome or alleviated by postulating that the sources are locally inhomogeneous. It seems that it is essential to take local inhomogeneities into account in order to reach a reasonable agreement between theory and observation.

* Proceedings of the Second CESRA Workshop on Particle Acceleration and Trapping in Solar Flares, held at Aubigny-sur-Nère (France), 23–26 June, 1986.

Solar Physics **111** (1987) 89–101.

The review is set out as follows. In Section 2 the generation of Langmuir turbulence is discussed, Section 3 contains an overview of nonlinear processes involving Langmuir and ion sound turbulence, and in Section 4 specific plasma emission mechanisms are summarized. In Section 5 some of the outstanding problems concerning the applications of plasma emission are discussed: these are the observed harmonic ratio being less than two, the lack of agreement between theory and observation on the polarization, the apparent source sizes and directivities, and the clumpiness of Langmuir waves in the solar wind.

2. Generation of Langmuir Turbulence

The first step in any plasma emission mechanism is the generation of Langmuir turbulence.

2.1. Plasma turbulence

A nonthermal distribution of waves in any wave mode constitutes plasma turbulence in that mode. In plasma emission the magnetic field plays only a minor role, and considerable simplification results by neglecting it. (Magnetic effects need to be included when discussing the polarization and they may be important in causing certain fine structures.) Once the magnetic field is neglected, only three waves need to be considered. These are transverse (T) waves with dispersion relation $\omega = (\omega_p^2 + k^2c^2)^{1/2}$, or refractive index $n = kc/\omega = (1 - \omega_p^2/\omega^2)^{1/2}$, Langmuir ($L$) waves with dispersion relation $\omega \cong \omega_p + 3k^2 V_e^2/2\omega_p$ and ion sound (S) waves with dispersion relation $\omega = kv_s$, where the ion sound speed v_s is related to the thermal speed of electrons by $v_s \cong V_e/43$. Only transverse waves (with $\omega > \omega_p$) can escape, and an emission mechanism may be defined as a process which produces transverse waves.

A thermal distribution of waves in mode M corresponds to an energy density $W_M(\mathbf{k})\, \mathrm{d}^3\mathbf{k}/(2\pi)^3$ in the range $\mathrm{d}^3\mathbf{k}$ of wavevectors such that $W_M(\mathbf{k})$ is equal to the thermal energy. It is convenient to set Boltzmann's constant esual to unity so that temperatures are expressed in energy units. Then a thermal level of waves corresponds to $W_M(\mathbf{k}) = T_e$, where T_e is the electron temperature. It is convenient to write $W_M(\mathbf{k}) = T_M(\mathbf{k})$ and to interpret $T_M(\mathbf{k})$ as the effective temperature of the turbulence. The turbulence may be anisotropic, implying that $T_M(\mathbf{k})$ is a strong function of $\mathbf{k}$. The evolution of the turbulent spectrum involves transfer of energy from one region of $\mathbf{k}$-space to another, say from $\mathbf{k}$ to $\mathbf{k}'$ with $T_M(\mathbf{k}') < T_M(\mathbf{k})$.

2.2. Bump-in-tail instability

The most important mechanism for the generation of Langmuir turbulence in solar radio bursts, particularly type III bursts, is a streaming instability. There are two versions of many instabilities and for a streaming instability these two versions are often called the bump-in-tail and the weak-beam instability, respectively. More generally these two versions may be called resistive and reactive instabilities, respectively, e.g., Melrose (1986a).

A resistive instability may be described in terms of negative absorption and attributed to a maser action. The saturation of a resistive instability is due to quasilinear relaxation when the particle distribution function is modified by the back-reaction of the waves on the particles as the free energy driving the instability drains away. For any resistive instability involving Langmuir waves and nonrelativistic electrons, the absorption coefficient $\gamma(\mathbf{k})$ is given by

$$\gamma(\mathbf{k}) = -\frac{\pi e^2 v_\phi^2}{\varepsilon_0 m \omega_p} \frac{\mathrm{d}F(v_\phi)}{\mathrm{d}v_\phi}, \tag{1}$$

with $v_\phi = \omega/k$, $\omega \cong \omega_p + 3k^2 V_e^2/2\omega_p$ and where $F(v_x)$ is the distribution function integrated over v_y and v_z, with the x-axis along $\mathbf{k}$ here. For semi-quantitative purposes (1) may be approximated by $\gamma \cong -\pi\omega_p(n_b/n_e)(v_b/\Delta v)^2$ for a stream with number density n_b, mean streaming velocity v_b and velocity spread Δv. Data on type III streams in the solar wind have been used to show that growth should occur according to (1) for the actual electron streams observed (Lin *et al.*, 1981).

Quasi-linear relaxation of the bump-in-tail instability involves formation of a plateau and its gradual spread to lower velocities, as illustrated in Figure 1. For type III streams advection is important; advection involves faster particles overtaking slower particles

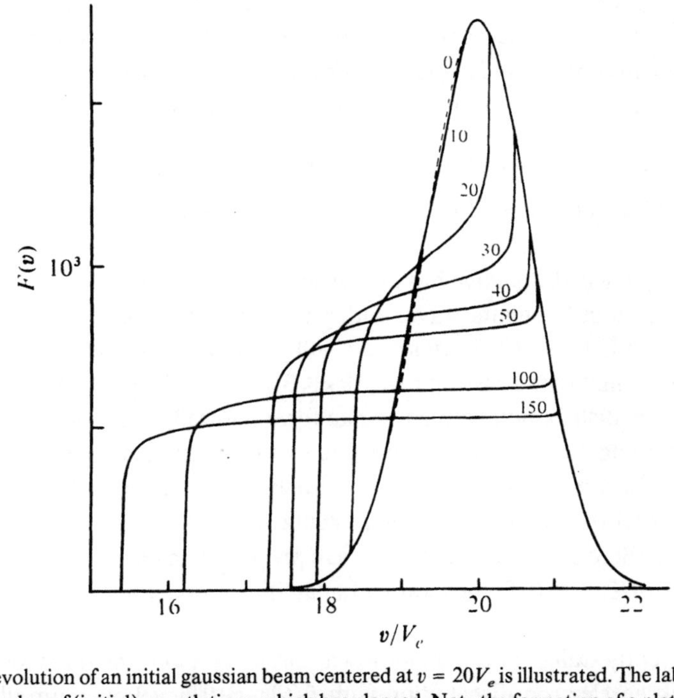

Fig. 1. The evolution of an initial gaussian beam centered at $v = 20V_e$ is illustrated. The labels on the curves refer to the number of (initial) growth times which has elapsed. Note the formation of a plateau which moves to lower velocities due to quasilinear relaxation (Grognard, 1975).

thereby tending to reform a region of positive slope $dF(v_\phi)/dv_\phi > 0$ in opposition to quasilinear relaxation tending to decrease this slope. A steady state with very specific properties is possible in principle due to a balance between these two effects. It is therefore highly significant that measured distributions of type III electrons in the solar wind exhibit these specific properties, as shown by Grognard (1982) by inserting the observed distributions in a numerical code that follows the evolution, and showing that essentially the same distribution emerges as the self-consistent solution. Evidently other effects are unimportant in influencing the evolution.

2.3. Weak-beam instability

The weak-beam instability is a reactive version of the bump-in-tail instability. This instability was first recognized by Bohm and Gross (1949), and it was invoked in Ginzburg and Zheleznyakov's (1958) original treatment of the plasma emission mechanisms. The weak-beam instability applies when the velocity spread is small. It is attributed to phase bunching of the streaming electrons. The growth rate is given approximately by $|\gamma| \cong \omega_p(\pi n_b/n_e)^{1/3}$; this growth rate applies for $\Delta v/v_b \lesssim (\pi n_b/n_e)^{1/3}$, with the bump-in-tail instability applying when this inequality is reversed. By way of illustration, parameters reported by Lin *et al.* (1986) for a type III event in the solar wind imply $\Delta v/v_b \cong 10^{-1}$ and $(n_b/n_e)^{1/3} > 10^{-2}$.

The weak-beam instability, like other reactive instabilities saturates due to wave trapping. Wave trapping is important only for phase-coherent waves; reactive instabilities involve phase-coherent processes, and resistive instabilities involve phase-random processes. The oscillatory or bouncing motion of the trapped particles can lead to the development of side bands, and this effect has been invoked in at least one theory for fine structures in solar radio bursts (Smith and de la Noë, 1976).

2.4. Alternative sources of Langmuir turbulence

Other proposed mechanisms for generation of Langmuir turbulence include the following.

(1) An isotropic *gap distribution* has no particles in a range $V_e \ll v \ll v_0$, and then the particles at $v \gtrsim v_0$ emit Langmuir waves with phase speeds in the gap but do not absorb them (Tidman and Dupree, 1965; Melrose, 1980b). A relativistic effect in the absorption (Robinson, 1978) limits the emission to $T_L(\mathbf{k}) \lesssim 3 \times 10^9$ K for phase speeds in the gap.

(2) A loss-cone distribution with a gap can form naturally (eg., Melrose and Brown, 1976) and then Langmuir waves can grow due to a loss cone instability, cf. Hewitt and Melrose (1985) for a review of the literature. Such a mechanism is favoured for type I emission (e.g., Melrose, 1980c; Benz and Wentzel, 1981).

(3) Recently, Winglee and Dulk (1986) have proposed an electron cyclotron mechanism (driven by $\partial f/\partial v_\perp > 0$) for generating Langmuir (or upper hybrid) waves for type V emission.

(4) Also, recently Cairns (1986) has shown that in a region (the foreshock) ahead of a curved shock an electron distribution with $\partial f/\partial v_\parallel > 0$ develops naturally, and may account for the continuum (non-herringbone) type II emission.

(5) A further alternative but controversial mechanism is turbulent bremsstrahlung (Tsytovich *et al.*, 1975), which I claim does not exist (Kuijpers and Melrose, 1985).

3. Nonlinear Evolution of Langmuir Turbulence

Nonlinear effects modify a spectrum of Langmuir turbulence by transferring energy from one range of wavevectors **k** to another. Relevant processes include three-wave interactions involving ion sound waves, scattering off thermal particles, scattering off density fluctuations, parametric instabilities and modulational instabilities.

3.1. Kinematic Restrictions on Three-Wave Interactions

Three waves, described by ω_i, $\mathbf{k}_i$ with $i = 1$, 2, and 3, can beat together when the conditions

$$\omega_1 + \omega_2 = \omega_3 , \qquad \mathbf{k}_1 + \mathbf{k}_2 = \mathbf{k}_3 , \tag{2a, b}$$

are satisfied. If we denote Langmuir waves, transverse waves and ion sound waves by L, T, and S, respectively, then the following processes are allowed: $L + S \to L$, $L + S \to T$, $T + S \to L$, $T + S \to T$, and $L + L \to T$. The first process is of interest here, the next two are of interest for fundamental plasma emission, the fourth is of interest for scattering of transverse waves and the last is of interest for second harmonic plasma emission.

Suppose that in (2a) one sets $\omega_1 = \omega_p + 3k^2 V_e^2/2\omega_p$ for an initial Langmuir wave, $\omega_3 = \omega_p + 3k'^2 V_e^2/2\omega_p$ for a final Langmuir wave and $\omega_2 = \pm k_s v_s$ for an ion sound wave, and that (2b) is then replaced by $\mathbf{k} + \mathbf{k}_s = \mathbf{k}'$. The following kinematic restrictions are implied:

$$k^2 - k'^2 = \mp k_0 k_s , \qquad \cos\theta = \frac{k_0 \mp k_s}{2k} , \tag{3a, b}$$

with

$$k_0 = \frac{2\omega_p v_s}{3V_e^2} = \frac{\omega_p}{65 V_e} , \tag{4}$$

and where θ is the angle between **k** and $\mathbf{k}_s$.

In applications to solar radio events the inequality $k > k_0$ usually applies. Then the three-wave interactions are possible for k_s ranging from arbitrarily small k_s to $k_s \cong 2k$. Interactions with $k_s \ll k$ cause a small change in the direction and magnitude of the wave vector **k**, and this effect may be described as a diffusion in **k** space (cf. Section 3.3). Interactions with $k_s \cong 2k$ lead to a backscatter with $\mathbf{k}'$ approximately antiparallel to **k** and with $k' - k \cong \pm k_0$. Interactions in the intermediate range $k_s \cong k$ lead to $\mathbf{k}'$ at an oblique angle to **k**.

3.2. Saturation of growth in three-wave interactions

Examination of the kinetic equations for a three-wave process leads to two simple conclusions (e.g., Melrose, 1980b, d, 1986a). Let $N_M(\mathbf{k}) = T_M(\mathbf{k})/\omega_M(\mathbf{k})$ be the wave action for any wave mode M. One conclusion is that in a coalescence process $A + B \rightarrow C$ in which waves in mode C are generated, the process saturates at

$$N_C(\mathbf{k}_C) = \frac{N_A(\mathbf{k}_A)N_B(\mathbf{k}_B)}{N_A(\mathbf{k}_A) + N_B(\mathbf{k}_B)}\,, \tag{5}$$

due to the absorption-like effect of the inverse process $C \rightarrow A + B$. This applies to the process $L + S \rightarrow L'$, which saturates at

$$T_L'(\mathbf{k}') = \frac{T_L(\mathbf{k})T_s(k_s)}{T_s(\mathbf{k}_s) + T_L(\mathbf{k})\omega_s(\mathbf{k}_s)/\omega_p}\,. \tag{6}$$

The other conclusion is that if modes A or B are generated by the decay process $C \rightarrow A + B$ then when the limit (5) implied by (2) on N_A or N_B is exceeded the absorption-like process is negative, implying growth of the waves A and B. This can occur for the secondary Langmuir waves L' in the decay process $L \rightarrow L' + S$. This kind of nonlinear wave growth is an example of a parametric instability.

3.3. Scattering processes

The conversion of Langmuir waves at one $\mathbf{k}$ into another $\mathbf{k}'$ is called scattering when it is attributed either to thermal particles or to density fluctuation. Both these can be treated as three-wave processes, with the ion-sound wave replaced by the relevant fluctuation.

For scattering off thermal ions the fluctuations have $k'' \gtrsim 1/\lambda_D = \omega_p/V_e$ and $\omega'' \lesssim k''V_i$. For $k'' = 1/\lambda_D$ and $\omega'' = k''V_i$ the relations (2a, b) apply with $k_0 = 2\omega_p V_i/3V_e^2$. The decay $L \rightarrow L' + F$, where F denotes the fluctuation ω'', $\mathbf{k}''$, leads to an instability called induced scattering (e.g., Kaplan and Tsytovich, 1968). This growth occurs only when a threshold analogous to (6) is exceeded, and the specific limit $T_L' \gtrsim T_i v_\phi/V_i$ corresponds to $T_L' \gtrsim$ few $\times\ 10^9$ K for type III bursts. The growth rate for this induced scattering instability is almost identical to that for the parametric decay $L \rightarrow L' + S$ mentioned above. For semiquantitative one may regard induced scattering off ions as being included in the parametric decay process $L \rightarrow L' + S$, and ignore it.

Scattering off density fluctuations may be treated as a limit of the three-wave interaction in which the low-frequency waves have negligible frequency and are described by a spectrum in k''. Suppose the fluctuations are described by an r.m.s. level δn_e and a mean wave number $\bar{k}''$. One may relate δn_e to $T_s(\mathbf{k}'') = k''v_s N_s(\mathbf{k}'')$ by writing

$$\int \frac{d^3\mathbf{k}''}{(2\pi)^3}\, T_s(\mathbf{k}'') = \left(\frac{\delta n_e}{n_e}\right)^2 n_e T_e\,, \tag{7}$$

$$D_\theta \cong \frac{\pi}{12}\,\frac{\omega_p}{(k\lambda_D)^2}\,\frac{k''}{k}\left(\frac{\delta n_e}{n_e}\right)^2. \tag{8}$$

This diffusion can suppress the bump-in-tail instability by removing the Langmuir waves from the range $\Delta\theta$ of forward angle where they grow to angles θ where they damp: suppression occurs when the scattering rate $\cong D_\theta(\Delta\theta)^2$ exceeds the growth rate of the bump-in-tail instability. Using observational data on type III events Muschietti *et al.* (1985) estimated that this suppression should be effective, thereby posing a serious problem for the conventional theory of type III bursts.

3.5. Parametric instabilities, modulational instabilities, and strong turbulence

Besides the processes mentioned so far, there is a rich and confusing variety of other nonlinear processes. The following is a brief summary of some relevant points.

(1) Parametric instabilities have both resistive and reactive versions (e.g., Melrose, 1986b).

(2) The resistive version of the parametric three-wave decay is the process most likely to be relevant in solar radio bursts. It has a nonlinear growth rate

$$\gamma_{NL} \cong \frac{\omega_p \omega_s}{4\Delta\omega} \frac{W_L}{n_e T_e}, \tag{9}$$

with $k_s \cong 2k_L$, and where the Langmuir waves have an energy density W_L over a frequency range $\Delta\omega \cong 3\omega_p \Delta v_\phi V_e^2/v_\phi^3$. A condition on the growth is that γ_{NL} exceed the damping rate $\cong \omega_s/50$ of the ion sound waves.

(3) Induced scattering off ions need not be distinguished from this decay process for semi-quantitative purposes.

(4) The parametric decay instability usually discussed in the literature (e.g., Liu and Kaw, 1976) is a reactive counterpart to this decay instability (Melrose, 1986b). The reactive version applies for $\Delta\omega < \gamma_{NL}$ and has a growth rate $\gamma_{PD} \cong [\gamma_{NL}\Delta\omega]^{1/2}$.

(5) Modulational instabilities are due to the ponderomotive force (e.g., Goldman, 1983, 1984). They cause a Langmuir wave packet to shrink and hence transfer some of the energy to larger k. The most familiar such instability is OTSI (e.g., Papadopoulos *et al.*, 1974, Bardwell and Goldman, 1976).

(6) Langmuir collapse occurs when this shrinking becomes dynamically unstable and the wave packet shrinks to much smaller dimensions. Collapse has yet to be explored fully other than in one dimension (e.g., Goldman, 1984). A relevant point is that even if collapse does occur, the transfer of the energy to larger k is not necessarily more important for the generation of plasma emission than the transfer to smaller k, which necessarily continues even during collapse.

4. Plasma Emission Mechanisms

The most likely forms for fundamental (F) and second harmonic (H) plasma emission involve three-wave processes. These are the only processes discussed here. Alternative forms of plasma emission were discussed by Melrose (1980b, d).

4.1. Fundamental emission

Fundamental plasma emission can occur due to the three-wave processes $L + S \rightarrow T$ and $L \rightarrow T + S$. In the presence of suitable ion sound turbulence both processes saturate at $T_T(\mathbf{k}_T) = T_L(\mathbf{k}_L)$. The requirements on the ion sound turbulence are twofold. First, the wave vectors $\mathbf{k}_s$ must satisfy $\mathbf{k}_L \pm \mathbf{k}_s = \mathbf{k}_T$, and $k_T \ll k_L$ then implies $\mathbf{k}_s \cong \mp \mathbf{k}_L$. Second, the level of the ion sound turbulence must be such that saturation occurs. If there is just one localized region where the ion sound and Langmuir turbulence coexist, then T_T must increase to reach T_L in less than the time it takes a transverse wave to propagate across the source. For a source with linear dimensions L, this condition reduces to (Melrose *et al.*, 1986)

$$\frac{v_g}{L} \lesssim \frac{\pi}{6}\, \omega_p \frac{v_\phi^2}{V_e^2} \frac{W_s}{n_e T_e}\,, \tag{10}$$

where $v_g = nc \cong \sqrt{3}\, V_e c/v_\phi$ is the group speed of the transverse waves, v_ϕ is the phase speed of the Langmuir waves, and W_s is the energy density in the (assumed isotropic) ion sound waves in a range $\Delta k_s \cong k_s$ about $k_s = \omega_p/v_\phi$.

The crucial question concerning the efficacy of this process is the presence of ion sound waves with $\mathbf{k}_s \cong \pm\mathbf{k}_L$. It may be that the broad spectrum of ion sound waves in the solar wind (e.g., Gurnett and Frank, 1978) includes the required waves, or it may be that they are generated through a decay instability, as suggested by Lin *et al.* (1986). It may also be that the required waves are not present, and that F emission is due to some other process.

An attractive feature of this mechanism for F emission is that it allows one to determine the value of the brightness temperature T_L at which saturation occurs. Saturation may be treated using (5) with $A = L$, $B = S$, and $C = T$, giving $T_T \cong T_L$ at saturation for $T_s \gg T_L v_s/v$ with $v_\phi = v_b$ for type III emission. For Langmuir waves confined to a range of angle $\Delta\theta$, the bump-in-tail instability starts to saturate when an energy $W_L \cong \frac{1}{2} n_b m v_b \Delta v$ is reached, and then the implied value of T_L and $T_T \lesssim T_L$ gives a saturation value

$$T_T \lesssim \frac{\frac{1}{2} n_b m v_b^2}{\pi(\Delta\theta)^2} \left(\frac{v_b}{f_p}\right)^3 . \tag{11}$$

4.2. Second harmonic emission

H emission is attributed to the coalescence $L + L \rightarrow T$, which also saturates at $T_T(\mathbf{k}_T) = T_L(\mathbf{k}_L)$. The kinematic conditions require $\mathbf{k}_L + \mathbf{k}_L' = k_T$ and $k_T = k_H \cong \sqrt{3}\, \omega_p/c$ for transverse waves at $\omega \cong 2\omega_p$. With $v_\phi \ll c/\sqrt{3}$ one has $k_L \gg k_T$, one requires $\mathbf{k}_L' \cong \mathbf{k}_L$ and, hence, the two Languir waves must coalesce nearly head on. This head-on condition would seem to be easily satisfied when a backscatter of the Langmuir waves has occurred. However, this is not necessarily the case: a backscatter changes k_L by k_0, cf. Equation (4), and for $k_0 > k_H$ the coalescence of k_L and $k_L' = K_L \pm k_0$ into $k_T = k_H$ is not possible (Melrose, 1983). Three possibilities arise:

(i) The condition $k_0 > k_H$ may not be satisfied. According to Equation (4), k_0 depends only on the electron temperature and for $T_e > 5 \times 10^5$ K the coalescence is possible. The condition $T_e > 5 \times 10^5$ K is satisfied in the solar corona but not in the solar wind.

(ii) The initial spectrum of Langmuir waves covers a relatively broad range Δk about k_L. The spectrum formed in one backscatter then covers a range Δk about $k_L - k_0$ in the backward direction. In this case, provided one has $\Delta k > k_H - k_0$, H emission is possible for $k_0 > k_H$.

(iii) Scattering of the Langmuir waves (Section 3.3) may produce a roughly isotropic distribution which contains nearly anti-parallel waves. The coalescence $L + L \rightarrow T$ is then possible between two such waves.

An important qualitative point is that the coalescence processes considered here for F and H emission saturate at the same value of T_T, which may be estimated using (11).

5. Some Outstanding Difficulties in Applications of Plasma Emission

Despite approximately three decades of theoretical work on plasma emission, there is no case where there is convincing quantitative agreement between theory and observation. Here several difficulties with the applications are discussed, concentrating on those with implications on local density structures in sources.

5.1. The Harmonic Ratio

Theory implies that the harmonic ratio for plasma emission should be very close to two, i.e., the ratio of the frequencies of H and F emission should differ insignificantly from 2 : 1. However, observations show that it is systematically less than 2 : 1. The only plausible explanation for this is that the F and H components have moderately broad bandwidths and that the lower part of the F band does not escape (Wild *et al.*, 1954). However, this bandwidth cannot be intrinsic to plasma emission itself, and must be due to a range of plasma frequencies in the source (Roberts, 1959). To prevent the lower half of the F band from escaping it may be necessary to assume that the source is in an underdense duct (Stewart, 1974). Thus to explain harmonic ratios different from two, it seems necessary to appeal to density structures in the source, with the F emission being able to escape only from the density peaks. This argument is independent of other arguments for ducting (Duncan, 1979), and combining the two arguments implies a model in which the source is an underdense duct, with local variations in the plasma density across the duct. The observed value of the harmonic ratio does not provide evidence on the sizes of the density irregularities, but it does imply a minimum value for the density variations. For example, to account for an harmonic ratio of 1.8 : 1 (Stewart, 1974) with only half the F bnd escaping requires a density variation of at least 20%. Observed fluctuations in the solar wind (Celnikier *et al.*, 1983) are considerably smaller than this.

Alternative mechanisms which could produce an harmonic ratio less than two seem implausible. The production of F emission could involve a coalescence with a low frequency wave with a higher frequency than the ion sound waves. The required frequency to produce a ratio 1.8 : 1, say, is $\omega = 0.1\omega_p$. However, all the candidates, such as electron cyclotron waves ($\omega \cong \Omega_e$ = cyclotron frequency) and electron acoustic waves (e.g., Gary and Tokar, 1985) seem quite implausible. Another possibility is that once the F emission is generated it is scattered e.g., off ion sound waves, many times before it escapes. Each such scattering changes the wave frequency by an amount equal to the frequency of the ion sound wave. The transverse waves then diffuse in frequency causing a spread in the bandwidth of the escaping F emission. However, the frequency of the ion sound waves involved is $\cong 10^{-4}\omega_p$ and, hence, $\cong 10^6$ scatterings would be required for the bandwidth to increase by 10% of the plasma frequency. These alternative mechanisms are much less plausible than that involving density variations.

5.2. Polarization of Plasma Emission

Theory implies that F emission should be essentially completely polarized in the sense of the o-mode for plausible values of the coronal magnetic field, and that H emission should be weakly polarized in the sense of the o-mode for streaming electrons (e.g., Melrose *et al.*, 1980). The argument for F emission is that the frequency of the Langmuir waves lies between the cutoff frequencies, ω_p for the o-mode and $\cong \omega_p + \frac{1}{2}\Omega_e$ for the x-mode, so that only o-mode emission can be generated. Although observed plasma emission is polarized in the sense of the o-mode, there are at least three difficulties relating to the magnitude of the polarization.

(i) The polarization of F emission in type III and type II bursts is always less than 100%. The degree of polarization does not exceed about 70% for type III bursts (Suzuki and Sheridan, 1977; Dulk and Suzuki, 1980).

(ii) The H emission in type III events is only modestly polarized (Suzuki and Sheridan, 1977), but the degree of polarization is higher than can be explained by simple theory (Melrose *et al.*, 1980).

(iii) Type I emission, which is accepted as being F emission, is typically 100% polarized. However, the degree of polarization can have a characteristic value less than 100%, especially for sources away from the central meridian (Zlobec, 1975).

There are two possibilities for explaining the $<100\%$ polarization of F emission in type III bursts: the lower polarization is intrinsic or the radiation is depolarized as a propagation effect. Lower intrinsic polarization can occur only if the frequency of the F component on generation exceeds the cutoff frequency for the x-mode, and this seems highly implausible. Depolarization due to propagation effects is more favorable. The evidence for type I bursts on variation of the degree of polarization with viewing angle (Zlobec, 1975) is strongly suggestive of a propagation effect. However, the details of the depolarizing mechanism remain unclear. There are three suggested mechanisms for depolarization, and in one sense all three are equivalent: these are mode coupling (Melrose, 1975), scattering by low-frequency waves (Wentzel, 1984) and reflection at density inhomogeneities (Hayes, 1985). All these mechanisms involve small scale lengths and, hence, large spatial Fourier coefficients in plasma inhomogeneities, allowing large angle changes for the scattered (or reflected) waves. After a large angle deflection an initial o-mode component is composed of a mixture of o-mode and x-mode components. Depolarization results whenever a large angle deflection occurs for waves above the cutoff frequency for the x-mode. The conditions under which the required degree of depolarization occurs are very specific (Hayes, 1985).

In summary, the observed polarization of plasma emission appears to have been strongly influenced by propagation effects. Models to account for this require relatively sharp gradients in plasma density involving scalelengths comparable with the wavelength of the radiation. Further investigation of detailed models is required.

5.3. Scattering and Directivity

It is thought that the apparent sizes of sources of plasma emission are scatter images of smaller actual sources (Steinberg *et al.*, 1971). Such scattering is attributed to density fluctuations in the source. The sources remain highly directive, however, and it has been assumed tha this is difficult to reconcile with the amount of scattering required to explain the apparent sizes. The directivity has been explained in terms of specific structures in the corona. For type I bursts Bougeret and Steinberg (1977) suggested a fibrous structure, involving parallel columns of overdense plasma. For type III bursts Duncan (1979) suggested that the sources are in underdense ducts. There is no other supporting evidence for either of these suggestions.

It is worth pointing out that the directivity may not be an actual problem, and the perceived difficulty in reconciling the scattering and the directivity may be an artefact of assumptions made in scattering models. The point is the following. Snell's law implies that radiation with refractive index $n \ll 1$ initially becomes confined to a cone with half angle $\theta \cong n/n'$ at another point where the refractive index is $n' \gg n$. It is assumed that scattering destroys this directivity. However, it is not clear that this is the case, and in fact it requires specific assumptions on the scattering for this directivity to be destroyed.

One way of seeing this point is to note the existence of a conserved quantity, called the 'generalized étendue' in geometric optics (e.g., Welford and Winston, 1978). Consider a collection of rays confined to direction cosines $\mathrm{d}L$ and $\mathrm{d}M$ about the z-axis and to an area $\mathrm{d}x\,\mathrm{d}y$ at some point P where the refractive index is n. At any other point P' where the refractive index is n', suppose the same rays are confined to ranges $\mathrm{d}L'$ and $\mathrm{d}M'$ of direction cosines about the new ray direction along the z'-axis, and to an area $\mathrm{d}x'\,\mathrm{d}y'$. Then one has

$$n^2\,\mathrm{d}x\,\mathrm{d}y\,\mathrm{d}L\,\mathrm{d}M = n'^2\,\mathrm{d}x\,\mathrm{d}y'\,\mathrm{d}L'\,\mathrm{d}M' \,. \tag{12}$$

Scattering, in the sense used here, involves only propagation in accord with the laws of geometric optics. The path from the actual to the apparent source can then be regarded as equivalent to an optical system to which (12) applies. The generalized étendue them implies that the range of angles to which the radiation is confined must shrink not only as the refractive index increases (as is well known) but also as the area increases. From this viewpoint the difficulty is in understanding why the radiation is not much more highly directive than it appears to be. To be compatible with the generalized étendue it appears that radiation filling a range of angles much greater than $\theta \cong n$ (for

$n' = 1$) must be such that rays at a particular angle arise only from a small fraction of the area of the apparent source. This has implications on the relation between the apparent brightness temperature and the value of T_T in the source.

These points require more detailed discussion. What the generalized étendue does is provide a different viewpoint on the problem of scattering by density inhomogeneities in the corona. Qualitatively, an implication is that it is much more difficult to destroy directivity by scattering than has been assumed hitherto.

5.4. Clumpy Langmuir Waves

Direct evidence on inhomogeneities in sources is available for type III sources in the solar wind. It is found that the Langmuir waves are distributed in a highly inhomogeneous way (Gurnett and Anderson, 1977; Lin *et al.*, 1986), referred to as 'spikes' or 'clumps' of Langmuir waves. A plausible explanation for the clumpy distribution is the effect of local density inhomogeneities (Smith and Sime, 1979). As mentioned in Section 3.3, scattering of Langmuir waves off density inhomogeneities appears to e strong enough to suppress growth entirely (Muschietti *et al.*, 1985) and specific assumptions are required to explain any Langmuir waves at all (Melrose *et al.*, 1986). Investigation of this problem is continuing.

The evidence from the solar wind provides qualitative support for the roles of inhomogeneities considered above for emission from the corona. However, a potentially serious problem is that the observed density fluctuations (Celnikier *et al.*, 1983) seem to be isotropic (Muschietti, 1986, private communication). How can one explain ducting without any anisotropic density structures? Either the anisotropic density structures are present and have escaped detection for some reason, or ducting effectively occurs without ducts. The remarks on the generalized étendue in Section 5.3 suggest that the latter possibility cannot be excluded.

Acknowledgement

I thank G. A. Dulk for presenting this paper on my behalf at the Workshop.

References

Bardwell, S. and Goldman, M. V.: 1976, *Astrophys. J.* **209**, 912.
Benz, A. O. and Wentzel, D. G.: 1981, *Astron. Astrophys.* **94**, 100.
Bohm, D. and Gross, E. P.: 1949, *Phys. Rev.* **75**, 1864.
Bougeret, J. L. and Steinberg, J. L.: 1977, *Astron. Astrophys.* **61**, 777.
Cairns, I. H.: 1986, *J. Geophys. Res.* **91**, 1507.
Celnikier, L. M., Harvey, C. C., Jegou, R., Kemp, M., and Moricet, P.: 1983, *Astron. Astrophys.* **126**, 293.
Dulk, G. A. and Suzuki, S.: 1980, *Astron. Astrophys.* **88**, 203.
Duncan, R. A.: 1979, *Solar Phys.* **63**, 389.
Gary, S. P. and Tokar, R. L.: 1985, *Phys. Fluids* **28**, 3691.
Ginzburg, V. L. and Zheleznyakov, V. V.: 1958, *Soviet Astron. AJ* **3**, 235.
Goldman, M. V.: 1983, *Solar Phys.* **89**, 403.
Goldman, M. V.: 1984, *Rev. Mod. Phys.* **56**, 709.
Grognard, R. J.-M.: 1975, *Australian J. Phys.* **28**, 731.

Grognard, R. J.-M.: 1982, *Solar Phys.* **81**, 173.
Gurnett, D. A. and Anderson, R. R.: 1977, *J. Geophys. Res.* **82**, 632.
Gurnett, D. A. and Frank, L. A.: 1978, *J. Geophys. Res.* **83**, 58.
Hayes, L. M.: 1985, *Australian J. Phys.* **38**, 705.
Hewitt, R. G. and Melrose, D. B.: 1985, *Solar Phys.* **96**, 157.
Kaplan, S. A., and Tsytovich, V. N.: 1968, *Soviet Astron. AJ* **11**, 834.
Kuijpers, J. and Melrose, D. B.: 1985, *Astrophys. J.* **294**, 28.
Lin, R. P., Levedahl, W. K., Lotko, W., Gurnett, D. A., and Scarf, F. L.: 1986, *Astrophys. J.* **308**, 954.
Lin, R. P., Potter, D. W., Gurnett, D. A., and Scarf, F. L.: 1981, *Astrophys. J.* **251**, 364.
Liu, C. S. and Kaw, P. D.: 1976, *Adv. Plasma Phys.* **6**, 83.
McLean, D. J. and Labrum, N. R.: 1985, *Solar Radiophysics*, Cambridge University Press, Cambridge.
Melrose, D. B.: 1975, *Solar Phys.* **43**, 79.
Melrose, D. B.: 1980a, *Plasma Astrophysics*, Vol. I, Gordon and Breach, New York.
Melrose, D. B.: 1980b, *Plasma Astrophysics*, Vol. II, Gordon and Breach, New York.
Melrose, D. B.: 1980c, *Solar Phys.* **67**, 357.
Melrose, D. B.: 1980d, *Space Sci. Rev.* **26**, 3.
Melrose, D. B.: 1983, *Solar Phys.* **87**, 359.
Melrose, D. B.: 1986a, *Instabilities in Space and Laboratory Plasmas*, Cambridge University Press, Cambridge.
Melrose, D. B.: 1986b, *J. Plasma Phys.* **36**, 269.
Melrose, D. B. and Brown, J. C.: 1976, *Monthly Notices Roy. Astron. Soc.* **176**, 15.
Melrose, D. B., Dulk, G. A., and Cairns, I. H.: 1986, *Astron. Astrophys.* **163**, 229.
Melrose, D. B., Dulk, G. A., and Gary, D. E.: 1980, *Proc. Astron. Soc. Australia* **4**, 50.
Muschietti, L., Goldman, M. V., and Newman, D.: 1985, *Solar Phys.* **96**, 181.
Papadopoulos, K., Goldstein, M. L., and Smith, R. A.: 1974, *Astrophys. J.* **190**, 175.
Roberts, J. A.: 1959, *Australian J. Phys.* **12**, 327.
Robinson, R. D.: 1978, *Astrophys. J.* **222**, 696.
Smith, D. F. and Sime, D.: 1979, *Astrophys. J.* **233**, 998.
Smith, R. A. and de la Noë, J.: 1976, *Astrophys. J.* **207**, 605.
Steinberg, J. L., Aubier-Girand, M., Leblanc, Y., and Boischot, A.: 1971, *Astron. Astrophys.* **10**, 362.
Stewart, R. T.: 1974, *Solar Phys.* **39**, 451.
Suzuki, S. and Sheridan, K. V.: 1977, *Radiophys. Quantum Electron.* **20**, 989.
Tidman, D. A. and Dupree, T. H.: 1965, *Phys. Fluids* **8**, 1860.
Tsytovich, V. N., Stenflo, L., and Wilhelmsson, H.: 1975, *Phys. Scripta* **11**, 251.
Welford, W. T. and Winston, R.: 1978, *The Optics of Nonimaging Concentrators*, Academic Press, New York.
Wentzel, D. G.: 1984, *Solar Phys.* **90**, 139.
Wild, J. P., Murray, J. D., and Rowe, J. D.: 1954, *Australian J. Phys.* **7**, 439.
Winglee, R. M. and Dulk, G. A.: 1986, 'The Electron-Cyclotron Maser Instability as the Source of Solar Type V Continuum', preprint.
Zlobec, P.: 1975, *Solar Phys.* **43**, 453.

SPOTLESS FLARES AND THE ASSOCIATED RADIO CONTINUUM EMISSION*

V. RUŽDJAK[1], M. MESSEROTTI[2], M. NONINO[3], A. SCHROLL[4], B. VRŠNAK[1], and P. ZLOBEC[5]

Abstract. We studied 24 spotless flares of Hα importance ≥ 1 which occurred during the 21st cycle of solar activity. The spotless flares could be grouped in three categories according to their location and time history of the associated active region. Our association of the flares with radio events was based on relative timing and on the flare importances. Weak microwave gradual rise and fall events were frequently recorded during the occurrence of the spotless flares. A few flares from our sample could be associated with impulsive and complex microwave bursts. Only in one case an association of a spotless flare with a significant metric type II/IV event seems to be justified.

1. Introduction

Spotless flares are a rather rare but well observed aspect of solar activity. We investigated spotless flares of Hα importance ≥ 1 which occurred in the period June 1976–December, 1985, i.e., during the 21st cycle of solar activity. The goal was to study continuum radio emission associated with this kind of flares which are believed to occur in regions of weak magnetic fields. A systematic study of the general characteristics of major flares occurring in centers of activity with small or no spots was published by Dodson and Hedeman (1970). We have to stress that our data set does not cover the full 24 hr period per day, but is rather limited to the daily observing schedules of our observatories which were operating roughly from 07 : 00 to 18 : 00 UT (April–September) and 08 : 00 to 15 : 00 UT (October–March). Our sample consists of 24 spotless flares; if we take into account also the lists of spotless flares published by Yunnan (Luo, 1982) and Udaipur (Ambashta and Bhatnagar, 1983) observatories, we can roughly estimate that about 60 spotless flares of Hα importance ≥ 1 occurred during the past cycle of Solar activity.

2. The Data

2.1. Optical data

We considered flares of Hα importance ≥ 1 and would like to point out that we took into account only truly spotless flares, i.e., that no spots were present within 15° at the

* Proceedings of the Second CESRA Workshop on Particle Acceleration and Trapping in Solar Flares, held at Aubigny-sur-Nère (France), 23–26 June, 1986.

[1] Hvar Observatory, 58450 Hvar, Yugoslavia.
[2] SISSA-ISAS, 34014 Trieste, Italy.
[3] Astronomical Department of Trieste University, 34131 Trieste, Italy.
[4] Sonnenobservatorium Kanzelhöhe, 9520 Sattendorf, Austria.
[5] Trieste Astronomical Observatory, 34131 Trieste, Italy.

Solar Physics **111** (1987) 103–111.

TABLE I

Data on the flares studied in the paper, grouped in categories

Date	Start	Max.	End	Position	Remarks	Importance Hα/X-ray	Radio emission						
							Metric	Freq.	Type	Microwave	Start	Max.	Duration (min)
Category I													
1978, Sept. 4	0808	0823 + 4	0927	S 15 E 52	G, U, S	2N M2	II/IV	3100	47	GB	0813	0818	19
	0800	0813	0818	S 31 E 15		1F		2950	21	GRF	0810	0828	29
1979, Jan. 5	1347	1351	1402	S 23 W 06	G, U	1N C4	–			–			
1982, Apr. 4	0800	0806	0910	S 23 W 51	G, F	1B C9	–	3750	21	GRF	0800	0815	35
								2000	45	C	0800	0807	10
								3750	5	S	0802	0806	3
1983, Apr. 14	0839E	0848 + 7	0929	S 14 E 49	G	2N C5	–	2950	20	GRF	0827	0935	213
	0817	0831	0929	S 06 E 37	S, U	2N							
	0850E	0855	0923D	N 06 E 90	H	1F							
Category II													
1977, Nov. 11	1235	1239 + 2	1309	N 25 E 02	G, U	1N	L.E.			–			
1978, May 30	1100	1131 + 6	1255	S 30 E 53	G, E	1N	–	2800	240	R	1120	1130	10
								2800	21	GRF	1135	1142	150
1979, June 28	1012	1014 + 9	1038	S 05 W 60	G, F	1F	–			–			
1979, Aug. 2	0845	0900	0945	N 24 E 74	G	1N	–			–			
1980, Sept. 17	0615	0620 + 5	0640	N 09 W 67	G	1N	–			–			
1981, Feb. 14	0828	0836	0906	S 15 W 69	G, E	1N	–	2950	20	GRF	0830	0833	12
	0752	0759 + 3	0840	N 12 E 30		1N							
1981, Apr. 1	1206	1214	1236	N 15 W 13	G, H, U	1N	–	2800	20	GRF	1150	1225	50
1981, May 7	1335	1409 + 2	1430	S 23 W 52	G, E, L	1N	–			–			
1981, Nov. 24	0956	1019	1114	S 15 E 33	G, E	1N	–			–			
1982, May 31	1356	1420	1501	S 10 E 33	G, F	1N C1	–			–			
1983, Dec. 5	1216	1229	1337	S 12 E 16	G, U	1N C2	–			–			
1984, June 8	0730	0737	0900	S 22 W 51	G, U	1N	–			–			

Category III

1980, Feb. 6	1020	1038	1227	S 18 W 37	G, S, U	2N	C5	L.E.	6100	24	R		1025	1033	
	1058	1102	1111	S 20 W 17	E	– F			6100	24	R		1034	1123	39
									6100	8	S		1033.7	1034	
1980, May 17	1006	1014	1119	N 26 E 41	G, H, U	2B	M1	L.E.	6100	27	RF		1000	1020	40
	1015E	1018	1058	N 17 E 29		– B			3100	1	S		1010	1011.5	10
									[Hard X-rays				1009.2	1012.4	8]
1980, July 20	0606	0610 + 5	0735	S 38 E 12	G, S, U	1N		–	3750	20	GRF		0603	0611	70
1983, Jan. 7	0841	0858	0912	S 14 W 01	G	1N	C5	–				–			
1983, Jan. 7	0930E	0935	1038	S 15 E 01	G, L, U	2N	C6	–				–			
1983, July 22	0746	0752 + 8	0841	S 14 E 26	G, U	1N	C2	–	2950	20	GRF		0742	0812	219
	0751	0756	0820	S 09 E 14		– F									
1983, Aug. 27	0735	0744	0839	S 10 E 33	G, F	1B	C2	–	3750	21	GRF		0740	0804	70
									2695	47	GB		0742.6	0742.8	4
									[Hard X-rays				0743.1	0744	2.4]

Explanation of Table I

Symbols as adopted in *Solar Geophysical Data* are used.

Remarks: E = two or more brilliant points, F = several eruptive centers, G = spotless flare, H = flare accompanied by a high-speed dark filament, L = existing filament shows signs of sudden activity, S = brightness follows disappearance of filament, U = two bright branches.

Radio emission: L.E. denotes low enhancement of the metric continuum, available data on two hard X-ray bursts are inserted in brackets.

If other flares occurred in spot groups during the radio bursts, they are listed below the spotless flare at a given date.

time of the flare occurrence. After discarding all flares occurring at CMD larger than 70°, as the recognition of small spots becomes difficult close to the limb, we were left with a sample of 24 flares (Table I). It appears that spotless flares can be grouped into three categories according to their location and the time history of the associated active regions.

Category I contains flares occurring during the spotless phases of active regions which either have been significant and flare productive in previous rotations or will become large and active in the following rotations. The maximal calcium plage area attained during the development of these active regions, which survived three or more rotations, was over 3000 millionths of the solar hemisphere. The area covered by spots was over 600 millionths of the hemisphere and the maximal spot number was over 30. Two of the flares in this category occurred after the eruption of a filament which was situated at the inversion line of the magnetic field in the plage, and were two-ribbon flares. We attribute the most significant recorded radio event (types II, III, and IV) to a flare from this category, the spotless flare of September 4, 1978.

Category II consists of flares which occurred in small active regions with low magnetic field strengths and had a maximal calcium plage area less than 2000 millionths of the hemisphere and survived only one or two rotations. The maximal area covered by spots was between 50 and 350 millionths of the hemisphere. In this category four flares showed the two-ribbon shape, the other ones were rather irregular. A typical flare from this category is shown in Figure 1(a).

Category III consists of flares occurring in the channel between plage regions, at the inversion line of the large-scale magnetic field of the Sun. Almost all flares from our sample in this category followed the classical two-ribbon scenario: eruption of the filament and formation of two-ribbons on both sides of the disappeared filament. We

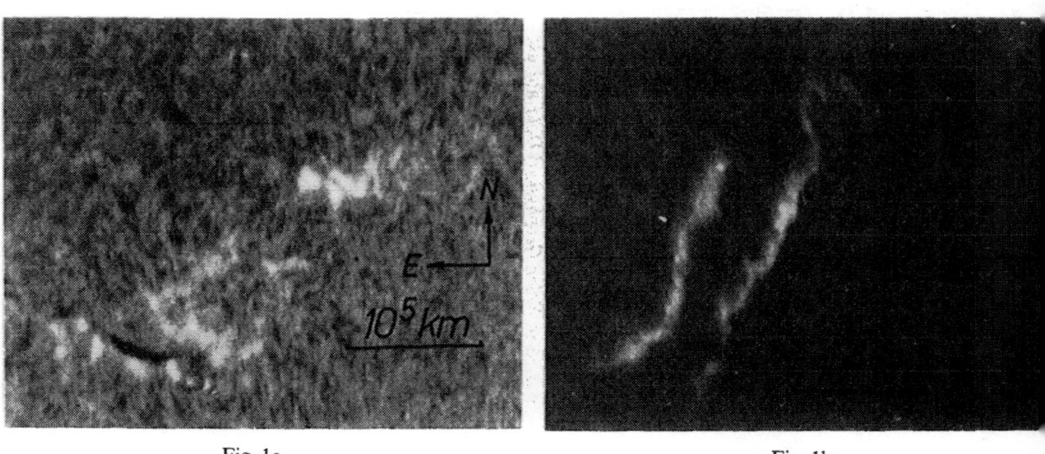

Fig. 1a. Fig. 1b.

Fig. 1. Hα images of the flares of November 11, 1977 (a) and December 9, 1980 (b) taken at 12 : 41 and 07 : 53 UT, respectively.

have to note that the ribbon separation was usually considerable from the beginning, typically it amounted to 30000 km. In several cases the length of the Hα ribbons was up to 300000 km. A flare from this category is shown in Figure 1(b). It is interesting, and perhaps surprising, that even in this category of flares occurring in the channel between plage regions, homologous flares are sometimes observed. This was the case with the 2N flare of May 17, 1980 which was preceded by a 1N flare at 01 : 15 UT (Luo, 1982) and the flares of January 7, 1983 from our sample.

It seems appropriate to include into this category disparition brusque (DB) events which occur in similar environment and produce only slight or no Hα brightening along the filament channel. In a few cases such DB events were associated with proton events (Sanahuja *et al.*, 1983) or with solar energetic particle events (Kahler *et al.*, 1986). We have to note that the flare from category III of February 6, 1980 could possibly be associated with the proton event registrated the same day with maximum at 18 : 50 UT as it was situated at the position W 37. No associated flare is reported in the list of proton events in *Solar Geophysical Data*, Comprehensive Reports No. 502, 1986.

2.2. Radio and X-ray data

Radio recordings obtained at Trieste Observatory at 237 MHz and 408 MHz stood at our disposal, while information on the microwave range and soft X-rays was taken from *Solar Geophysical Data*. Data on hard X-rays were taken from Dennis *et al.* (1986). The association of the radio bursts with a particular spotless flare was based on the timing of the radio and Hα events. However, we have to note that such an association based only on timing is not always unambigous. See Table I where all other flares occurring in regions with spots during the development of the radio events are also listed. We skipped from Table I the spotless two-ribbon 2N, M1 flare of our category III (1980, December 9, max at 07 : 03 UT, S 19 W40) as four subflares occurred during the development of this flare and the radio events (L.E., g.r.f., and a 4 S/F burst).

The only outstanding radio event with metric type II, III, and IV emission which was recorded during the occurrence of the flares from our sample, we associated with the category I flare from September 4, 1978. Two other flares from this category we associated with the occurrence of weak microwave g.r.f. events with peak flux values of some 10 solar flux units (s.f.u.). One of them, the flare of April 4, 1982 was also accompanied by a complex microwave burst (according to Krüger, 1979, a precise distinction between complex and type IVμ bursts seems to be difficult to achieve). The soft X-ray importance ranged from C5 to M2 in this category.

The flares in category II had no X-ray emission above C2 level and were practically radio quiet. We associated the flare of November 11, 1977 with a weak, short duration enhancement of the metric continuum, and in three cases very low intensity microwave g.r.f. events (peak flux values of some 5 s.f.u.) have been recorded during the occurrence of the spotless flares (Table I). So, these seem to be the least energetic events in our sample, which is not surprising as the flares from this category occurred in regions with very weak magnetic fields.

The flares of category III, although of considerable Hα importance, do not suffer from the 'big flare syndrom' which tends to screen many interesting effects. Weak microwave g.r.f. events (peak flux values of about 15 s.f.u.) were recorded during the occurrence of all the flares from this category. In three cases also a low (peak flux of about 10 s.f.u.) and short duration (10 min) enhancement of the metric continuum was recorded.

It is interesting to note that there are indications for the occurrence of an impulsive phase during three flares of this category. Four minutes before the Hα maximum of the spotless flare of February 6, 1980 a microwave spike and a spike like soft X-ray burst, importance M2, which preceded the more gradual C5 soft X-ray burst, were recorded. An impulsive microwave and a hard X-ray burst peaked two minutes before the Hα maximum of the May 17, 1980 spotless flare. The flare of August 27, 1983 was accompanied by a microwave type IV burst and a hard X-ray burst (see Table I).

3. The Event of September 4, 1978

Two flares were observed during the radio continuum event on September 4, 1978: a spotless flare Hα importance 2N and a flare occurring in a spot group, Hα importance 1F (Table I). We associate the radio event with the spotless flare because we deal with a metric type II/IV event, usually connected with large dynamic two-ribbon flares.

The spotless flare occurred in Hale plage region No. 15521 during its first rotation. The region survived four rotations and was most flare productive in the second rotation (Hale No. 15570) when it produced 58 subflares, 3 importance 1 flares, and one importance 2 flare which was also associated with a metric type IV event.

On September 3, three small spots were present and the active region was classified as type BXO according to the modified Zürich classification. Next day the spots had disappeared. The filament which was situated at the inversion line of the magnetic field in the plage region showed first signs of activity at 06 : 40 UT, and at 08 : 06 UT was almost not visible any more in the center of the Hα line. Two minutes later the first Hα brightenings showed up close to the position of the disappeared filament. Two ribbons were fully developed at 08 : 14 UT and showed a rapid lateral separation at the beginning, which later slowed down – a typical behaviour for two-ribbon flares. The Hα maximum was reached at 08 : 23 UT (Figure 2), the flare ended at 09 : 30 UT. The separation of the ribbons amounted to 50 000 km at the end of the event.

A group of type III bursts was recorded at 08 : 13 UT, at the same time the microwave type IV burst started. A type II metric burst with harmonic structure started at 08 : 16 UT. The metric type IV event began at 08 : 17 UT (at 237 MHz), the maximum flux density was about 450 s.f.u. at 08 : 27 UT (Figure 3). The polarization was essentially zero at the beginning, then it increased and amounted to 20% Rh polarized near the end of the event. The whole event was rather low and short, the most prominent characteristic was the oscillatory phase which began at 08 : 18 UT and lasted for about 5 min. At 08 : 23 UT the sudden absorption phase began and lasted for 1 min. The duration of this oscillatory – sudden absorption phase was exceptionally long, almost 7 min, and its polarization percentage was very low (see Zlobec *et al.*, 1987). The

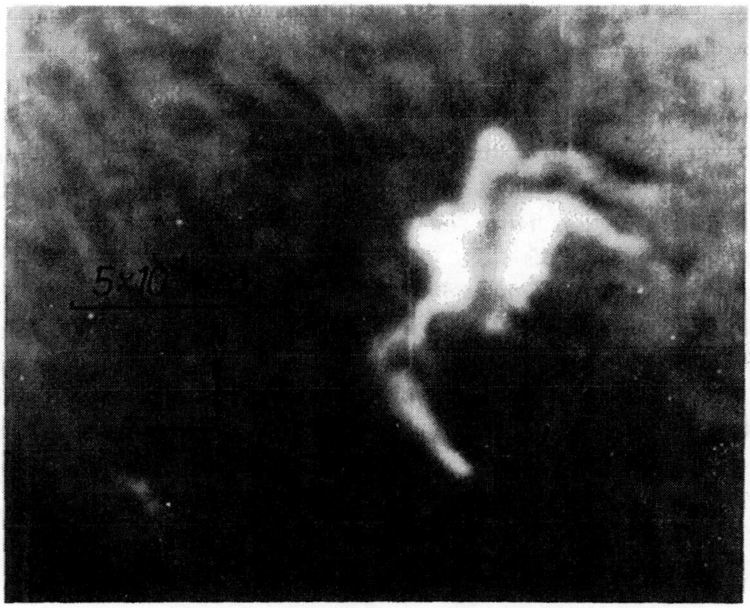

Fig. 2. Hα image of the flare of September 4, 1978 taken at 08 : 23 UT.

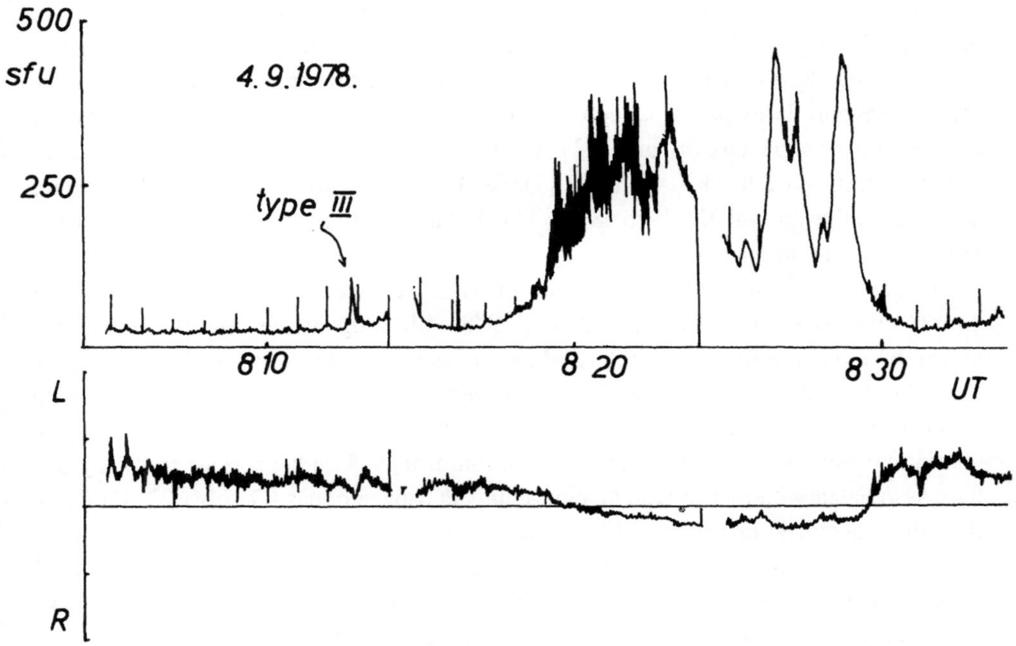

Fig. 3. Recording of the type IV event at 237 MHz. The polarization behaviour is indicated below; note that the polarization is slightly changing during the event.

polarization of the background and of the oscillations was changing contemporaneously and identically. This means that the background source should be the same as the one which was provoking the oscillations. Such a result was reported also by Trottet *et al.* (1981) for another event observed by the Nancay radioheliograph. The general trend of the metric type IV event was rather smooth if we except the oscillatory phase.

What were the characteristics of this plage region after the occurrence of the September 4, 1978 spotless flare? The calcium plage area showed a pronounced drop (from 1800 to 1000 millionths of the hemisphere) and no further spots were observed during the first rotation. Only one subflare (– F) was observed on September 13. This seems to be indicative for a decrease of flux in the active region after the occurrence of a major flare. Švestka (1976) proposed that one should try to measure magnetic fields prior and after spotless flares as according to the expected low fluxes it should be most easy to detect a decrease after the flare. However, no definitive results have been obtained so far as usually detailed magnetic field maps are not being performed for regions where spots are absent.

4. Discussion

Investigating the 24 spotless flares from our sample and their relation to radio continuum radiation it was found that the flares frequently could be associated with weak microwave g.r.f. events. It seems that the detection of such events is only a question of the g.r.f.'s peak flux and the limit of detectability of single-frequency patrol observations. Similarly, Švestka (1976) notes that almost every Hα flare would be found to be accompanied by a soft X-ray burst should the X-ray recorder be sensitive enough.

There are indications for the presence of a weak impulsive phase during the development of several flares of our category III (flares occurring between plage regions). A good example is the flare of August 27, 1983 which was accompanied by a microwave type IV burst and a hard X-ray burst (Table I). This spotless flare was the only flare observed in the interval 01 : 28 to 08 : 56 UT that day, so the association with the bursts seems rather unambiguous.

The association of spotless flares with metric type IV events was very low and we connected only the flare of September 4, 1978 with a significant type IV event. This seems to be in contradiction with the findings of Dodson and Hedeman (1970) who state that 13 out of 38 major Hα flares (Hα importance ≥ 2) which occurred in regions with small or no spots were accompanied by metric type IV (6 cases) or significant metric continuum radiation. However, an examination of their flare list reveals that only one flare associated with metric type IV emission was truly spotless (October 23, 1962), all the others occurred in regions where small spots were present. Another spotless flare associated with a metric type IV event is the well studied Skylab flare of July 29, 1973 (Michalitsanos and Kupferman, 1974; Hejna, 1975; Švestka *et al.*, 1982). We would like to add in proof that from our preliminary list of spotless flares associated with metric type IV events, after careful examination, we had to discard three cases as small spots have been present at the flaring site. So we can conclude that spotless flares showing a well developed explosive phase, accompanied by metric type II and type IV bursts are quite rare. It appears that although not exclusively necessary, the presence of even small spots is highly favouring the occurrence of an explosive phase and the associated metric continuum radiation.

Acknowledgements

We would like to thank Dr A. Zappalà (Catania Astrophysical Observatory) for informations on several spotless flares and Dr M. Pick (D.A.S.O.P. Observatoire de Meudon) for critical reading of the manuscript. This work was partially supported by CNR (Italian Research Council) MPI (Ministry of Education). The radio data reduction was carried out at the Computing Centre of the Astronet node at the Trieste Observatory.

References

Ambashta, A. and Bhatnagar, A.: 1983, *Photographic Atlas of the Solar Chromosphere*, Udaipur Solar Observatory, Udaipur, India.

Dennis, B. R., Orwig, L. E., Kipplinger, A. L., Gibson, B. R., Kennard, G. S., and Tolbert, A. K.: 1986, *Hard X-ray Burst Sepectrometer Event Listing 1980–1985*, NASA Technical Memorandum No. 86236.

Dodson, H. W. and Hedeman, E. R.: 1970, *Solar Phys.* **13**, 401.

Hejna, L.: 1975, *Bull. Astron. Inst. Czech.* **26**, 119.

Kahler, S. W., Cliver, E. W., Cane, H. V., McGuire, R. E., Stone, R. G., and Sheely, N. R., Jr.: 1986, *Astrophys. J.* **302**, 504.

Krüger, A.: 1979, *Introduction to Solar Radio Astronomy and Radio Physics*, D. Reidel Publ. Co., Dordrecht, Holland, Ch. II.

Luo, Bao-rong: 1982, *Chin. Astron. Astrophys.* **6**, 186.

Michalitsanos, A. G. and Kupferman, P.: 1974, *Solar Phys.* **36**, 403.

Sanahuja, B., Domingo, V., Wenzel, K.-P., Joselyn, J. A., and Keppler, E.: 1983, *Solar Phys.* **84**, 321.

Švestka, Z.: 1976, *Solar Flares*, D. Reidel Publ. Co., Dordrecht, Holland.

Švestka, Z., Dodson-Prince, H. W., Martin, S. F., Mohler, O. G., Moore, R. L., Nolte, J. T., and Petrasso, R. D.: 1982, *Solar Phys.* **78**, 271.

Trottet, G., Kedraon, A., Benz, A. O., and Treumann, R.: 1981, *Astron. Astrophys.* **93**, 129.

Zlobec, P., Messerotti, M., Li Hong-wei, Barry, R. and Comari, M.: *Solar Phys.* **111**, 000 (this issue).

THEORY OF RADIO PULSATIONS IN CORONAL LOOPS*

(Invited Review)

M. J. ASCHWANDEN

Institute of Astronomy, ETH, Zürich, Switzerland

Abstract. 'Pulsations' include a wide range of phenomena from strictly sinusoidal oscillations up to quasi-periodic fine structures, observed in the radio, microwave and X-ray frequency range. The various versions of pulsation models are reviewed and classified in three groups according to their driver mechanisms: (1) Magnetic flux tube oscillations (the emissivity of trapped particles is modulated by a standing or propagating MHD wave), (2) cyclic self-organizing systems of plasma instabilities (wave-particle, wave-wave interactions), and (3) modulation of acceleration (acceleration/injection of particles into the source). Observational references illustrate the applicability of the models. In conclusion, discrimination criteria of models are discussed, in order to give a key for interpretation of observations.

1. Introduction

Why is the phenomenon called 'pulsations', a marginal by-product of some solar flares, subject of numerous observational and theoretical investigations? What kind of physical information is inherent in an oscillating or pulsating system? The harmonic motion of the classical mechanical oscillator results from a restoring force proportional to the displacement. Any elastic body can be excited to oscillate in eigen-modes. In presence of a magnetic field the characteristic eigen-frequencies of an oscillating system are determined by the magnetic field strength, density and geometry of the hydrodynamic fluid. Hence, the only observation of some oscillation parameters (e.g., time period, quality factor, modulation depth) reduces the number of free parameters. Also in complex systems such as wave-particle interactions the observable parameters (temporal and spectral characteristics) are related to relevant parameters of the resonance condition (resonance frequency, resonance bandwidth, coherence time, stability). Consequently, the establishment of a well-chosen set of parameters would enable us to discriminate between alternative pulsation models. In analogy to the study of global oscillations (helioseismology) providing powerful results concerning the solar interior, the exploration of local coronal oscillations may provide prospective enlightening of coronal plasma processes.

For over two decades radio astronomers have been observing pulsating structures in the solar corona. 'Pulsations' include a wide range of phenomena from strictly sinusoidal oscillations up to rather quasi-periodic patterns or fine structures, observed in the metric/decimetric/microwave or in the X-ray frequency range. The richness of structural details requires different approaches of theoretical modeling. It is still an open question which pulsation mechanisms are relevant for the interpretation of the observed events,

* Proceedings of the Second CESRA Workshop on Particle Acceleration and Trapping in Solar Flares, held at Aubigny-sur-Nère (France), 23–26 June, 1986.

Solar Physics **111** (1987) 113–136.

and how the proposed models work in detail. An extensive collection of pulsation models is reviewed here. They form a very inhomogeneous group with respect to the degree of elaboration. Some models explain the complex pulsation phenomenon rather completely. On the other hand, some models only explain selected sophisticated details ignoring important steps of the puzzle such as the driving mechanism, the link of the driver to the emission, the emission mechanism, or the conversion into observable radio waves.

Pulsation models can be classified by their basic driver mechanism; a further subclassification may regard the involved emission mechanism (cf. Table I). The aim of this article is to cover the range of possible solar pulsation models as completely as possible, supplementing previous attempts (e.g., Krüger, 1979). In this review we describe the different proposed pulsation models following the classification scheme as summarized in Table I (Sections 2, 3, 4). In Section 5 we confront specific observations with the various theories. From the compilation of the proposed models discrimination criteria are extracted providing a preliminary key for the interpretation of observations.

TABLE I

Classification of pulsation models (the chapters are indicated by the numbers in parentheses)

(1) MHD oscillations (eigenmodes):	Gyrosynchrotron modulation (Section 2.4) Plasma emissivity modulation (Section 2.5) Bounce motion of wave trains (Section 2.6)
(2) Cyclic self-organizing systems:	Wave-wave interactions (Section 3.2) Wave-particle interactions (Section 3.3) (e.g., stream or loss-cone instability)
(3) Modulation of acceleration:	Periodic injection (Section 4.1) Periodic magnetic reconnection (Section 4.2)

2. MHD Pulsations

The first established and most elaborated model of coronal pulsations considers MHD oscillations (also called Fast Magnetosonic or FMS oscillations). Since the birth of this idea (Rosenberg, 1970) many modifications and theoretical improvements have succeeded. Theoretical aspects of MHD oscillations in different geometries of flux tubes were most thoroughly investigated by Roberts *et al.* (1984) whose notation we follow.

2.1. MHD WAVES

In a magnetically structured atmosphere, with density inhomogeneities, the surface of such regions forms natural boundary conditions for standing MHD waves, or acts as a barrier resp. waveguide for propagating waves. According to the wave propagation characteristics, two kinds of waves may be distinguished: the body wave and the surface wave. The body wave resonates with the (homogeneous) volume as a whole, the surface wave has a nonvanishing amplitude only near the well-defined magnetic surface.

The corona is structured by magnetic loops, the photosphere by flux tubes. Both structures show cylindrical geometry in first approximation if the curvature of the homogeneous magnetic field is neglected. Systematic studies of the propagation of MHD waves have been done for different geometries: (a) unbounded homogeneous medium (b) single magnetic interface (Wentzel, 1979a, Roberts, 1981a), (c) slender magnetic slab (Roberts, 1981b; Edwin and Roberts, 1982; Rae and Roberts, 1982), and (d) cylindrical symmetry (Wentzel, 1979b; Edwin and Roberts, 1983).

An unbounded homogeneous medium permeated by a uniform magnetic field can support the Alfvén wave and the fast and slow magnetoacoustic waves. For cylindrical symmetry of a flux tube or coronal loop, the dispersion relation has been derived using the following assumptions: (1) pressure balance across the boundary, (2) Fourier form of the disturbance, and (3) neglecting of gravitational effects (e.g., Edwin and Roberts, 1983). The solution for standing waves implies that there is no propagation of energy outside the cylinder. Hence, the oscillations of the magnetic flux tube are confined to the tube and penetrate only a short distance into their surroundings. These are the *free modes* of the oscillation (ω and k real).

Two free modes, the symmetric (sausage or pulsational) mode given by $n = 0$, and the asymmetric (kink or taut-wire) mode given by $n = 1$ were studied extensively by Edwin and Roberts (1983). Symmetric oscillations have their analogy in oceanography (Pekeris waves), asymmetrical disturbances are mathematically related to Love waves in seismology. The phase speeds of these modes are shown in Figure 1 for coronal conditions. There are two branches of MHD waves, the slow body waves on an acoustic scale, and the fast body waves on an Alfvénic scale.

2.2. MHD WAVES IN CORONAL LOOPS

The nature of the occurring wave modes depends on the relative values of the sound speed $c = \sqrt{\gamma p/\rho}$ and Alfvén speed $v = B/\sqrt{\mu\rho}$ inside and outside the cylinder. In the photosphere the plasma parameter β is of order unity or larger, whereas in the corona it is generally small. Using pressure balance across the boundary of the fluxtube we get typical conditions for the relative speed values in the corona and photosphere summarized in Table II (parameters inside the flux tube are labeled by the subscript 'o', outside by 'e').

In Rosenberg's (1970) theory the important influence of the loop's magnetic environment was ignored. He considered radial MHD oscillations. Meerson *et al.* (1978) developed a more refined model that involves the emissive damping of radial modes. The damping rate ν was found to be roughly proportional to the ratio of the densities outside ρ_e and inside ρ_i the fluxtube. Thus the requirement of a large scale ratio (ρ_i/ρ_e) is essential for the existence of free oscillations. The authors propose to compensate this radiative damping by additional amplification, e.g., due to resonant interaction with trapped energetic particles. They invoke the bounce-resonance instability of protons with velocity $\beta_p \geq 0.2$ for compact sources (density ratio of $\rho_e/\rho_i \approx 0.02$). However, their model is confined to damped modes, overlooking the free oscillation modes presented by Roberts *et al.* (1984).

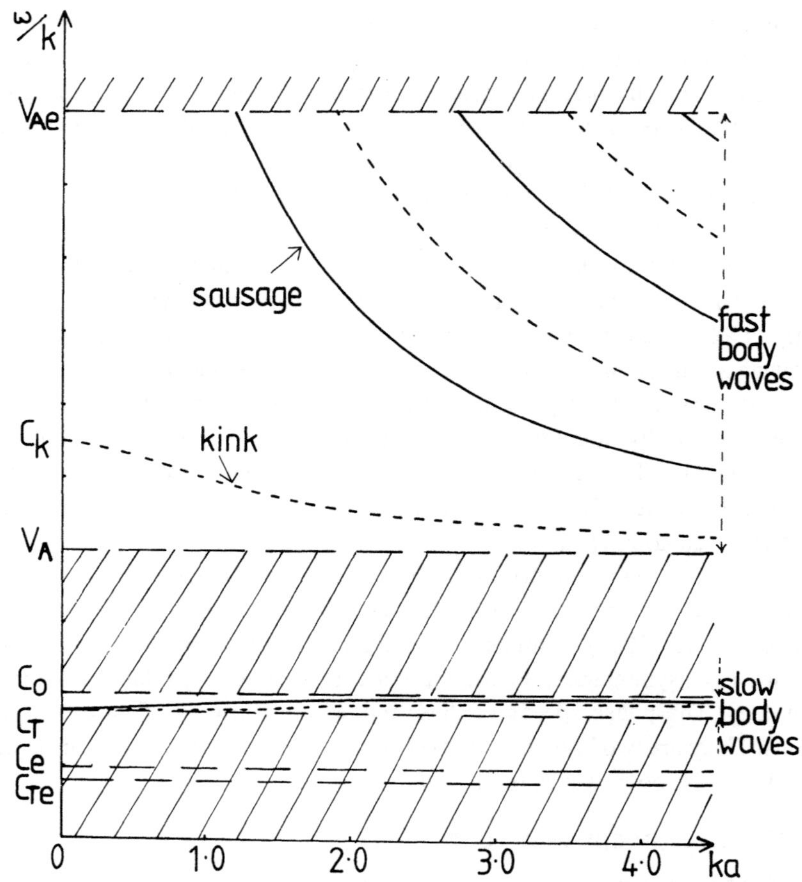

Fig. 1. The phase-speed ω/k as function of the longitudinal wavenumber k for the fast and slow magneto-acoustic waves in a coronal flux tube under typical conditions $v_{Ae} > v_A > c_0 > c_T > c_e$. Sausage mode (solid curves) and kink modes (dashed curves). (Courtesy of Edwin and Roberts, 1983.)

TABLE II

Coronal and chromospheric conditions

	Corona	Photosphere
Temperature $c_{th} = 2kT_e/m_e$	$T_e < T_0 \rightarrow c_e < c_0$	$T_0 < T_e \rightarrow c_0 < c_e$
Plasma parameter $\beta = (c_{th}/v_A)^2$	$\beta_0 \ll 1 \rightarrow c_0 \ll v_A$ $\beta_e \ll 1 \rightarrow c_e \ll v_{Ae}$	$\beta_0 < 1 \rightarrow c_0 < v_A$ $\beta_e \geq 1 \rightarrow c_e \geq v_{Ae}$
Density $\dfrac{\rho_e}{\rho_0} \approx \dfrac{v_A^2}{v_{Ae}^2}\left(\dfrac{1+\beta_0}{1+\beta_e}\right)$	$\rho_e \ll \rho_0 \rightarrow v_A \ll v_{Ae}$	$\rho_e > \rho_0 \rightarrow v_A > v_{Ae}$
Typical conditions	$c_e < c_0 < v_A < v_{Ae}$	$v_{Ae} < c_0 < c_e < v_A$

Standing waves result if the whole magnetic loop is synchroneously excited. If disturbances are impulsively generated, *propagating* waves are produced. From the study of impulsively generated Pekeris-waves in oceanography Roberts *et al.* (1984) derived the behaviour of such propagating disturbances in coronal loops which correspond closely to the fast sausage mode. The evolution of the fast sausage wave in the low-β corona passes a periodic, a quasi-periodic and a decay phase. During the quasi-periodic phase, starting at the time $t = z/v_A$ (when high-frequency information arrives at the source), and ending at the time $t = z/v_g^{min}$ (when the group velocity passes through a minimum), the amplitude of the propagating MHD wave is strongly amplified. In this quasi-periodic phase the shortest periods of $\tau^{min} = 2\pi/\omega^{min}$ occur. This model may be tested by the following observed quantities: (i) the oscillation period τ, (ii) the predicted duration $\tau_{duration}$ of the quasi-periodic phase,

$$\tau_{duration} = h\left(\frac{1}{c_g^{min}} - \frac{1}{v_A}\right), \tag{2.1}$$

and (iii) its typical phases of the time evolution (Figure 2).

Wave propagation in flux tubes may also provide the formation of *solitons*, if non-linearities are accounted and weak dispersion occurs, mainly under photospheric conditions (Roberts and Mangeney, 1982).

2.3. Pulse periods of MHD oscillations

An observational test for the existence of these MHD waves in the corona may be the comparison of the time periods. Assuming a standing wave, with its knots at the footpoints of a loop with length L, the wavenumbers are $k = j \times \pi/L$ for integers $j = 1, 2, \ldots$ Pulsation periods $\tau = 2\pi/\omega$ may be derived by inserting the phase speed $v_{ph} = \omega/k$, hence, we obtain $\tau = 2L/jv_{ph}$. In Table III the periodicities of the relevant modes are compiled with estimates using typical coronal parameters (Roberts *et al.*, 1984).

The expression of the period τ_c for propagating waves in Table III is similar to that given by Rosenberg (1970) derived from radial flux tube oscillations. Many of the observed short-period pulsations can be explained in terms of an impulsively generated fast sausage wave propagating in a dense coronal inhomogeneity (open field region, or dense loop).

Time periods of the order of 0.1–5 s have been frequently found, and of the order of ≈ 50 s in rare cases. In the range of > 500 s no significant observation is available. The reported periodicities in the radio frequency range are listed in Table V. However, additional criteria are needed to decide whether the MHD pulsation model explains these observations.

Substructures with shorter time periods were observed that may be interpreted as subharmonics. Rosenberg (1970) observed subharmonics with a period of $\tau/3$ well fitting the ratio of the Bessel harmonic numbers 1.8 : 5.3. Other observations with 1 to 6 substructures were reported without explanation (Pick and Trottet, 1978).

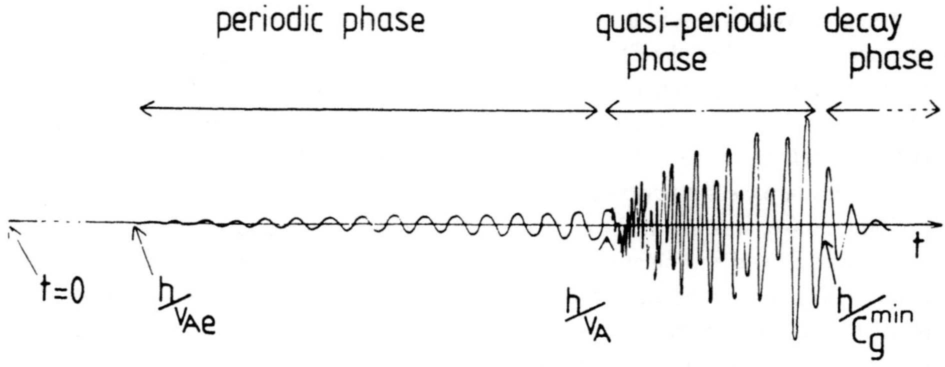

FIG. *a*

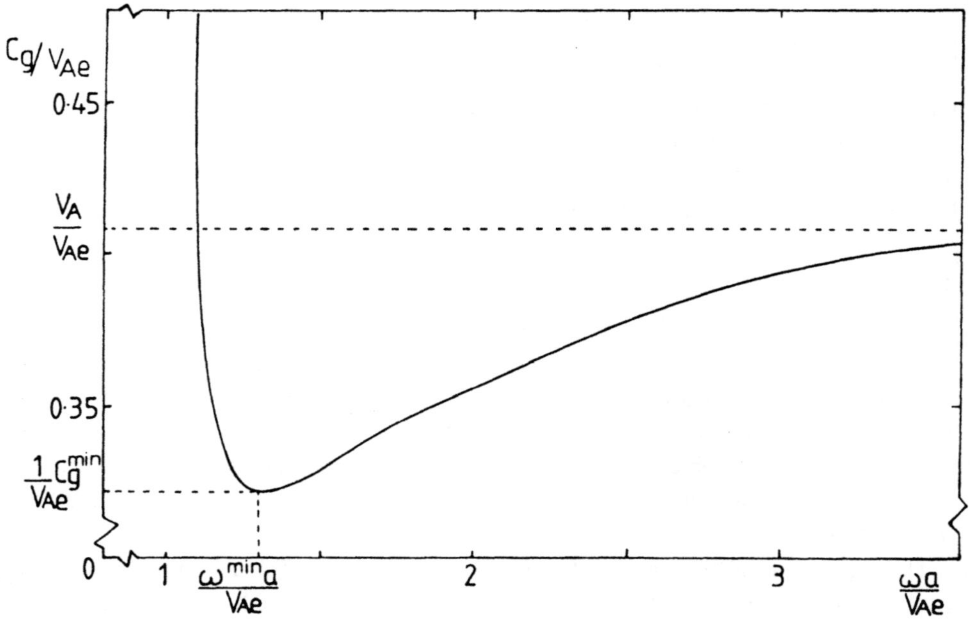

FIG. *b*

Fig. 2. (a) The evolution of the propagating fast sausage wave under coronal conditions. The subsequent phases in the disturbance are shown as observed at height $z = h$ away from an impulsive source at $z = 0$. (b) The group velocity c_g in units of the external Alfvén speed v_{Ae} as function of the dimensionless frequency $\omega a/v_{Ae}$ for the sausage mode. (Courtesy of Roberts *et al.*, 1984.)

2.4. Modulation of synchrotron emission

Both standing and propagating MHD waves are confined to the oscillating flux tube which cannot be directly observed. The oscillating MHD wave drives a periodic variation in the magnetic field. The most immediate link to the observer may be realized by an electromagnetic radiation whose emissivity is directly related to the magnetic field,

TABLE III

Pulse periods of MHD waves

Standing MHD waves		
Slow modes	$T_0 = 2 \times 10^6$ K	$\tau_s \approx 850$ s
$\tau_s = \dfrac{2L}{kc_T} = \dfrac{1.2 \times 10^{-4} L}{jT_0^{1/2}} \left(1 + \left(\dfrac{c_0}{v_A}\right)^2\right)^{1/2}$	$L = 10^{10}$ cm $j = 1$, $c_0 \ll v_A$	
Fast kink mode	$N_0 = 10^9$ cm^{-3}	$\tau_s \approx 50$ s
$\tau_f = \dfrac{2L}{kc_k} = \dfrac{4\pi^{1/2} L}{j} \left(\dfrac{\rho_0 + \rho_e}{B_0^2 + B_e^2}\right)^{1/2}$	$L = 10^{10}$ cm $B_0 = 40$ G, $\rho_0 \gg \rho_e$	
Fast sausage mode	$N_0 = 10^9$ cm^{-3}	$\tau_s \approx 1.5$ s
$\tau_{f'} = \dfrac{2\pi a}{c_k} = 4\pi^{3/2} a \left(\dfrac{\rho_0 + \rho_e}{B_0^2 + B_e^2}\right)^{1/2}$	$a/L = 10^{-2}$ $B_0 = 40$ G, $\rho_0 \gg \rho_e$	
Propagating MHD waves		
$\tau_c = \dfrac{2\pi a}{j_{0,e} v_A} \left(1 - \dfrac{\rho_e}{\rho_0}\right)^{1/2} \approx 2.6 \left(\dfrac{a}{v_A}\right)$	$a = 2 \times 10^8$ cm, $s = 1$ $N_0 = 10^9$ cm^{-3} $B_0 = 40$ G, $\beta \ll 1$	$\tau_c \approx 2$ s

i.e., the gyrosynchrotron emission. The first interpretation of MHD pulsations in terms of synchrotron radiation emitted by trapped electrons was given by Rosenberg (1970). The evidence for synchrotron radiation was based on the analysis of the radio spectrum and the high energetic electron flux observed simultaneously by spacecraft.

Fluctuations in the synchrotron emission due to small variations in the magnetic field were systematically analyzed by Gotwols (1973). He explored four cutoff mechanisms and their effects on the pulsations: (i) synchrotron self-absorption, (ii) Razin effect, (iii) gyrosynchrotron absorption by thermal electrons, and (iv) collisional (free-free) absorption. For a pulsating self-absorbed source he predicted two bands of pulsations out of phase (180°) on both sides of the spectral peak. In contrast, a Razin-suppressed source (relevant for ultra-relativistic particles) would produce only a single band of pulsations around the spectral peak, since the synchrotron emissivity and the absorption coefficient are depressed below the critical frequency. The free-free absorption coefficient is not a function of the magnetic field and modifies the synchrotron emissivity by a frequency-dependent weighting factor. Although these four cutoff mechanisms clearly predict the spectral features of a MHD-modulated synchrotron source, no observations have confirmed their appearance up to now.

Modulation of the gyrosynchrotron emissivity may also be affected by the modulated energy spectrum during radial oscillations (Gaizauskas and Tapping, 1980). The high-energy electrons trapped in a radially oscillating flux tube change their momentum according to the betatron process. The first adiabatic constant (magnetic momentum) describes the increase of transverse electron momentum during the compression phase of the oscillation. If part of it is transferred to the parallel momentum before the tube expands (e.g., by collisions), the electron gains net energy. Consequently, the betatron process is pumping energy to the oscillating system and may enhance considerably the duration which is limited by radiational and collisional damping. This energy support is necessary especially for microwave oscillations which take place in the lower and denser regions of the corona.

Another unsolved problem is the origin of pulsations, respectively, their excitation. McLean *et al.* (1971) observed a simultaneous shock wave (type II) and concluded the excitation of radial oscillations at the top of the arch when the shock front crosses them. The acceleration of the energetic electrons responsible for the synchrotron emission was assumed by Fermi-process set up by reflected Alfvén wave trains between the rising shock front and the top of the arch.

The discovery of asymmetric pulse shapes casted new doubts on the reliability of the MHD pulsation model in terms of the modulated synchrotron emissivity. Rosenbergs radial MHD pulsations were expected to produce symmetrical sinusoidal pulse shapes. McLean and Sheridan (1973) observed asymmetric pulse profiles with longer rise times than decay times suggesting nonlinear terms in the MHD theory. Abrami (1972) found frequent pulsations in absorptions and sawtooth time profiles.

A pulsation model based upon torsional waves is discussed by Tapping (1983). The strong damping of radial oscillation at the boundary of the flux tube, due to radiation of acoustic waves into the surrounding plasma, is assumed to inhibit long-lived, stable, deeply modulated pulsations. As an alternative, Tapping suggests torsional waves of small amplitude in a flux tube firmly anchored in the photosphere. The synchrotron emission from electrons in the tube is modulated by changing orientation of the magnetic field lines with respect to the observer. Large modulation depths in pulsations could be observed when the twisting flux tube is viewed end-on. The pulse period of standing waves is given by the bounce time of trapped torsional waves

$$\tau_{\text{torsional}} = n \frac{2L}{n\overline{v_A}} \tag{2.2}$$

thus requiring high harmonical numbers n to fit typical observed pulsation times in the order of 1 s.

Torsional waves can not be generated by compression (at least in linear theory) and, hence, cannot be excited by motions in the external plasma. They would have to originate in the disturbed tube, either above or below the photosphere (Tapping, 1983). Magnetic reconnection during flares in twisted flux tubes would generate torsional pulsations. Since the observed pulse periods of typically 1 s require very high orders of torsional wave harmonic numbers ($n > 100$). Tapping proposed disturbances in the chromosphere as the most likely exciter of torsional pulsations. The steep density gradient would then act as a high-pass filter and transmit only the high-order pulsation modes.

Another modulation mechanism situated outside the synchrotron source was offered by Zaitsev and Stepanov (1982). If the synchrotron emission traverses an MHD resonator, such as a pulsating flux tube or loop, the original radio emission may undergo a modulation, since the absorption coefficient of the MHD resonator varies periodically, if the density varies. The varying free-free absorption coefficient (for $\omega \gg \omega_p$) was found to produce an intensity modulation.

All these proposed models suggest synchrotron emission as the radiation mechanism. The linear theory of MHD oscillations yields time periods (see Table III) based on the knowledge of some physical parameters (B_0, N_0, T_0, a, L) and predicts symmetric sinusoidal time profiles. The spectral features of synchrotron emission depends on the involved particle energy spectrum. However, although the interpretation of MHD pulsations in terms of gyrosynchrotron emission is established for microwaves, some discrepancies arose from metric observations, i.e., from the interferometric measurements of pulsating source sizes. Either the derived source extent did not fit the time period relation $\tau \approx 2.6a/v_A$ (Pick and Trottet, 1978; Trottet *et al.*, 1979) or it delivered too high brightness temperatures (Trottet *et al.*, 1981). The authors concluded that incoherent as well as coherent gyrosynchrotron radiation are not compatible with their observations in the metric case. Further puzzling problems (asymmetric pulse shape, narrow frequency bandwidth, questionable exciter mechanisms, energy support of radiative modes, suppression of low-order modes for torsional pulsations, etc.) led to alternative suggestions, abandoning synchrotron radiation as emission process, and invoking plasma emission modulated by MHD waves.

2.5. Modulation of plasma emissivity

Since 1975, several papers have been published based on the suggestion that plasma emission plays the role of an important emission process in many kinds of decimetric radio bursts. Benz and Tarnstrom (1976) investigated the question 'Synchrotron or plasma process emission in narrow-band type IVdm bursts'. Their conclusion confirmed the existence of plasma emission for the considered bursts, and this result is also valid for pulsations with a frequency bandwidth of the order of $\Delta \nu / \nu \lesssim 1$. Further support of this idea is given in Roberts *et al.* (1983, 1984).

MHD pulsations, radial as well as torsional ones, modulate the magnetic field. Since radial pulsations (sausage mode) are compressive, the radius of the flux tube varies periodically, and the constancy of the magnetic flux $\Phi \approx B_z(t)\,(\pi r^2(t))$ requires also a periodic variation of the axial magnetic field $B_z(t)$. (For noncompressive modes, e.g., torsional waves or kink mode of radial oscillations, the axial magnetic field is not perturbed in linear theory.) The time dependence of the magnetic field changes the perpendicular component of the velocity $u_\perp$ of the trapped electrons according to the first adiabatic invariant (constancy of magnetic momentum) that is for the nonrelativistic case $\mu = B(t)/u_\perp^2(t)$. If the MHD period is longer than the collision time of the trapped particles ($\tau > \tau_{ei}$), transfer of momentum from the perpendicular to the parallel component will occur, thus tending to restore the isotropy in momentum. On the other hand, if the MHD pulsation period is shorter than the collision time, the perpendicular momentum of the trapped electrons will vary in the same way, and the parallel momentum remains constant. Consequently, the pitch angle will vary according to

$$\frac{\tan \alpha(t_1)}{\tan \alpha(t_2)} = \frac{u_\perp(t_1)}{u_\perp(t_2)} = \sqrt{\frac{B(t_1)}{B(t_2)}} = \frac{r(t_2)}{r(t_1)} . \tag{2.3}$$

In this way, MHD pulsations may modulate the magnetic field and the pitch angle distribution of the trapped electrons simultaneously. Therefore, any plasma instability which is sensitive to the pitch angle, especially loss-cone instabilities, may be dramatically affected by MHD pulsations.

Several plasma emission mechanisms driven by loss-cone instabilities have been proposed to explain fine structures and type IV radio continua: loss-cone instabilities either interrupted by injected particles leading to sudden reductions (Kuijpers, 1974, 1975; Zaitsev and Stepanov, 1975; Benz and Kuijpers, 1976), or modulating the excitation of waves (Stepanov, 1974, 1978). Berney and Benz (1978) have shown for whistler and upper hybrid waves that the linear growth rate and the stability threshold is highly sensitive to the loss-cone (pitch) angle. For 'marginal stable' particle-wave systems a modulation of the pitch angle would switch the instability on and off. The final fluctuation in radio emission would be a nonlinear response to the modulated magnetic field.

In the scenario of Zaitsev and Stepanov (1983) the periodic variation of the mirror ratio by MHD pulsations in the flare release region leads to a modulation of the loss-cone instability growth rate (upper hybrid waves) and, hence, of the resulting microwave emission. This model includes pulsations in hard X-ray, microwave and decimetric emissions. Rosenberg (1972) proposed nonlinear coupling of upper hybrid plasma waves and Bernstein waves as possible explanation for 'Zebra'-like pulsating fine structures. However, a critical analysis of the polarization mode by Sy (1974) revealed some difficulties of this model. Another instability sensitive to the pitch angle is the growth of whistlers. Tapping (1978) described a possible generation mechanism. In his scenario radial MHD oscillations of the flux tube near the legs of the loop modulate the magnetic field, the pitch angle distribution and, hence, the growth rate of whistler waves packets. These whistlers rise upwards in the loop along the magnetic field lines and modulate the metric radio emission at the top of the loop.

The enumerated instabilities exhibit a high sensitivity on the pitch angle distribution thus producing a nonlinear response to the oscillating magnetic field. The modulation of the observed radio emission is interlocked with the on/off-switching of the involved instability. The MHD pulsation plays the role of a timing clock (pulse period), whereas the driven instability determines the pulse shape (amplitude, rise time, decay time). The identification of the involved instability has to be done by analysis of the observed (mostly asymmetric) pulse profiles. The rise time contains information on the linear growth rate, and the decay time which is governed by thermal, collisional damping, thermal conductivity or diffusion processes.

2.6. Periodic bounce motion of wave trains

Several pulsation models were proposed wherein the observed pulse period of radio, microwave or hard X-ray emission is related to the bounce period within the loop. A traveling wave such as Alfvén or whistler waves is assumed to travel forth and back between two mirror points of a loop, where the propagation conditions change the direction of the wave vector. The interaction of these periodically reflected waves with the emission region of radio waves is believed to cause a kind of pulsating signatures.

Chiu (1970) presented a detailed model of bouncing whistlers in a large-dimensioned loop. Whistler wave trains are assumed to be generated by the pitch angle instability. The resultant ≈ 10 s bounce motion modulates (i) the synchrotron emission (microwave) via the resonant electrons at some harmonics of the gyrofrequency, (ii) the hard X-ray emission due to pitch-angle scattering of MeV-electrons, and (iii) type III emissions by scattering with cyclotron plasma waves.

Yao (1983) suggested the interaction of bouncing whistlers with the loss-cone distribution of the electrons at the top of the loop, thus enhancing the diffusion into the loss-cone. The subsequent precipitation and simultaneous gyrosynchrotron emission would be modulated with the same period.

A disturbance traveling with Alfvén speed, bouncing in a magnetic loop, is assumed to drive the quasi-periodic modulation of microwave emission during an adiabatic compression phase of a flare in a model of Wiehl and Mätzler (1980).

3. Cyclic Self-Organizing Systems

In contrast to the models of MHD pulsations (Section 2) which correspond to the physical picture of a classical oscillator characterized by discrete eigen-modes, we present in this chapter an entirely different approach: quasi-stabilized systems of non-linear plasmas governed by an oscillatory phase of wave-wave or wave-particle interactions. These systems share the principle of self-organization, which enjoys an increasing attraction in astrophysical applications.

3.1. Lotka–Volterra-like systems

The properties of self-organizing processes are characterized by (i) cyclic processes (closed trajectories in phase space) and (ii) connection of the considered system with the environment by other processes (energy dissipation or transfer). There is a big difference with isolated systems which tend to reach equilibrium position (converging trajectory in phase space), whereafter the evolutionary dynamics disappears. Dissipative self-organizing systems, however, communicate with the environment. These interactions enable the system to persist in a cyclic state *far* from the equilibrium.

To illustrate the typical behaviour of a self-organizing system we remember the classical Lotka–Volterra equation. The classical mathematical description includes two time-dependent functions $[X(t), Y(t)]$ that interact mutually in the following way:

$$\frac{dX}{dt} = k_1 X - k_2 XY,$$
$$\frac{dY}{dt} = k_2 XY - k_3 Y. \tag{3.1}$$

This type of system dynamics represents a key instrument and is used in a great number of modifications also in auroral and solar plasma physics (Trakhtengerts, 1968; Zaitsev, 1971; Stepanov, 1974; Bespalov and Trakhtengerts, 1977a, b; Kuijpers, 1978; Bardakov and Stepanov, 1979; Aschwanden and Benz, 1987). For plama physical applications the two parameters may be represented, e.g., by wave $N(\mathbf{k}, t)$ and/or particle distribution functions $f(\mathbf{p}, t)$. The implementation of distributions in the coupled system of integro-differential equations complicates the analytical solution considerably. However, in many cases the dynamics can be studied by reduction to a classical Lotka–Volterra equation system. For small perturbations the asymptotic cycle period is

$$\tau_0 = \frac{2\pi}{\sqrt{k_1 k_3}} . \tag{3.2}$$

The system rotates around the stationary point in the phase space ('limit cycle'). For strong perturbations the oscillation period is amplitude-dependent and the (closed) phase trajectory is determined by the initial boundary conditions. The development of a limit cycle occurs only for a restricted set of initial values, whereas the system evolves towards nonlinear oscillations for other initial values (e.g., Kuin and Martens, 1982).

3.2. Oscillatory wave-wave interactions (resonant/nonresonant regimes)

In the scope of the present review we are interested in applications to solar plasma instabilities. Some promising approaches have been made, mainly through the constructive cross-fertilization between terrestrial (laboratory and ionospheric) and solar plama physics. Up to now two classes of instabilities have been applied (beam instabilities and loss-cone instabilities) to study nonlinear wave-wave interactions in the oscillating saturation phase.

The nonlinear interaction of plasma waves is subject of numerous theoretical investigations, especially in studying the evolution of an explosive instability. The results depend on the assumed coherent-phase or random-phase description. The stabilization of an explosive instability due to nonlinearity may be caused by frequency shifts which detune the resonance condition, and may quench the instability due to the finite width of the resonance condition (Fukai *et al.*, 1969; Weiland and Wilhelmsson, 1973). Saturation effects produce two kinds of analytical solutions: (1) soliton-like in time or (2) oscillatory dependence in time. The soliton-type solution in time, realized in the case where all interacting waves have the same amplitude, changes into an oscillatory type of structure if dissipative effects are included (Oraevskii *et al.*, 1973; Weiland and Wilhelmsson, 1977). The oscillatory type of stabilization is found to exist due to the combined effect of nonlinear frequency shifts and effective nonlinear dissipation (Weiland and Wilhelmsson, 1973). A numerical example of this repetitive nature of explosive instabilities is shown in Figure 3 (from Weiland and Wilhelmsson, 1973). A similar behaviour for nonlinear wave-particle interaction in the case of exponentially growing electrostatic (uper hybrid) instabilities has been investigated by Benz (1980).

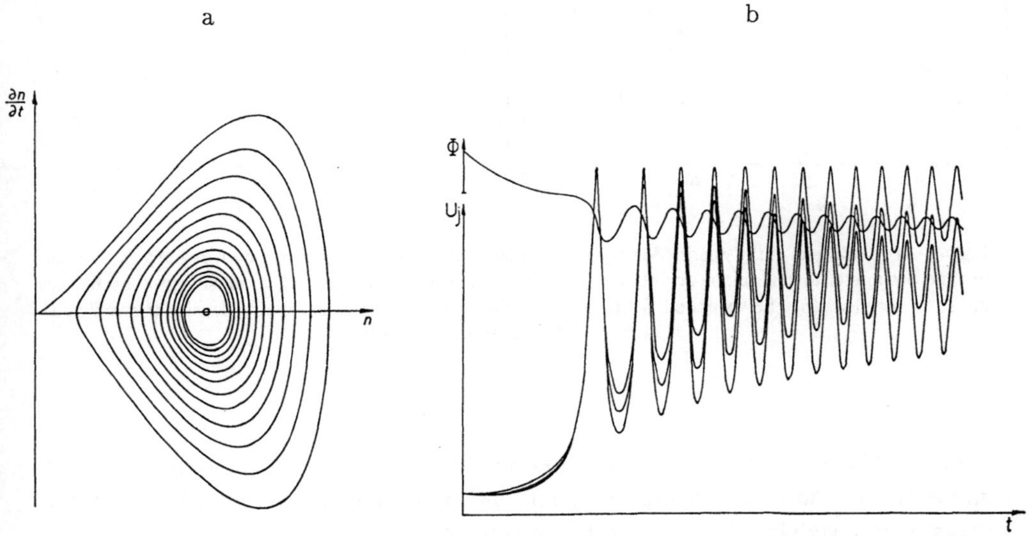

Fig. 3. (a) Phase space diagram in the presence of dissipation. (b) Wave amplitude u_j and phase Φ developments in the presence of mutually different dissipation of the explosive instability. (Courtesy of Weiland and Wilhelmsson, 1973.)

The nonlinear interaction of plasma waves on a stream-plasma system driven by the two-stream instability has been analyzed by, e.g., Smith and Fung (1971) and Heyvaerts and Verdier de Genouillac (1974). The nonlinear wave-particle interaction leads to a stabilization of the stream; the Langmuir waves of the 'resonant' region are scattered into the 'nonresonant' region and *vice versa*. During the quasi-steady phase the wave energies show oscillatory time profiles. Numerical calculations (Smith and Fung, 1971) show the initial unstabilized phase, followed by a quasi-steady oscillatory phase (Figure 4). The oscillating wave energies of Langmuir waves may produce radio bursts with periodic fine structure.

Interpretations of fine structures invoking periodical regimes of plasma instabilities were proposed by Gotwols (1972) for broad-band pulsations upon an underlying type IV continuum, by Elgarøy and Sveen (1973) for wide band quasi-periodic bursts with periods of 0.1–0.2 s, by Zaitsev (1971) for 'rain'-type bursts (succession of regularly repeated short-lived broad-band bursts).

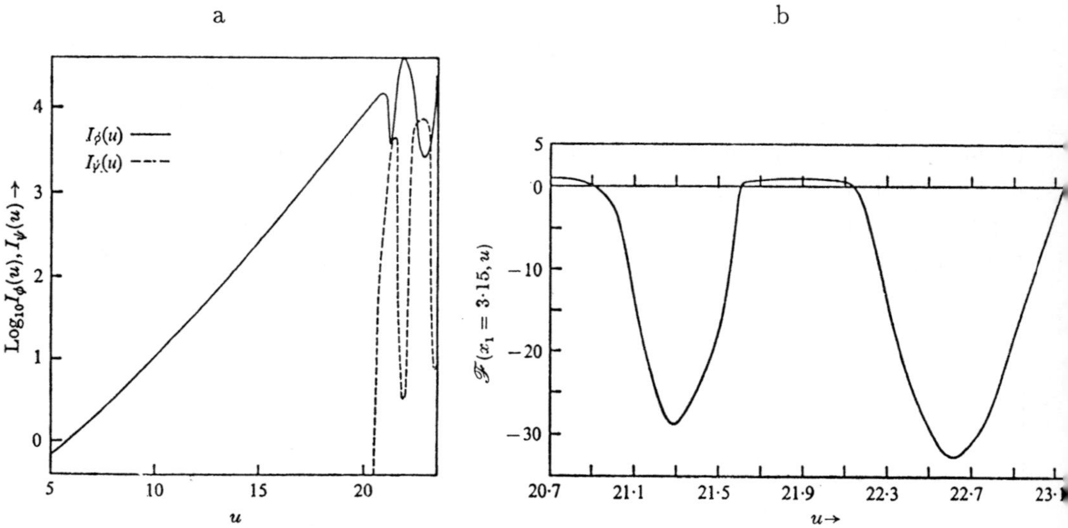

Fig. 4. (a) Evolution of the energy density I_Φ (resonant region) and I_Ψ (nonresonant region) as function of time parameter u. The initial evolution reveals a linear growth, after $u = 20$ a quasi-steady oscillatory phase is reached. (b) The extended figure displays the rate of change in energy density (normalized). (Courtesy of Smith and Fung, 1971.)

Zaitsev (1971) presented a dynamic analysis of the quasi-steady oscillating phase. The interaction of the waves in the 'resonant' and 'nonresonant' regions (driven by the two-stream instability) is described by an equation system similar to Lotka–Volterra equation (3.1):

$$\frac{\partial W}{\partial t} + V_{gr} \frac{\partial W}{\partial x} = (\gamma a - \nu)W - \zeta W W^* ,$$
$$\frac{\partial W^*}{\partial t} + V^*_{gr} \frac{\partial W^*}{\partial x} = \zeta W W^* - \nu W^* , \tag{3.3}$$

where γa is the growth rate, ν is the collision rate, ζ is the nonlinear scattering rate between the 'resonant' (wave energy = W) and the 'nonresonant' (wave energy = W^*) regions. The feedback upon the particle distribution, a diffusion equation describing the spreading of the stream distribution function, is noted as

$$\frac{\partial a}{\partial t} + V_s \frac{\partial a}{\partial x} = -Da^{5/2} W , \tag{3.4}$$

for a one-dimensional ion stream distribution function f_s (where $a(x, t) = \partial f_s / \partial V$ denotes the positive gradient of the distribution function).

From the 'limit cycle' (Equation (3.2)) we get for small amplitude modulation a time period in the order of $\tau_0 = 2\pi/\sqrt{\gamma \nu}$. The spatial dependence of the plasma included in Equation (3.3) plays the role of nonuniform initial conditions. Zaitsev (1971) found, as a criterion for the region of instability to pulsate as a whole with a single definite frequency (small dispersion of oscillation period), the following relation:

$$\frac{l_0^2 \nu \gamma}{V_{gr}^2} \gg 1\,, \tag{3.5}$$

where l_0^2 represents the size of the initial disturbance that is apparently comparable to the stream dimensions or the trap dimensions. Zaitsev's (1971) model, taken into account by several authors for the interpretation of radio pulsations, has been ruled out by Zaitsev *et al.* (1984) in the application to solar flares. Their argument was based on the beam stabilization condition which requires proton beams due to their much longer life time than that of electrons.

The nonlinear wave-wave interactions driven by two-stream instabilities as described above may work in a similar way for loss-cone instabilities. Trakhtengerts (1968) proposed a model based on the loss-cone instability exciting upper hybrid waves in terrestrial magnetic loops. Using the Lotka–Volterra equation (3.1) he estimated the modulation of the precipitating auroral flux due to the oscillatory solution of the nonlinear wave-wave interactions between resonant and nonresonant regimes. Experimental evidence was given by measurements of the precipitating electron flux.

A definitive answer to the significance of periodic wave-wave interactions in explaining radio pulsations, which works for auroral pulsations, but has been disproved in the solar context, may be established by exploration of additional parameters, e.g., the detailed shape of the pulse time profile.

3.3. Oscillatory wave-particle interactions (relaxational oscillations)

The positive growth of an instability dissipates energy from the resonant particles. A stationary operation of the instability, e.g., the quasi-stabilized oscillatory saturation phase will modify the particle distribution considerably. This process is attributed to quasi-linear diffusion. The interaction of the unstable waves with the resonant particles may also be approximated by a Lotka–Volterra-like equation. The periodic solution predicts not only an oscillatory wave energy amplitude, but also a phase-shifted oscillatory particle distribution (relaxational oscillation).

Relaxational oscillations were studied by Stepanov (1974), Bespalov and Trakhtengerts (1977a, b), Bardakov and Stepanov (1979), Stepanov (1981), and Aschwanden and Benz (1987). Bespalov and Trakhtengerts (1977a, b) considered the cyclotron instability for whistlers in the Earth's radiation belt, Bardakov and Stepanov (1979) invoked the two-stream instability for solar magnetic traps, and Aschwanden and Benz (1987) applied the electron-cyclotron maser driven by the loss-cone in coronal loops.

The dynamics of the wave-particle interaction is described in all cases cited above by the following system of coupled equations:

$$\frac{\partial N(\mathbf{k}, t)}{\partial t} + \mathbf{v}_g(\mathbf{k}) \frac{\partial N(\mathbf{k}, t)}{\partial \mathbf{r}} = \Gamma(\mathbf{p}, \mathbf{k}, f) N(\mathbf{k}, t) - \gamma(\mathbf{p}, \mathbf{k}, f) N(\mathbf{k}, t),$$

$$\frac{\partial f(\mathbf{p}, t)}{\partial t} + \mathbf{v}(\mathbf{p}) \frac{\partial f(\mathbf{p}, t)}{\partial \mathbf{r}} = \frac{\partial}{\partial p_j} \hat{D}_{ji}(\mathbf{p}, \mathbf{k}, N) \frac{\partial f(\mathbf{p}, t)}{\partial p_i} + \left. + \frac{\partial f(\mathbf{p}, t)}{\partial t} \right|_{\text{source}} + \left. \frac{\partial f(\mathbf{p}, t)}{\partial t} \right|_{\text{loss}} . \tag{3.6}$$

The waves are represented by the photon number density $N(\mathbf{k}, t)$ in the $\mathbf{k}$-space, the particle system is described by its density distribution $f(\mathbf{p}, t)$ in momentum space, $\Gamma(\mathbf{p}, \mathbf{k}, f)$ is the growth rate, $\gamma(\mathbf{p}, \mathbf{k}, f)$ is the damping rate, $\mathbf{v}_g(\mathbf{k})$ is the group velocity of the emitted waves, and $\hat{D}_{ji}(\mathbf{p}, \mathbf{k}, N)$ is the diffusion tensor. The wave equation describes the balance between emission and growth/damping rate. Since the growth rate depends generally on the particle distribution, a coupling between the wave and particle equation results. The change in the particle distribution due to the interaction with resonant waves is described by the quasi-linear diffusion. Additional modifications of the particle distribution result from the precipitation out of the loss-cone (loss term). Consequently, a quasi-stationary model requires a particle supply (source term). In the stationary limit, the source rate counterbalances the diffusion and loss rate.

A rigorous solution of the coupled integro-differential equation has not been tackled up to now. However, a recent analysis of the problem revealed a consistent solution for the 'limit-cycle' applied to the electron–cyclotron maser (Aschwanden and Benz, 1987). The pulse period was determined as product average of growth and diffusion time

$$\tau = 2\pi \sqrt{\tau_{\text{growth}} \, \tau_{\text{diff}}} \, . \tag{3.7}$$

The variation of the pulse amplitude was found to be very sensitive to fluctuations of the growth rate. As consequence, the pulse amplitude varies exponentially with the growth rate ($I \sim \exp(\Gamma R/v_g)$), and the pulse period varies with the square root of the amplitude ($\tau \sim$ I), herewith well explaining the strong fluctuations of peak intensities and pulse periods observed for decimetric solar pulsations.

Interpretation of radio pulsations implying oscillatory wave-particle interactions were given by Sastry *et al.* (1981) for decametric pulsations, by Bardakov and Stepanov (1979), Stepanov (1981) for type IV pulsations. The oscillatory particle distribution in a loss-cone modulates the precipitating flux of particles towards the legs of the magnetic loop. This modulated particle precipitation may be observed (i) directly, e.g., for auroral events in the Earth's magnetic field (Trakhtengerts, 1968; Bespalov and Trakhtengerts, 1977a, b; Bardakov and Stepanov, 1979) or (ii) indirectly by modulated hard X-ray flux for coronal loops (Zaitsev *et al.*, 1984).

To identify this class of pulsation models involving oscillatory saturation of nonlinear wave-particle interactions of the stream or loss-cone instability type, refined theoretical work has to be done basing on reliable parameters of the solar environment (magnetic topology, source function of trapped particles). Especially the effect of the quasi-linear diffusion has to be solved selfconsistently.

4. Modulation of Acceleration

Investigating all possible periodical exciters of solar radio emissions, not only mechanisms possessing an internal pace-maker, but also mechanisms driven by external repetitive forces have to be considered. The common aspects of this residual third group of pulsation models is two-fold: (a) they do not fit into one of the previous two categories and (b) the resonant region where the pulsating radio emission originates is driven by external dynamics. This external driver may be represented by a modulated acceleration process: repetitive injection of particles (Section 4.1), whose driver may be a pulsed acceleration, e.g., a quasi-periodic magnetic reconnection (Section 4.2). The theoretical framework of such mechanisms is rather fragmentary; neither the dominant acceleration process nor its efficiency is well known for the solar environment.

4.1. Periodic Injection

Quasi-periodic fine structures during type IV are observed in a variety of pulse shapes, from emissive bursts up to clear absorption features upon a smooth continuum. Assuming plasma emission for narrow-banded decimetric type IV continua, absorption features have been interpreted in terms of injected particles into the magnetic trap (Zaitsev and Stepanov, 1975; Benz and Kuijpers, 1976). The injection of particles into the magnetic trap modifies the quasi-stationary loss-cone distribution of the particles, e.g., it flattens the gradient in the perpendicular velocity distribution which is responsible for the growth rate of the loss-cone instability, hence, the loss-cone instability becomes quenched and the radio emission is reduced. Calculations of this model confirm the time profile and observed flux reduction of the wedge-shaped absorption bursts. The periodicity of sudden reductions is commonly not very regular. The origin of quasi-periodic injection may be attributed to a quasi-periodic acceleration process. Interpretation of observations in terms of the 'sudden reduction'-model has been suggested, e.g., by Trottet *et al.* (1981) for quasi-periodic modulations of moving type IV bursts.

In addition, the reported hard X-ray pulsations (e.g., Takakura *et al.*, 1983; Kaufmann *et al.*, 1984; Costa and Kaufmann, 1986) are currently proposed to be associated with multiple injections of particles into the flare energy region. The interpretation of quasi-periodic radio emission due to injected particles may have the same origin.

The production of runaway electrons in pulses by a constant electric field, as observed in laboratories, may also take place in coronal loops (Kuijpers *et al.*, 1981). The runaway process becomes repetitive, respectively, pulsating, if the growth times of the slow and fast plasma waves are less than the collision time in the case of $\omega_C/\omega_p \gg 1$.

4.2. Periodic magnetic reconnection

The electric fields resulting from unstable coalescent current sheets are able to accelerate particles. Studying the current instability leading to anomalous resistivity in current sheets in which magnetic reconnection occurs, three cases are derived, depending on the relative characteristic lengths of inertial and resistive effects (Smith, 1977). We mention only the case of a pulsating regime of reconnection.

The pulsating regime originates in the following way: under specific conditions (half diffusion region smaller than the characteristic length of anomalous resistivity due to ion acoustic turbulence, but larger than the inertial characteristic length) a pulsating reconnection occurs due to the mutual balance of current instability growth/decay and current sheet collapse/expansion around a critical value. In this regime, 'reconnection is intrinsically a dynamical phenomenon, since anomalous resistivity is switching on and off and steady-state reconnection theory is inapplicable' (Smith, 1977).

In loop models of solar flares basing on magnetic reconnection such pulsating quasi-periodic phenomena have been currently overlooked. Spicer (1981) mentioned relaxation processes by the tearing mode, where marginal stability may cause the process to repeat itself in a quasi-periodic way leading to quasi-periodic bursts. Nakajima *et al.* (1984) reported observations of quasi-periodic pulses with double peak structure, seen in hard X-ray, γ-ray, and microwave emissions of two intense solar flares, and could explain them consistently in terms of coalescence instability of two current loops by computer simulations.

5. Observational Constraints

The pulsation models are classified in this review according to their intrinsic periodic driver: (1) free oscillations in eigen-modes, (2) nonlinear dynamics in saturated state, and (3) modulated acceleration. A subdivision is given, either by the emission mechanism (gyrosynchrotron, plasma emission, bremsstrahlung), or by the kind of coupling (wave-wave, particle-wave, acceleration-emission). This classification is based on physically different mechanisms.

The inherent dissimilarities of the reviewed models put constraints on the interpretation of observations. In this conclusion we summarize differing properties which may serve as discrimination criteria for the interpretation of observations. However, the key criteria presented in this synthesis, are not always sufficient to evaluate unambigeously the correct theoretical model for a given observation. Much more refined theoretical research is needed to fulfill this task, therefore the presented approach may serve as preliminary guideline.

5.1. The pulse period

The pulse period, the parameter which is most accessible by observations even with low time resolution, is not a appropriate tool to discriminate among the various theories. Nearly all of the models cited above are able to match the typical observed pulse period of ≈ 1 s. The pulse period relations (see Tables III and IV) are better suited to rule out a presumed model. For instance, in the case of MHD pulsations, interferometric observations determining the source size and brightness temperature have been used to disprove the interpretation in terms of MHD pulsations (Pick and Trottet, 1978; Trottet *et al.*, 1979, 1981).

The main use of the pulse period consists in the determination of the relevant mode for a given pulsation mechanism. For standing MHD pulsations the various modes (slow, fast kink, and fast sausage) differ each by an order of magnitude (Table III). In the case of the electron–cyclotron maser, where the pulse period (determined by the growth and diffusion time) is highly sensitive to the magneto-ionic mode and harmonic number, the numerical values differ by many orders of magnitude.

5.2. The Periodicity

As the term 'quasi-periodic' suggests, the constancy of the pulse period may be heavily disturbed in some cases. For systems oscillating in eigen-modes, the period is determined by geometry and elasticity (modulus) of the resonant volume and remains generally constant. In the case of MHD pulsations, the period can be expected as highly constant, since the matching parameters such as the axial magnetic field, the overall electron density and the flux tube topology are slowly varying parameters on the time scale of a typical event duration (≈ 1 min).

In contrast, cyclic self-organizing systems which communicate with the environment may exhibit variable pulse periods if the ambient parameters are only slightly time-dependent. Wave-particle interactions are very sensitive to fluctuations of the particle distribution: e.g., if a source of particles modifies the pitch angle distribution in the loss-cone, growth and diffusion time and consequently the pulse period are strongly affected. Moreover, instabilities operating in a loss-cone may be in marginal stability, threshold effects may then suppress significant growth sporadically. The variability of the pulse period is presumed to be much larger for coherent emission mechanisms than for others.

Also for the third class of pulsation models (modulation of acceleration processes) the variability of the observed pulse period may be determined by the kind of the accelerator, which may be driven by eigen-modes or self-organization (e.g., runaway electron production, relaxational magnetic reconnection).

The variability of the pulse period may serve as important criterion to distinct between MHD pulsations and self-organizing systems. The period may be assumed to be constant for radial and torsional MHD modes, and to be slowly varying for impulsively generated propagating MHD waves (Figure 2). Table IV contains a systematic overview of the time characteristics.

5.3. The Variability of the Pulse Amplitude

The observed intensity flux amplitude is the response of the resonant volume to the on-going emission mechanism. The response may be linear for small amplitudes, thus keeping the amplitude constant for a regular modulation. For this reason, the modula-

TABLE IV

Time characteristics of pulsation models

Pulsation model	Pulse period	Variability of pulse period	Variability of amplitude	Frequency bandwidth	Time profile
(1a) MHD-oscillation (Gyro)-synchrotron emission	(a) Radial (fast sausage mode) (Rosenberg, 1970) $\tau_{f'} \approx 4.4(a/v_A)$	τ = constant	I = constant or exponentially damped	$\Delta\nu/\nu > 1$	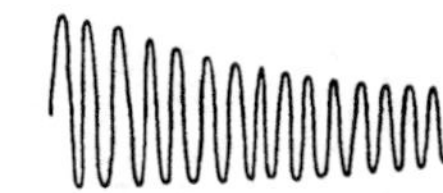
	(b) Torsional mode (Tapping, 1983) $\tau_t \approx 2L/nv_A$	τ = constant	I = constant	$\Delta\nu/\nu > 1$	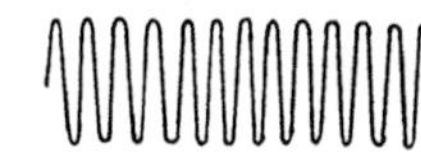
(1c) MHD-oscillation Plasma emission	(c) Impulsively generated propagating MHD-wave (Roberts *et al.*, 1984) $\tau_c \approx 2.6(a/v_A)$	τ = slowly variable	I = rapidly variable	$\Delta\nu/\nu < 1$	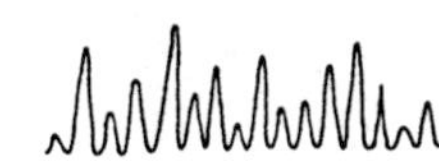
(2a) Cyclic self-organizing system (wave-wave-interactions)	(a) (E.g., Langmuir waves) (Zaitsev, 1971) $\tau = 2\pi\sqrt{\tau_{growth}\,\tau_{diff}}$	τ = rapidly variable	I = rapidly variable	$\Delta\nu/\nu < 1$	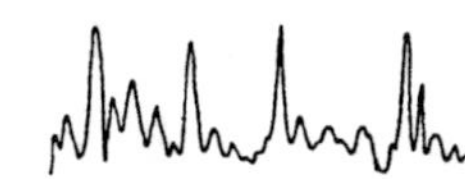
(2b) Cyclic self-organizing system (wave-particle-interactions)	(b) (E.g., loss-cone driven electron–cyclotron maser) (Aschwanden *et al.*, 1987) $\tau = 2\pi\sqrt{\tau_{growth}\,\tau_{diff}}$	τ = rapidly variable	I = rapidly variable	$\Delta\nu/\nu < 1$	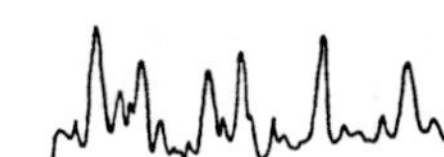
(3) Modulation of acceleration (repetitive injection into the loss-cone)	(e.g., sudden reduction model) (Kuijpers *et al.*, 1976) $\tau = \tau_{injection}$	τ = rapidly variable	I = modulated by absorption	$\Delta\nu/\nu < 1$	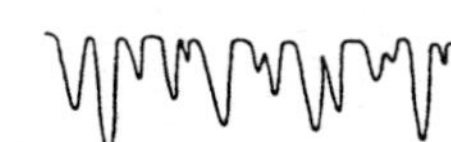

tion of the synchrotron emission by regular MHD oscillations may produce a constant height of the pulse amplitude except for damping. For the case of torsional MHD oscillations Tapping (1983) predicted a long, undamped lifetime.

In contrast, the maximum of a wave amplitude produced by an instability is highly nonlinear. The maximum is governed by saturation of the wave energy, by resonance frequency shifts, by flattening of the resonant particle distribution, or by other nonlinear mechanisms which quench the instability. Because of the (initial) exponential growth of the wave amplitude the obtained maximum value is very sensitive to small fluctuations of environmental parameters. Hence strong variability of the pulse amplitude is a straightforward consequence for pulsations governed by any instability. Moreover, in the case of self-organizing plasma-instabilities where the pulse period varies losely, the amplitude is highly unstable.

The pulse period and pulse amplitude are correlated for self-organizing systems, especially in Lotka–Volterra systems. Prospective investigations have to quantify this relation for specific plasma instabilities. It may serve as important key criterion to identify the involved instability from pulsations recorded with high-time-resolution pulsations. Meanwhile we note the remarkable difference between plasma instabilities generated by MHD oscillations or by self-organizing systems. The variability of the pulse amplitude is independent of the pulse period in the former case, but dependent in the second case.

The pulse amplitude occurs as absorption feature upon a slowly varying background for repetitive injection of particles into the loss-cone (Benz and Kuijpers, 1976) independent of any periodicity.

5.4. Spectral characteristics

In the current state of knowledge, the analysis of spectral characteristics has not been used extensively to distinguish between competitive pulsation models. From the theoretical point of view, the frequency bandwidth should reveal whether synchrotron emission or plasma emission is responsible. According to Benz and Tarnstrom (1976) synchrotron emission requires a bandwith of $\Delta\nu/\nu \geq 1$. The double-band feature of a pulsating synchrotron source predicted by Gotwols (1973) has never been observed. Plasma emission exhibits an intrinsic narrower bandwidth due to the finite resonance width, however, several propagation effects may broaden the original bandwidth. Most of the previous observations of pulsations were carried out by single-frequency instruments. The lack of good spectral data and of comprehensive models hamper spectral parameters for diagnostic purposes until now.

An observational fact is the predominance of regular MHD oscillations in the metric domain, whereas the oscillations in the decimetric range $\nu > 500$ MHz are irregular as a rule. This suggests that favourable conditions for MHD oscillations occur mainly in the upper part of the corona (metric emissions), respectively, in large loops.

The bandwidth criterion for synchrotron emission is generally not fulfilled in the metric/decimetric range, thus suggesting plasma emission or maser emission. In contrast, microwave observations exhibit often a bandwith of $\Delta\nu/\nu \geq 1$, thus giving support for synchrotron emission.

5.5. Interpretation of previous observations

Gathering the time characteristics of the mentioned pulsation models (see Table IV) we obtain preliminary guidelines to classify the reported observations within some uncertainities. Most of the reported radio pulsations are listed in Table V. Using only the criteria of pulse periodicity (column 4) and the pulse amplitude variability (column 5) we suggest a most likely fitting model (column 6) which may not necessarily be identical with the proposed one from the observer.

TABLE V

Observations of pulse periods (column 3), periodicity (column 4), pulse amplitude variability (column 5), and estimated best fitting model (column 6). The symbols characterizing the pulse period and amplitude are: C = constant, S = slowly varying, R = rapidly varying, E = exponentially damped, A = in absorption. The symbols of the suggested models (1a, 1b, 1c, 2a, 2b, 3) are given in Table IV.

Observer	Observed frequency range ν [MHz]	Pulse periods τ (s)	Pulse periodicity	Pulse amplitude	Best fitting model
Dröge (1967)	240, 460	0.2–1.2	C, S	S, R	1c
Abrami (1970)	239	1.7–3.1	C, S	S, R	1c
Rosenberg (1970)	220–320	1.0–3.0	C	S	1a
De Groot (1970)	250–320	2.2–3.5	R	R	2
McLean *et al.* (1971)	100–200	2.5–2.7	C	S	1a
Gotwols (1972)	565–1000	0.5	R	R	2
Rosenberg (1972)	220–320	0.7–0.8	S	R	1c
McLean and Sheridan (1973)	200–300	4.20 ± 0.01	C	E	1a
Elgarøy and Sveen (1973)	500–550	0.1–0.3	S, R	S, R	1c, 2
Achong (1974)	18–28	4	C	S	1a, 1c
Abrami and Koren (1978)	237	–	C, S	S, R, A	1c, 3
Pick and Trottet (1978)	169	0.37, 1.7	R	R	2
Trottet *et al.* (1979)	169	60	–	–	–
Tapping (1978)	140	0.06–5	C, S	R	1c
Elgarøy (1980)	310–340	1.1	C, S, R	S, R	1c, 2
Bernold (1980)	100–1000	0.5–5	S, R	R	1c, 2
Trottet *et al.* (1981)	140–259	1.7 ± 0.5	C, S	R, A	3
Slottje (1981)	160–320	0.2–5.5	R	R, A	2, 3
Sastry *et al.* (1981)	25–35	2–5	S, R	R	1c, 2
Wiehl *et al.* (1985)	300–1000	1–2	S, R	R	1c, 2
Aschwanden (1985)	300–1100	0.4–1.4	S, R	R	1c, 2
Aschwanden and Benz (1986)	237, 650	0.5–1.5	R	R	2

A more detailed interpretation may succeed from the analysis of pulse shapes and period-amplitude relations requiring further theoretical effort. Future work, using realistic source functions in order to compute the streaming distributions in the coronal environment (e.g., in a loss-cone trap) would be able to simulate specific pulse shapes and spectra.

Recommendations for observations include interferometric observations (spatial dimension), higher time resolution (detailed pulse shape profile), reliable measurement of magnetic field and electron densities, and determination of polarization (magneto-ionic mode). Observations of pulsations should also be extended to the stellar domain. Recent high time resolution observations (e.g., Lang, 1986) have discovered quasi-periodic trains of pulses, opening a new field for model applications.

Acknowledgements

I am gratefully indepted A. O. Benz for constructive discussions and corrections, L. Vlahos, G. Dulk, and D. Spicer for valuable comments, Mrs Helen Studer and M. Güdel for assistance in preparation of the paper. This work was supported by grant No. 2.211–0.84 of the Swiss National Science Foundation.

References

Abrami, A.: 1970, *Solar Phys.* **11**, 104.
Abrami, A.: 1972, *Nature Phys. Sci.* **238**(80), 25.
Abrami, A. and Koren, U: 1978, *Astron. Astrophys. Suppl.* **34**, 165.
Achong, A.: 1974, *Solar Phys.* **37**, 477.
Aschwanden, M. J.: 1985, *Solar Phys.* **104**, 57.
Aschwanden, M. J. and Benz, A. O.: 1986, *Astron. Astrophys.* **158**, 102.
Aschwanden, M. J. and Benz, A. O.: 1987, *Astrophys. J.*, submitted.
Bardakov, V. M. and Stepanov, A. V.: 1979, *Soviet Astron. Letters* **5**, 247.
Benz, A. O.: 1980, *Astrophys. J.* **240**, 892.
Benz, A. O. and Kuijpers, J.: 1976, *Solar Phys.* **46**, 275.
Benz, A. O. and Tarnstrom, G. L.: 1976, *Astrophys. J* **204**, 597.
Berney, M. and Benz, A. O.: 1978, *Astron. Astrophys.* **65**, 369.
Bernold, T. E. X.: 1980, *Astron. Astrophys. Suppl.* **42**, 43.
Bespalov, P. A. and Trakhtengerts, V. Yu.: 1977a, *Soviet J. Plasma Phys.* **2**, 215.
Bespalov, P. A. and Trakhtengerts, V. Yu.: 1977b, *Radiophys. Quantum Electron.* **19**, 567.
Chiu, Y. T.: 1970, *Solar Phys.* **13**, 420.
Costa, J. E. R. and Kaufmann, P.: 1986, *Solar Phys.* **104**, 253.
De Groot, T.: 1970, *Solar Phys.* **14**, 176.
Dröge, F.: 1967, *Z. Astrophys.* **66**, 200.
Edwin, P. M. and Roberts, B.: 1982, *Solar Phys.* **76**, 239 (Paper III).
Edwin, P. M. and Roberts, B.: 1983, *Solar Phys.* **88**, 179.
Elgarøy, Ø.: 1980, *Astron. Astrophys.* **82**, 308.
Elgarøy, Ø. and Sveen, O. P.: 1973, *Solar Phys.* **32**, 231.
Fukai, J., Krishan, S., and Harris, E. G.: 1969, *Phys. Rev. Letters* **23**, 910.
Gaizauskas, V. and Tapping, K. F.: 1980, *Astrophys. J.* **241**, 804.
Gotwols, B. L.: 1972, *Solar Phys.* **25**, 232.
Gotwols, B. L.: 1973, *Solar Phys.* **33**, 475.
Heyvaerts, J. and Verdier de Genouillac: 1974, *Astron. Astrophys.* **30**, 211.
Kaufmann, P., Correia, E., Costa, J. E. R., Dennis, B. R., Hurford, G. J., and Brown: 1984, *Solar Phys.* **91**, 359.
Krüger, Albrecht: 1979, *Introduction to Solar Radio Astronomy and Radio Physics*, D. Reidel Publ. Co., Dordrecht, Holland.
Kuijpers, J.: 1974, *Solar Phys.* **36**, 157.
Kuijpers, J.: 1975, *Solar Phys.* **44**, 173.
Kuijpers, J.: 1978, *Astron. Astrophys.* **69**, L9.

Kuijpers, J., van der Post, P., and Slottje, C.: 1981, *Astron. Astrophys.* **103**, 331.
Kuin, N. P. M. and Martens, P. C. M.: 1982, *Astron. Astrophys.* **108**, L1.
Lang, K. R.: 1986, *Solar Phys.* **104**, 227.
McLean, D. J. and Sheridan, K. V.: 1973, *Solar Phys.* **32**, 485.
McLean, D. J., Sheridan, K. V., Stewart, R. T., and Wild, J. P.: 1971, *Nature* **234**, 140.
Meerson, B. I., Sasorov, P. V., and Stepanov, A. V.: 1978, *Solar Phys.* **58**, 165.
Nakajima, H., Tajima, T., Brunel, F., and Sakai, J.: 1984, *ESA Special Publ., ESA* **SP-207**, 193.
Oravskii, V. N., Wilhelmsson, H., Kogan, E. Ya., and Pavlenko, V. P.: 1973, *Phys. Scripta* **7**, 217.
Pick, M. and Trottet, G.: 1978, *Solar Phys.* **60**, 353.
Rae, I. C. and Roberts, B.: 1982, *Astrophys. J.* **256**, 761.
Roberts, B.: 1981a, *Solar Phys.* **69**, 27 (Paper I).
Roberts, B.: 1981b, *Solar Phys.* **69**, 39 (Paper II).
Roberts, B. and Mangeney, A.: 1982, *Monthly Notices Roy. Astron. Soc.* **198**, 7.
Roberts, B., Edwin, P. M., and Benz, A. O.: 1983, *Nature* **305**, 688.
Roberts, B., Edwin, P. M., and Benz, A. O.: 1984, *Astrophys. J.* **279**, 857.
Rosenberg, H.: 1970, *Astron. Astrophys.* **9**, 159.
Rosenberg, H.: 1972, *Solar Phys.* **25**, 188.
Sastry, Ch. V., Krishan, V., and Subramanian, K. R.: 1981, *J. Astrophys. Astron.* **2**, 59.
Slottje, C.: 1981, *Atlas of Fine Structures of Dynamic Spectra of Solar Type IV-dm*, Dwingeloo.
Smith, D. F.: 1977, *J. Geophys. Res.* **82**, 704.
Smith, D. F. and Fung, P. C. W.: 1971, *J. Plasma Phys.* **5**, 1.
Spicer, D. S.: 1981, *Solar Phys.* **70**, 149.
Stepanov, A. V.: 1974, *Soviet Astron.* **17**(6), 781.
Stepanov, A. V.: 1978, *Soviet Astron. Letters* **4**(2), 103.
Stepanov, A. V.: 1981, SibIZMIR Preprint 17–81, Irkutsk.
Sy, W. N.-C.: 1974, *Solar Phys.* **34**, 427.
Takakura, T., Kaufmann, P., Costa, J. E. R., Degaonkar, S. S., Ohki, K., and Nitta, N.: 1983, *Nature* **302**, 317.
Tapping, K. F.: 1978, *Solar Phys.* **59**, 145.
Tapping, K. F.: 1983, *Solar Phys.* **87**, 177.
Trakhtengerts, V. Yu.: 1968, *Geomagnetizm i Aeronomiya* **8**, 776.
Trottet, G., Kerdraon, A., Benz, A. O., and Treumann, R.: 1981, *Astron. Astrophys.* **93**, 129.
Trottet, G., Pick, M., and Heyvaerts, J.: 1979, *Astron. Astrophys.* **79**, 164.
Wentzel, D. G.: 1979a, *Astron. Astrophys.* **76**, 20.
Wentzel, D. G.: 1979b, *Astrophys. J.* **227**, 319.
Weiland, J. and Wilhelmsson, H.: 1973, *Phys. Scripta* **7**, 222.
Weiland, J. and Wilhelmsson, H.: 1977, *Coherent Nonlinear Interactions of Waves in Plasmas*, Pergamon Press, Oxford.
Wiehl, H. J. and Maetzler, C.: 1980, *Astron. Astrophys.* **82**, 93.
Wiehl, H. J., Benz, A. O., and Aschwanden, M. J.: 1985, *Solar Phys.* **95**, 167.
Yao Jinxing: 1983, in C. de Jager and Chen Biao (eds.), 'Solar Physics and Interplanetary Travelling Phenomena', *Proceeding of Workshop Kunming/China*, p. 588.
Zaitsev, V. V.: 1971, *Solar Phys.* **20**, 95.
Zaitsev, V. V. and Stepanov, A. V.: 1975, *Astron. Astrophys.* **45**, 135.
Zaitsev, V. V. and Stepanov, A. V.: 1982, *Soviet Astron.* **26**(3), 340.
Zaitsev, V. V. and Stepanov, A. V.: 1983, *Solar Phys.* **88**, 297.
Zaitsev, V. V., Stepanov, A. V., and Chernov, G. P.: 1984, *Solar Phys.* **93**, 363.

BEAT STRUCTURE IN PULSATING TYPE IV SOLAR RADIO BURSTS AND A POSSIBLE MECHANISM*

H.-W. LI[1,3], M. MESSEROTTI[1,2], and P. ZLOBEC[1]

Abstract. Pulsating type IV solar radio bursts with beat structure are presented and analysed in this paper. Based upon the theory of whistler soliton emission we interpret the beat structure by the combination of two components with different pulsation frequencies due to radial oscillations of two legs of the magnetic loop. The large depth of pulsation is also explained in this model.

1. Introduction

Periodic or quasi-periodic pulsations are interesting and not rare phenomena in type IV solar radio bursts at meter and decimeter wavelengths. A model that radial oscillations of magnetic flux tubes may modulate synchrotron emission from trapped energetic electrons was investigated extensively (e.g., Rosenberg, 1970; McLean and Sheridan, 1973; Tapping, 1978; Roberts *et al.*, 1983, etc.).

In recent years, the radiopolarimeter at Trieste Observatory recorded some interesting pulsating type IV bursts at frequencies 237, 327, 408, 505, and 610 MHz. In these events the pulsating flux densities are modulated, and appear like what we call 'beat structure'. In this paper we will interpret this beat structure by the combination of two components of periodic whistler soliton emissions with different pulsation frequencies.

2. Observations and Power Spectra

Many pulsating type IV solar radio bursts have been observed at Trieste Observatory since 1968 (Abrami, 1968). By means of high scan frequency (about 40 Hz) it is possible to detect rapid variations of the flux density. Some of these events reveal a peculiar feature that the depth of pulsation varies with time, i.e., what we call 'beat structure' (Figure 1(a)). Usually, this structure occurs during the main phase of the bursts, and appears evident in some relatively short time intervals compared with the duration of the bursts. Using Fourier analysis (Messerotti, 1985), we obtained the power spectra of these events, which show two bigger peaks in the higher frequency range. These peaks can be seen more clearly in Figure 1(b), where we filtered the data with a high-pass filter (cut-off frequency = 0.5 Hz). It is well known that the combination of two oscillations with slightly different frequencies and fixed phase difference will produce a modulation of the amplitude, or beat. Based upon this idea we propose a model in the next Section

* Proceedings if the Second CESRA Workshop on Particle Acceleration and Trapping in Solar Flares, held at Aubigny-sur-Nère (France), 23–26 June, 1986.

[1] Astronomical Observatory, Trieste, Italy.
[2] SISSA-ISAS, International School for Advanced Studies, Trieste, Italy.
[3] On leave from the Department of Astronomy, Nanjing University, Nanjing, The People's Republic of China.

Solar Physics **111** (1987) 137–142.

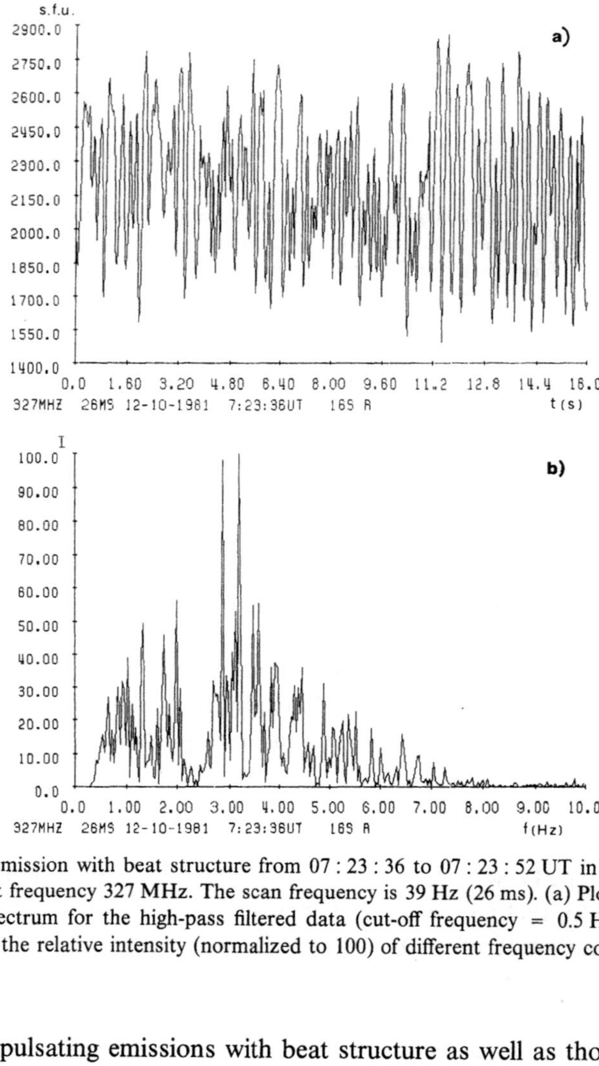

Fig. 1. Pulsating emission with beat structure from 07 : 23 : 36 to 07 : 23 : 52 UT in a type IV burst on 12 October, 1981 at frequency 327 MHz. The scan frequency is 39 Hz (26 ms). (a) Plot of flux density vs time; (b) power spectrum for the high-pass filtered data (cut-off frequency = 0.5 Hz). The ordinate I represents the relative intensity (normalized to 100) of different frequency components.

to interpret the pulsating emissions with beat structure as well as those without beat structure.

3. Model and Interpretation of the Beat Structure

We adopt a similar model proposed by Bernold and Treumann (1983; see also Figure 2), in which the energetic electrons with energy of a few tens of keV are trapped in a coronal loop in solar flares. The distribution of these electrons will develop temperature anisotropy ($T_\perp > T_\parallel$, where $\perp$ and $\parallel$ represent perpendicular and parallel directions to the magnetic field) or a loss cone, and consequently excite two kinds of waves, upper hybrid waves and whistler waves, at the top region of the loop.

Under solar conditions we have $\omega_p \gg \Omega_e > \omega_w > \Omega_i$, where ω_w is the frequency of whistler waves, ω_p is the plasma frequency, Ω_e and Ω_i are the gyrofrequencies of

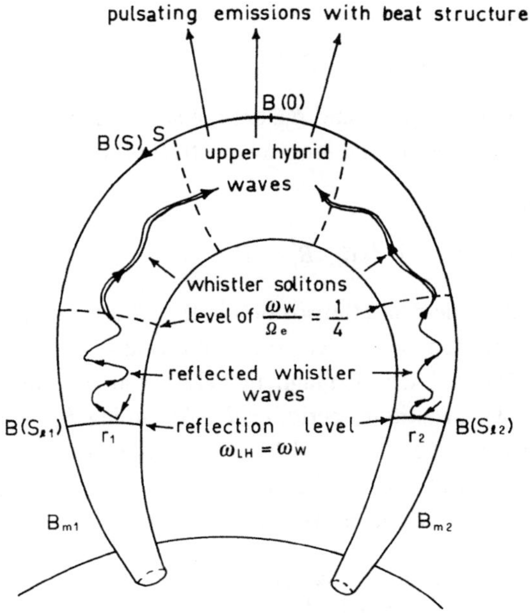

Fig. 2. Proposed model for pulsating emission with beat structure.

electrons and ions, respectively. The dispersion relation for whistler waves is (Thorne and Kennel, 1967)

$$k_w^2 c^2/\omega_w^2 = \omega_p^2/(\omega_w \Omega_e \cos\theta - \omega_w^2 + \Omega_e \Omega_i)\,, \tag{1}$$

where θ is the angle between the wave vector of whistler, $\mathbf{K}_w$, and the magnetic field $\mathbf{B}$. The group velocity of whistler waves is then

$$V_g = \frac{\partial \omega_w}{\partial k_w} = 2c\,\frac{(\omega_w \Omega_e \cos\theta - \omega_w^2 + \Omega_e \Omega_i)^{3/2}}{\omega_p \Omega_e(\omega_w \cos\theta + 2\Omega_i)} \tag{2}$$

Equations (1) and (2) are valid for nearly parallel propagating whistler waves ($\theta \ll 1$) and are still a good approximation for large θ only if $\cos\theta \gg \omega_w/\Omega_e - \Omega_i/\omega_w$ (Thorne and Kennel, 1967).

At its major region of generation whistler waves have the frequency ω_w well above the local lower hybrid frequency $\omega_{\mathrm{LH}} = (\Omega_e \Omega_i)^{1/2}$. The magnetic field $B(s)$ and the electron number density of plasma $n_e(s)$ [hence, $\omega_p(s)$, $\Omega_e(s)$, $\Omega_i(s)$, and $\omega_{\mathrm{LH}}(s)$] increase with s, where s is the distance from the top along the loop (see Figure 2). Therefore, as the whistler waves propagate down to a level, where $\omega_{\mathrm{LH}}(s) = \omega_w$, the whistler will be reflected (Thorne and Kennel, 1967; Bernold and Treumann, 1983). It can be seen from (1) and (2) that the reflected whistler has very small group velocity at the reflection level. Therefore, a large amount of whistler wave energy accumulates and modulation instability sets in, leading to the formation of whistler solitons (Karpman and Washimi, 1977).

For simplicity, assuming

$$B(s) = B_0[1 + (s/L)^2(B_m/B_0 - 1)]\,, \tag{3}$$

where L is the half length of the loop, B_0 and B_m are the strengths of the magnetic field at the top and at the mirror points of the loop, respectively (cf. Figure 2), we have

$$B_l/B_0 = 1/2(m_i/m_e)^{1/2}\,, \tag{4}$$

$$s_l = L[1/2(m_i/m_e)^{1/2} - 1]^{1/2}(B_m/B_o - 1)^{-1/2}\,, \tag{5}$$

where B_l and s_l are the magnetic field and the position of the reflection level, respectively. In some cases, both legs of the loop may undergo oscillations as described by Rosenberg (1970) and Roberts *et al.* (1983). Then the magnetic field B will vary and the reflection level will move up and down in altitude periodically. Assuming $B_m = B_{m0} + \delta B_m(t)$, where B_{m0} is the unperturbed value of B_m and $\delta B_m(t)$ is the periodic perturbation, we have $\dot{B}_m = \delta B_m \omega$, where the dot means the time derivative and ω is the angular frequency of the tube oscillation. Then the velocity of the reflection level is

$$\dot{s}_l = -0.5L\left[0.5\left(\frac{m_i}{m_e}\right)^{1/2} - 1\right]^{1/2}\left(\frac{B_m}{B_0} - 1\right)^{-3/2}\frac{\delta B_m}{B_0}\,\omega\,. \tag{6}$$

During the half period of the down motion of the reflection level if $|\dot{s}|$ is larger than the group velocity of whistler waves, i.e.

$$\omega\frac{L}{2}\left(\frac{1}{2}\left(\frac{m_i}{m_e}\right)^{1/2} - 1\right)^{1/2}\left(\frac{B_0}{B_{m0}}\right)^{1/2}\frac{\delta B_m}{B_{m0}} > V_g = \frac{c}{2}\,\frac{\Omega_e(0)}{\omega_p(0)}\,, \tag{7}$$

there will be no whistler reflected. In (7) $B_{m0}/B_0 \gg 1$ has been assumed. During the other half period, whistler solitons will be produced when the reflected whistler propagates upward above a level where the condition $\omega_w/\Omega_e > \frac{1}{4}$ is satisfied (Bernold and Treumann, 1983). Therefore, the whistler solitons will be generated periodically with a frequency which is equal to the frequency of the radial oscillation of the magnetic tube and hence is $\omega = 3.9\, V_A/r$ (Tapping, 1978), where V_A is Alfvén velocity and r is the radius of the tube at the mirror points.

In general, the configuration of the magnetic field in an active region is complicated. Then we could expect that the parameters for the two legs of a magnetic loop, B_m, n_e, and r, are somewhat different rather than the same. Consequently, the frequency of the generation of whistler solitons at two legs are different. For illustration, taking $\omega_w/\Omega_e(0) = \frac{1}{2}$, $B(0) = 4$ G, $n_e(0) = 1.3 \times 10^9$ cm^{-3}, $n_{e1} = 10^{10}$ cm^{-3}, $n_{e2} = 8 \times 10^9$ cm^{-3}, $L = 7 \times 10^9$ cm, $r_1 = 10^{-2} L$, $r_2 = 1.2 \times 10^{-2} L$, $B_{m01} = 150$ G, $B_{m02} = 180$ G, and $\delta B_m/B_{m0} = 0.02$, we have $V_g = 5.2 \times 10^8$ cm s^{-1}, $|\dot{s}_{l1}| = 9.4 \times 10^8$ cm s^{-1}, $|\dot{s}_{l2}| = 9.6 \times 10^8$ cm s^{-1}, $\omega_1 = 2\pi \times 2.9$ s^{-1}, and $\omega_2 = 2\pi \times 3.3$ s^{-1}, so the condition (7) for the periodic generation of whistler solitons is fulfilled. These whistler soliton chains propagate upward and interact with upper hybrid waves at the top region and produce electromagnetic emission (Bernold and

Treumann, 1983). Because the pulsation frequencies for the two legs are somewhat different, the combination of these two components will lead to the occurrence of beat structure.

It can be seen from the parameters chosen above that condition (7) can be satisfied even if the perturbation is small ($\delta B_m / B_{m0} \ll 1$). In other words, with a small periodic perturbation on the magnetic field we can get quite deep pulsations of whistler soliton emission, say larger than 50%. This can cover the difficulty of the theory of small perturbations in the application for some deep pulsating emissions.

4. Summary and Discussions

Pulsating type IV solar radio bursts with beat structure observed at Trieste Observatory are presented and analysed in this paper (Figure 1). The power spectra of these emissions contain basically two higher frequency components, appearing as two bigger peaks in the spectra. Based upon the model by Bernold and Treumann (1983), strong emission will be produced due to the interaction between upper hybrid waves and whistler solitons. We found out that if both legs of the magnetic loop undergo radial oscillations, whistler solitons will be generated periodically, leading to a pulsating emission. Because the two legs of the magnetic loop are usually characterized by different parameters, the pulsation frequencies differ from each other. The combination of these two pulsating emissions will give rise to a beat structure in the combined emission.

It can be expected reasonably that usually more loops participate in the emission. The beat structure is evident only when the contribution coming from one loop dominates the emission. If the contribution coming from other loops becomes stronger, or the oscillations are distorted, or the magnetic fields at the footpoints vary irregularly, the beat structure will become unidentified or disappear. Therefore, the beat structure is probably not so rare as looked apparently, but it is more difficult to be recognized.

In some cases oscillations occur only in one leg of the loop, then no beat appears in the pulsating emission and the spectra contain mainly one significant peak; such events have been observed.

A prominent feature of this model is that for reasonable parameters selected for the solar corona condition (7) can be fulfilled even if the perturbation is small, so that the depth of pulsation can be very large.

Acknowledgements

H.-W. Li is grateful for the hospitality to Trieste Observatory (especially the Solar Radio Group), where this work was carried out during his stay as a visiting scholar (CNR grant). This work was partially supported by CNR (Italian Research Council) and MPI (Ministry of Education). The computations were carried out at the Computing Centre of the Astronet at Trieste Observatory.

References

Abrami, A.: 1968, *Mem. Soc. Astron. Ital.* **39**, 371.
Bernold, T. X. and Treumann, R. A.: 1983, *Astrophys. J.* **264**, 677.
Karpman, V. I. and Washimi, H.: 1977, *J. Plasma Phys.* **18**, 1373.
McLean, D. J. and Sheridan, K. V.: 1973, *Solar Phys.* **32**, 485.
Messerotti, M.: 1985, Internal Report No. 1011, Trieste Astronomical Observatory.
Roberts, B., Edwin, P. M., and Benz, A. O.: 1983, *Nature* **305**, 688.
Rosenberg, H.: 1970, *Astron. Astrophys.* **9**, 159.
Tapping, K. F.: 1978, *Solar Phys.* **59**, 145.
Thorne, R. M. and Kennel, C. F.: 1967, *J. Geophys. Res.* **72**, 857.

REPETITION RATES OF FAST PULSES IN A SOLAR BURST OBSERVED AT MM-WAVES AND HARD X-RAYS*

E. CORREIA and P. KAUFMANN

INPE: Instituto de Pesquisas Espaciais, MCT, C.P. 515, 12.200, São José dos Campos, SP, Brazil

Abstract. The solar burst of 21 May, 1984, 13 : 26 UT, showed radio spectral emission with a turnover frequency above 90 GHz, well correlated in time with the hard X-ray emission. It consisted of seven major time structures (1–3 s in duration), of which each was composed of several fast pulses with rise times between 30 and 60 ms. The spectral indices of the millimeter and hard X-ray emission exhibited sudden changes during each major time structure. The subsecond pulses were nearly in phase at 30 and 90 GHz, but their relative amplitude at 90 GHz ($\approx 50\%$) were considerably larger than at 30 GHz ($< 5\%$). It was also found that the 90 GHz and the 100 keV X-rays fluxes were proportional to the repetition rate of the subsecond pulses, and that the hard X-ray power law index hardens with increasing repetition rate.

1. Introduction

There are some solar bursts reported in the literature that display a spectral flattening, or a peak, above 30 GHz (Croom and Powell, 1969; Croom, 1970; Feix, 1970; Shimabukuro, 1970, 1972; Cogdell, 1972; Akabane *et al.*, 1973). These events, however, were recorded with poor time resolution and sensitivity.

A peculiar event was observed on May 21, 1984 at various radio frequencies and at hard X-rays (Kaufmann *et al.*, 1985). It was possibly of the same nature of the events reported above. It exhibited a turnover frequency above or around 90 GHz and consisted of seven major structures (duration of seconds) built up by fast pulses (duration of tens of milliseconds) with relative amplitude of about 50% and $< 5\%$, at 90 and 30 GHz, respectively. The 90 GHz burst was extremely well correlated in time (better than 128 ms) with the hard X-ray emission.

The properties of the pulses (repetition rate and spectral indices) are analysed, as well as their relationship with the radio and X-ray fluxes. These results impose new boundary conditions which may be relevant to further theoretical work on the energy release mechanisms in solar flares.

2. Observational Characteristics

2.1. Observations

The solar burst of May 21, 1984 was observed in a broad spectral range, from hard X-rays to metric waves, by different instruments in space and at the ground.

At mm-waves it was observed by the Itapetinga Radio Observatory at 30 and 90 GHz with high time resolution (1 ms) and high sensitivity (≈ 0.03 s.f.u.) and at 7 GHz by a

* Proceedings of the Second CESRA Workshop on Particle Acceleration and Trapping in Solar Flares, held at Aubigny-sur-Nère (France), 23–26 June, 1986.

Solar Physics **111** (1987) 143–154.

patrol instrument (Kaufmann *et al.*, 1985, 1986a; Correia *et al.*, 1986). Radio spectral data was obtained by Sagamore Hill Radio Observatory, AFGL, at 245, 410, 606, 1415, 2695, 4995, 8800, and 15400 MHz, with 1 s time resolution (Cliver, 1984). The same burst was also observed in the course of a VLBI experiment between Westerbork and Onsala at 5 GHz with high time resolution (< 1 ms) and high sensitivity ($\ll 0.01$ s.f.u.) (van Nïeuwkoop, 1985).

Hard X-ray data were recorded by HXRBS on board of the SMM satellite, with 128 ms time resolution.

At Nançay Radio Observatory it was recorded at metric waves, at 169 MHz in a compressed scale (Raoult, 1984), and by a solar spectrograph at 468–162 MHz range (Bougeret, 1984). The Trieste group observed it at 610, 408, 327, and 237 MHz (Zlobec, 1986) with 20 ms time resolution. The association of mm-wave/X-ray emissions to metric and decimetric observations is being studied separately.

2.2. Time profiles and spectra

At 90 GHz the time profile shows seven major structures (of duration ≈ 1–4 s) well correlated with hard X-ray data (Figure 1). The comparison with the 30 GHz data shows that a good correlation is only present for the three last structures. For frequencies smaller than 15 GHz (using AFGL spectral data) all emission seem to be close or below instrumental limit, which is typically 5 s.f.u. This was confirmed by the 5 GHz Westerbork high sensitivity observations. The flux peak emission at 5 GHz was

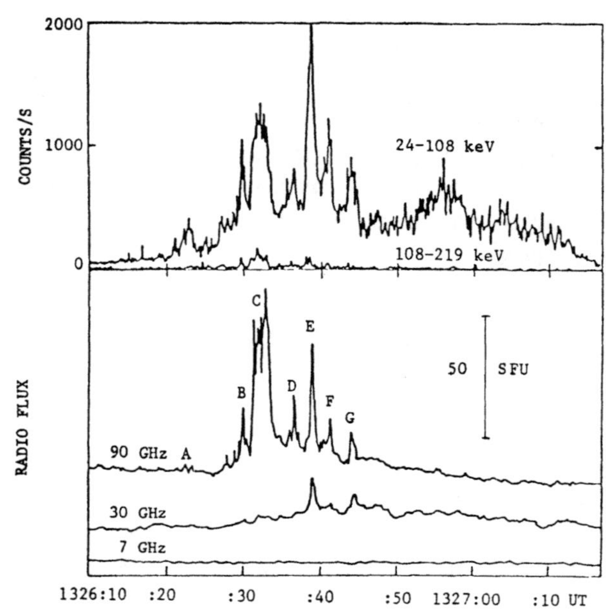

Fig. 1. The solar burst of May 21, 1984, 13:26 UT, observed at 30 and 90 GHz (Itapetinga Radio Observatory) and at hard X-rays (HXRBS–SMM). The seven major time structures are labeled A–G.

≈ 5 s.f.u., measured at a temporal resolution of 40 ms. The VLBI analysis of this event is still in progress (van Nieuwkoop, 1985).

All seven main structures show an increasing spectrum between 30 and 90 GHz (Correia *et al.*, 1986). The simplest suggestion of this trend is that the spectral turnover frequency is above or around 90 GHz, as it has been assumed in a number of interpretations (Kaufmann *et al.*, 1985, 1986a, b; Correia *et al.*, 1986; McClements and Brown, 1986; de Jager *et al.*, 1987). However, since there is no more spectral data available, a possible spectral peak between 30 and 90 GHz can not be excluded. In this case, however, the spectral index for the structures presenting the larger flux ratio between 90 and 30 GHz may well exceed 3, which is unlikely according to various emission mechanisms discussed in the literature (Guidice and Castelli, 1975).

The main burst structures are shown in Figure 2 in high time resolution. Each one of the structures can be considered as a packet of several millisecond pulses, which are

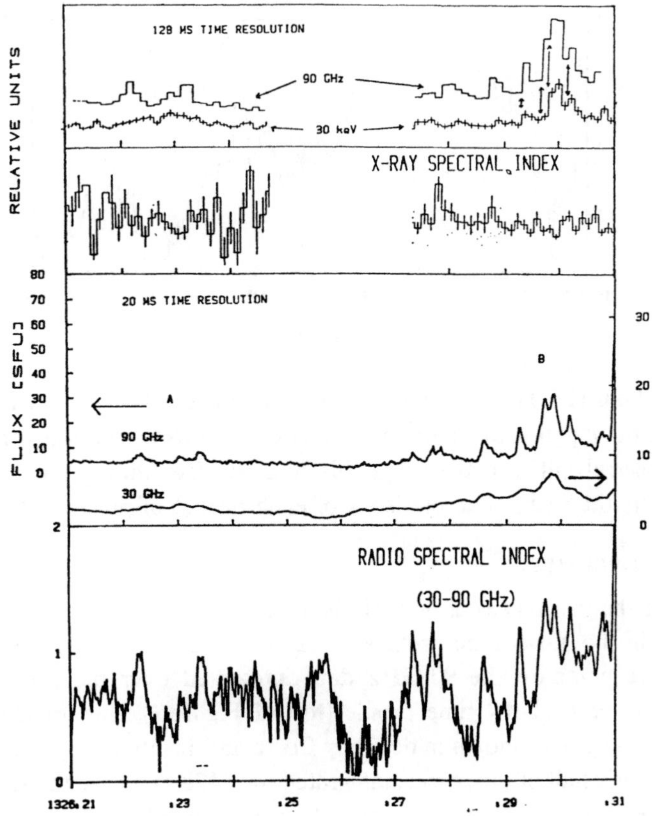

Fig. 2a.

Fig. 2(a–c). A comparison of flux and spectral indices at 90 and 30 GHz, and hard X-rays (> 30 keV) for all seven structures as indicated in Figure 1. The time resolution for the hard X-ray and radio emission was 128 and 4 ms, respectively. In the upper plots the 90 GHz data was time integrated over 128 ms, in order to allow the direct comparison with the X-ray time profiles.

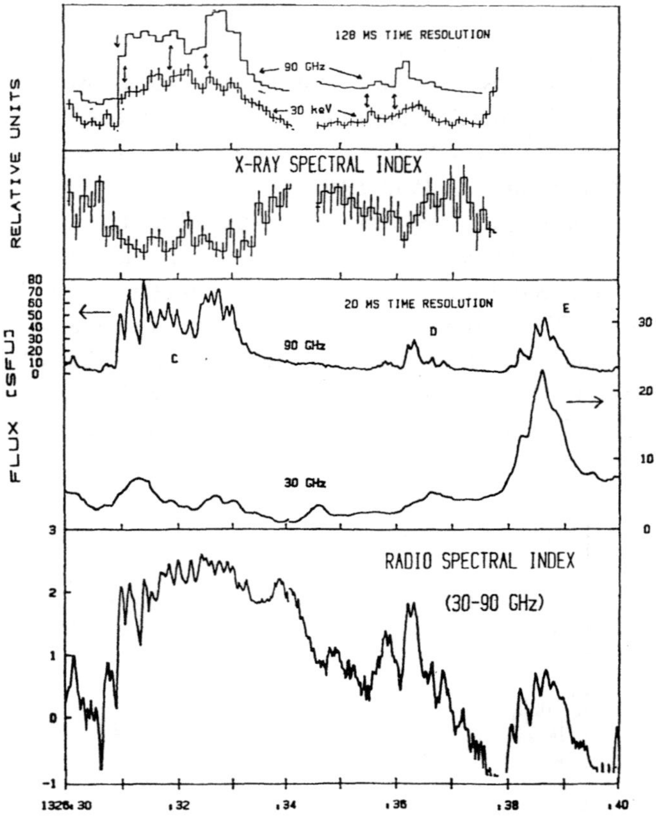

Fig. 2b.

well defined at 90 GHz. The hard X-ray power law indices, shown at the top of Figure 2, indicate a hardening for the most prominent structures. On the other hand, the microwave spectral index in the range 30–90 GHz rises during most of the pulses. Table I contains the main characteristics of each packet of pulses.

2.3. Pulse properties

The lifetime of the fast pulses at 90 GHz is approximately 50 ms. It is not possible to observe them in the X-ray time profiles, because of the low time resolution (128 ms). However, if we compare the 90 GHz data integrated over the same 128 ms time segments to the hard X-ray time profile (top of Figure 2), the correlation is highly suggestive. This was also shown in detail by Costa and Kaufmann (1986) for the major structure *C*. Nearly half of the pulses, integrated over 128 ms, have the same start times (arrows in the Figure 2). Fast pulses at hard X-rays obtained with higher time resolution were reported in the literature (Takakura *et al.*, 1983; Kiplinger *et al.*, 1983). However, a definitive confirmation that fast pulses are also present at hard X-rays in the present event would require time resolution and sensitivity higher than the available ones. The delays between hard X-rays and 90 GHz, if any, are smaller than 128 ms.

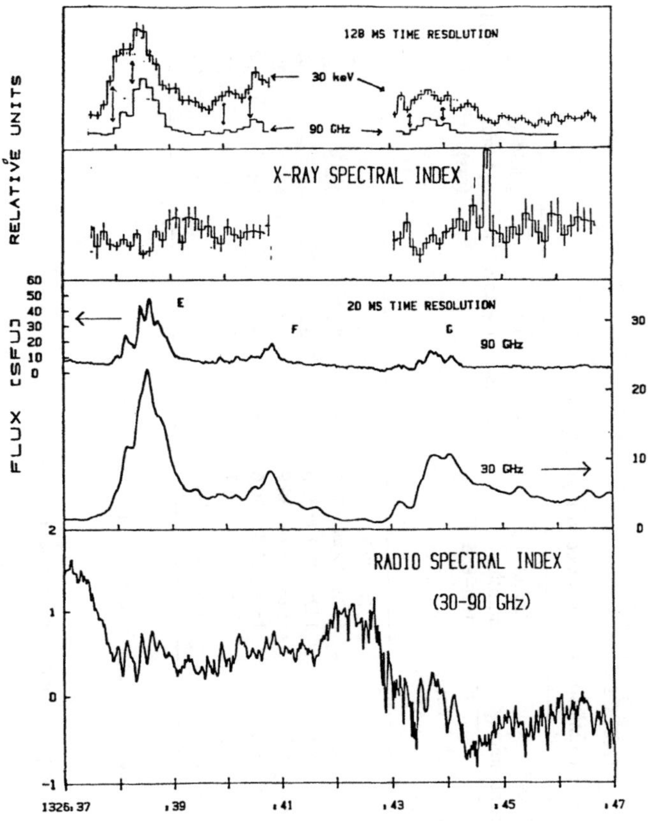

Fig. 2c.

The relative amplitude of the fast pulses, defined as $\Delta F/F$, where ΔF is the pulse amplitude and F the mean flux, was about 50% at 90 GHz, while at 30 GHz it was less than 5%. This confirms the earlier results obtained at 7, 10.7, 22, and 44 GHz (Figure 3), that the sub-second pulses appear with increasing relative amplitude at shorter mm-waves. The sub-second pulses cannot be caused by atmospheric or instrumental effects, as it has been discussed extensively elsewhere (Kaufmann *et al.*, 1984; Zodi Vaz *et al.*, 1984). It has been pointed out that clear sky turbulences may produce fluctuations in rare occasions, but that they are of a much slower time-scale (> seconds). The antenna structure cannot vibrate on sub-second time-scales, while the Cassegrain subreflector displacements are negligible, according to laser beam determinations. The rise of pulse repetition rate for larger fluxes, cannot be of experimental origin anyway.

The 30 GHz pulses, when measurable, are generally in phase with their counterpart at 90 GHz, and the delays are not bigger than 40 ms. It is important to note that if the 90 GHz burst data was not available, one could get the erroneus impression that the 30 GHz data were 'delayed' with respect to the hard X-rays.

TABLE I

Parameters of the burst major time structures (labeled in Figure 1)

Structure	Flux[a] 90 GHz (s.f.u.)	Flux[a] 30 GHz (s.f.u.)	Flux X-ray[b]		$\bar{q}$: X-ray[c] spectral indices		Packet duration start–end (UT)	No. pulses N	R[e] ($\pm 0.5\ s^{-1}$)	e-folding rise time (ms)	$\Delta F/F$[f] 90 GHz (%)
			> 30 keV	> 100 keV	burst (± 0.3)	slow comp. (± 0.5)					
A	8	3	0.6 (313)	0.03 (19)	2.2	4.0	~1326 : 22[d]	–	–	–	–
B	34	5	1.8 (1042)	0.09 (110)	2.5	4.0	1326 : 28.4– : 30.1	5	2.9	70	20
C	83	7	2.1 (1400)	0.15 (190)	1.9	3.0	1326 : 30.7– : 33.5	14	5.0	40	50
D	40	9	1.2 (730)	0.06 (37)	2.7	3.4	1326 : 35.7– : 36.9	4	3.3	60	20
E	58	24	3.3 (1980)	0.12 (110)	2.4	3.4	1326 : 37.8– : 39.0	5	4.0	50	30
F	27	12	1.8 (1150)	0.06 (37)	2.8	3.4	1326 : 39.7– : 41.1	4	2.8	60	50
G	23	16	1.5 (820)	0.09 (37)	2.7	3.5	1326 : 42.8– : 44.2	4	2.7	60	40

[a] (1 s.f.u. = $10^{-22}\ W\ m^{-2}\ Hz^{-1}$) the absolute flux error is 10 s.f.u.

[b] The flux units is $10^{-25}\ erg\ cm^{-2}\ s^{-1}\ Hz^{-1}$. The term in brackets is in counts s^{-1} and the error is the Poisson deviation.

[c] $\bar{q}$ is the averaged X-ray spectral index.

[d] Pulses not defined.

[e] $R = N/\Delta t$ is repetition rate of fast pulses, with Δt the duration of the pulse packet and N the number of pulses.

[f] $\Delta F/F$ is the mean relative amplitude of the pulses.

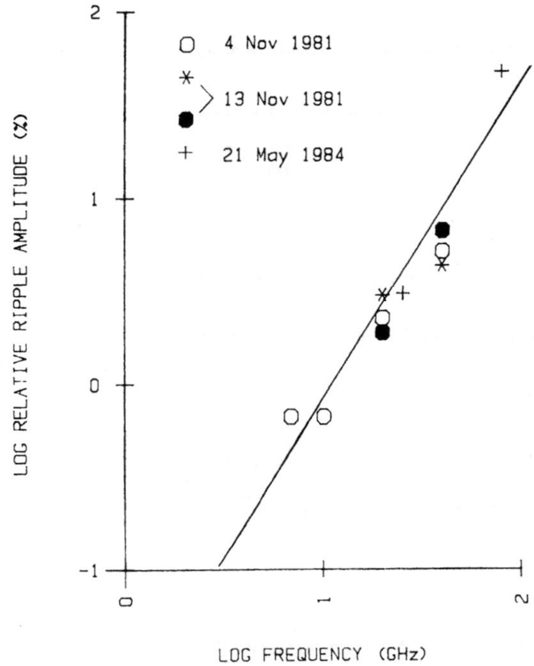

Fig. 3. Relative amplitude of the ultrafast pulses versus radio frequency.

The time profiles of the fast pulses at 30 and 90 GHz during the burst evolution are shown in Figure 4. The running mean with 220 ms integration time was used to remove the slow component and another of 60 ms to smooth the remaining signal. The pulses appeared in packets and were well above the noise level previous to the burst. Figure 4 also identifies the individual pulses and the time segments of each packet used for the determination of the pulse rate. Table I summarizes the most important properties of the pulses.

Scatter diagrams relating the different pulse packet characteristics were taken from Table I. The first one relating the repetition rate R and the associated peak flux F at 90 GHz shows an excellent linear correlation (Figure 5):

$$F \sim KR ,$$

with $K \simeq 25 \times 10^{-19}$ erg cm^{-2} Hz^{-1}.

The above relation could be explained if the solar bursts are a superposition of fast pulses at different repetition rates with quasi-quantized energy (Kaufmann *et al.*, 1980, 1984; Correia, 1983; Sturrock *et al.*, 1984; Loran *et al.*, 1985).

The flux scatter diagram between 90 GHz and > 30 and > 100 keV X-ray shows a better correlation for the last case (Figure 6) as expected.

Finally, a new interesting result was obtained from the scatter diagram between the X-ray spectral index and the repetition rate and the radio pulses (Figure 7). It shows

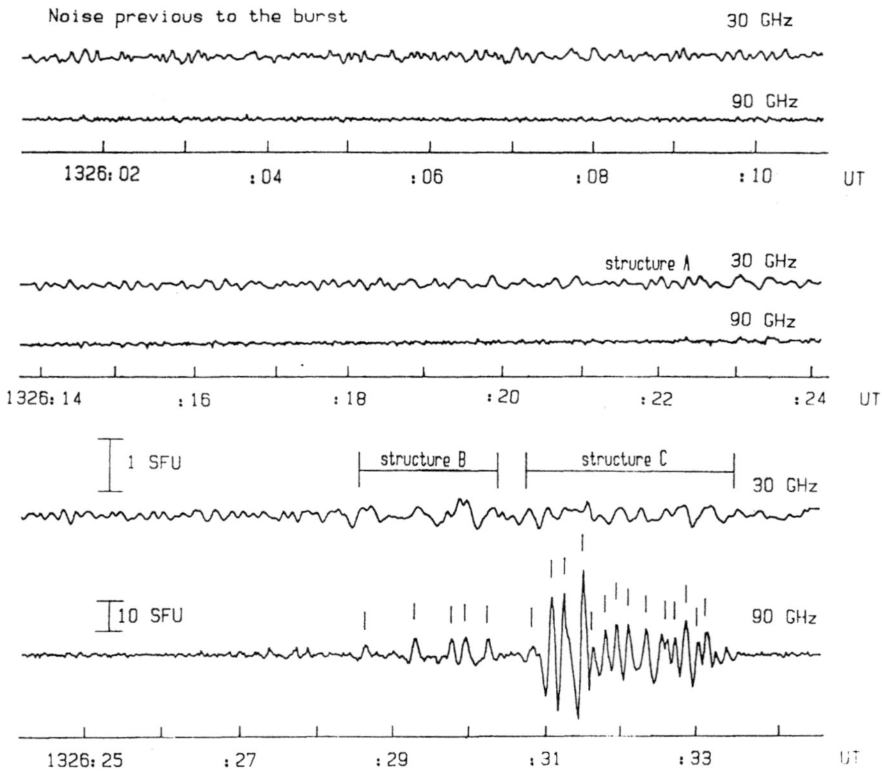

Fig. 4a.

Fig. 4(a–b). Detailed time profiles of the ultrafast pulses after the subtraction of the slow component by the running mean technique (220 ms). The flux scales are indicated in (a) and kept the same for all the other 10 s sections.

that the harder X-ray spectrum is associated with a higher repetition rate. This result could be a simple consequence of the good correlation between the hard X-ray spectral index and the hard X-ray flux (Figure 2) and the good correlation between hard X-ray and mm-wave flux the latter being correlated with the repetition rate (Figure 5) but has an important evidence, the hardening of the X-ray spectrum is related with an increased repetition frequency of elementary energy releases.

3. Fast Pulses and Primary Energy Release

The observational evidence for harder X-ray spectra during higher pulse repetition rates implies that the efficiency of the particle acceleration mechanism is associated to the conversion rate of magnetic energy into particle kinetic and/or thermal energy. The mechanism needs to explain the 10^{27} erg energy released by the smallest bursts with durations of less than 0.1 s (Kaufmann *et al.*, 1984).

One of the first mechanisms used qualitatively to interpret energy conversion in solar flares was the tearing mode reconnection. However, the time-scale of this process

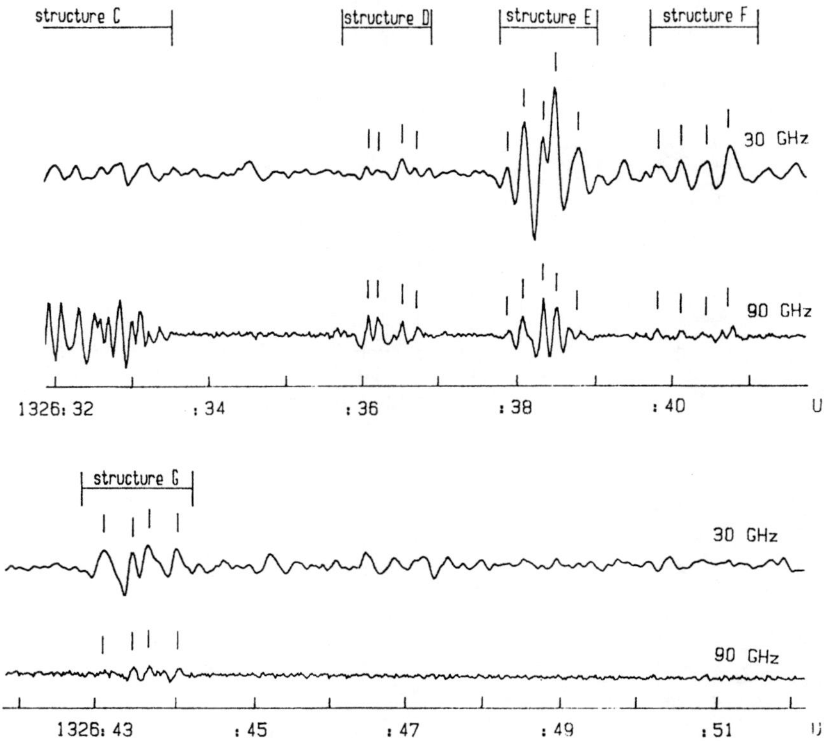

Fig. 4b.

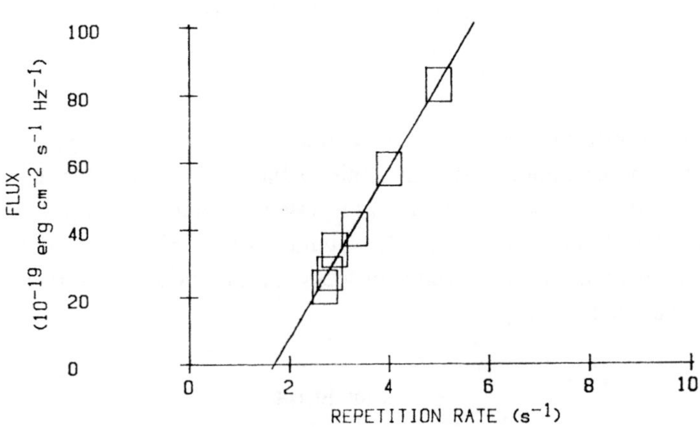

Fig. 5. Scatter diagram of pulse repetition rate and flux at 90 GHz for the six structures which include 36 fast pulses. The correlation coefficient is 99%. The boxes represent the limits of accuracy.

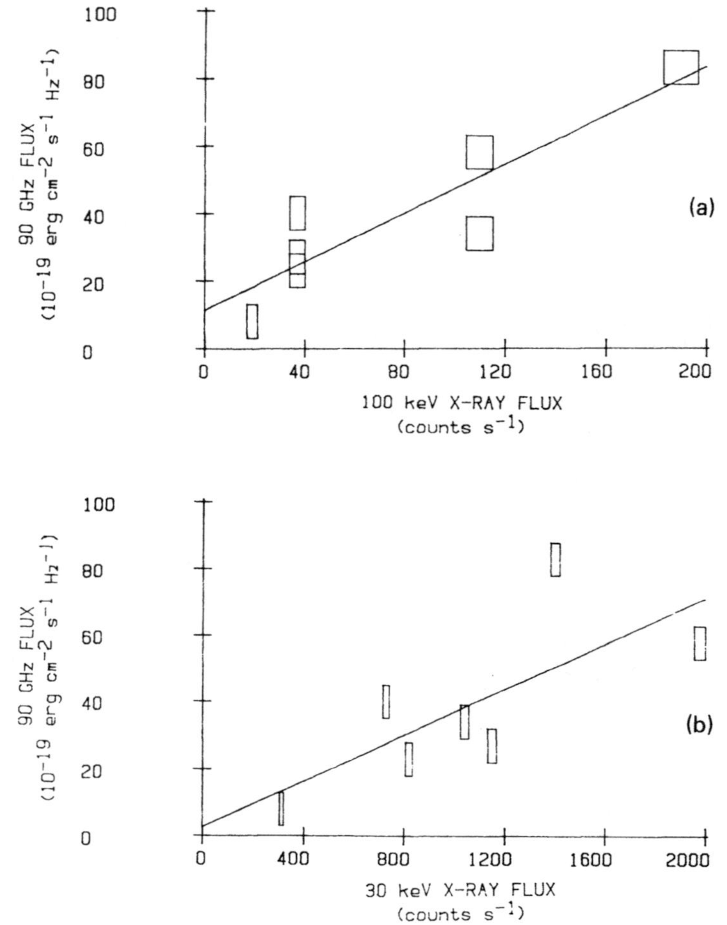

Fig. 6. Scatter diagram of the flux at 90 GHz and: (a) 100 keV X-rays and (b) 30 keV X-rays. The boxes represent the limits of accuracy.

cannot account for the fast conversion of magnetic into particle energy, even in its fastest mode. On the other hand, if the dynamic behaviour is considered, the nonlinear development of the coalescence instability is able to explain the energy released within short time-scales (Leboeuf *et al.*, 1982; Tajima *et al.*, 1982; Sakai *et al.*, 1985). The time-scales involved in the coalescence process correspond to the Alfvén time that can explain the observed pulses.

4. Conclusion

We believe that the emission at high radio frequencies and hard X-rays are the closest responses to the primary energy release mechanism. Therefore the May 21, 1984 solar

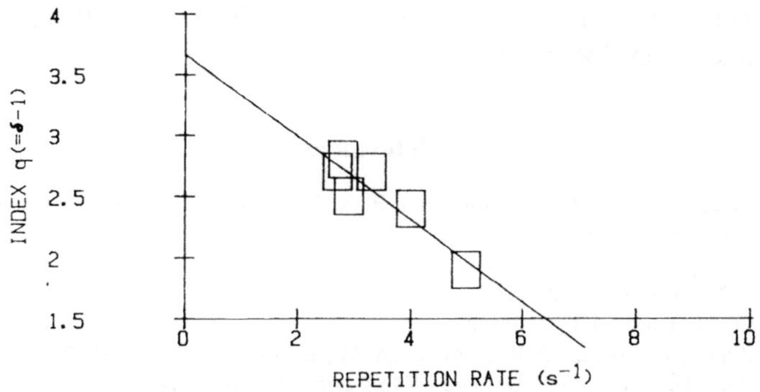

Fig. 7. Scatter diagram of pulse repetition rate and X-ray spectral index for the six burst structures. The correlation coefficient is 92%. The boxes represent the limits of accuracy.

burst, observed at 90 GHz and hard X-rays, could be a valuable source of information about the operating mechanisms.

The event time profile showed major time structures (time-scales of 1–3 s) which could be associated to the Elementary Flare Bursts (de Jager and de Jonge, 1978) and which may be generated by interloop interactions (Emslie, 1981). These structures are explained by groups of pulses, of tens of milliseconds, quasi-quantized in energy, and produced at different repetition rates.

Some new interesting properties were obtained from the fast pulses: the delays between X-rays and the mm-wave emission, if any, were smaller than 128 ms. The repetition rate of the fast pulses was associated to the hard X-ray spectral index which hardens for higher rates. At microwave and mm-wave frequencies the relative amplitude of the fast 'ripple' increased with frequency, being less than about 1% for 30 GHz, and over 50% at 90 GHz. Care must be taken in the determination of delays between microwave and X-ray emission, since a higher radio frequency spectral component can be present, exhibiting no delays.

The observations suggest that each pulse was produced by a different electron beam accelerated by some discontinuous process, and that the high energy conversion rate was associated to an efficient particle acceleration mechanism, capable of increasing the electron energies to relativistic levels during the impulsive phase.

For improvements in the understanding of the solar flare energy release process further observations are necessary at various radio frequencies, specially above 100 GHz, in the sub-millimetric and infrared ranges and at hard X-rays above 100 keV, obtained with high sensitivity and millisecond time resolution.

Acknowledgements

We are grateful to Dr B. R. Dennis, Mr S. Kennard, and Mr B. R. Gibson for preparing and providing the SMM-HXRBS X-ray data, and to Dr E. W. Cliver, Dr J. van Nieuwkoop, in providing unpublished burst radio data. This research was partially supported by Brazilian research agency FINEP. One of the authors (P. K.) is Guest Investigator on SMM-Project.

References

Akabane, K., Nakajima, H., Ohki, K., Moriyama, F., and Miyaji, T.: 1973, *Solar Phys.* **33**, 431.
Bougeret, J.: 1984, private communication.
Cliver, E. W.: 1984, private communication.
Cogdell, J. R.: 1972, *Solar Phys.* **22**, 147.
Correia, E.: 1983, Ms.Thesis, INPE, Brazil.
Correia, E., Kaufmann, P., Costa, J. E. R., Zodi Vaz, A. M., and Dennis, B. R.: 1986, *Proc. of the SMM Topical Workshop on Rapid Fluctuations in Solar Flares, NASA Lanham, MD, September 30–October 4, 1985*, NASA Conf. Publ. 2449, 161.
Costa, J. E. R. and Kaufmann, P.: 1985, *Solar Phys.* **102**, 97.
Croom, D. L.: 1970, *Solar Phys.* **15**, 414.
Croom, D. L. and Powell, R. J.: 1969, *Nature* **221**, 945.
De Jager, C. and de Jonge, G.: 1978, *Solar Phys.* **58**, 127.
De Jager, C., Kuijpers, J., Correia, E., and Kaufmann, P.: 1987, *Solar Phys.* **110**, 317.
Emslie, A. G.: 1981, *Astrophys. Letters* **22**, 171.
Guidice, D. A. and Castelli, J. P.: 1975, *Solar Phys.* **44**, 155.
Feix, G. J.: 1970, *Geophys. Res.* **75**, 211.
Kaufmann, P., Correia, E., Costa, J. E. R., Dennis, V. R., Hurford, G. J., and Brown, J. C.: 1984, *Solar Phys.* **91**, 359.
Kaufmann, P., Correia, E., Costa, J. E. R., Zodi Vaz, A. M., and Dennis, B. R.: 1985, *Nature* **313**, 380.
Kaufmann, P., Correia, E., Costa, J. E. R., and Zodi Vaz, A. M.: 1986a, in E. Tandberg-Hanssen and H. S. Hudson (eds.), *Solar Flares and Coronal Physics using P/OF as a Research Tool*, NASA Conf. Publ. 2421.
Kaufmann, P., Correia, E., Costa, J. E. R., and Zodi Vaz, A. M.: 1986b, *Astron. & Astrophys.* **157**, 11.
Kaufmann, P., Strauss, F. M., Raffaelli, J. C., and Opher, R.: 1980, in R. F. Donnelly (ed.), *Solar Terrestrial Proc.*, NOAA, Boulder, Colorado, Vol. III, p.c.-115.
Kiplinger, A. L., Dennis, B. R., Emslie, A. G., Frost, K. J., and Orwig, L. E.: 1983, *Astrophys. J.* **265**, L99.
Leboeuf, J. N., Tajima, T., and Dawson, J. M.: 1982, *Phys. Fluids* **25**, 784.
Loran, J. M., Brown, J. C., Correia, E., and Kaufmann, P.: 1985, *Solar Phys.* **97**, 363.
MacKinnon, A. L., Costa, J. E. R., Kaufmann, P., and Dennis, B. R.: 1986, *Solar Phys.* **104**, 191.
McClements, K. G. and Brown, J. C.: 1986, *Astron. Astrophys.* **165**, 235.
Raoult, A.: 1984, private communication.
Sakai, J., Zaidman, E., Tajima, T., Nakajima, H., Kosugi, T., and Brunel, F.: 1986, *Proc. of the SMM Topical Workshop on Rapid Fluctuations in Solar Flares, NASA, Lanham, MD, September 30–October 4, 1985*, NASA Conf. Publ. 2449, 393.
Shimabukuro, F. I.: 1970, *Solar Phys.* **15**, 424.
Shimabukuro, F. I.: 1972, *Solar Phys.* **23**, 169.
Sturrock, P. A., Kaufmann, P., Moore, R. L., and Smith, D. F.: 1984, *Solar Phys.* **94**, 341.
Tajima, T., Brunel, F., and Sakai, J.: 1982, *Astrophys. J.* **258**, L45.
Takakura, T., Kaufmann, P., Costa, J. E. R., Degaonkar, S. S., Ohki, K., and Nitta, N.: 1983, *Nature* **302**, 317.
van Nieuwkoop, J.: 1985, private communication.
Zlobec, P.: 1986, private communication.
Zodi Vaz, A. M., Kaufmann, P., and Zirin, H.: 1984, *Solar Phys.* **92**, 283.

ELECTRON CYCLOTRON MASER EMISSION FROM SOLAR FLARES*

(Invited Review)

LOUKAS VLAHOS

Department of Physics, University of Thessaloniki, 54006 Thessaloniki, Greece

Abstract. Energetic electrons, with energies 10–100 keV, accelerated during the impulsive phase of solar flares, sometimes encounter increasing magnetic fields as they stream towards the chromosphere. A consequence of the conservation of their magnetic moment is that the electrons with large initial pitch angle will be reflected at different heights from the atmosphere. Energetic electrons reflected below the transition zone will lose most of their energy to collisions and will never return to the corona. Thus, electrons reflected above the transition zone form a loss-cone velocity distribution which can be unstable to *Electron Cyclotron Maser* (*ECM*). The interaction of quasi-perpendicular shocks with the ambient coronal plasma will form a 'ring' or 'hollow beam' velocity distribution upstream of the shock. 'Ring' velocity distributions are also unstable to the ECM instability. A review of the recent results on the theory of ECM will be presented. We will focus our discussion on the questions: (a) What are the characteristics of the linear growth rate of the ECM during solar flares? (b) How does the ECM saturate and what is its efficiency? (c) How does the ECM generated radiation modify the flare environment? Finally we will review the outstanding questions in the theory of ECM and we will relate the theoretical predictions to current observations.

1. Introduction

Over the past years there has been great intrest in solar physics in the excitation of direct electromagnetic emission from unstable electron distributions (Twiss and Roberts, 1958; Fung and Yip, 1966; Ramaty, 1969; Mangeney and Veltri, 1976; Stepanov, 1978). Unlike plasma emission (Melrose, 1987), which is a multi-stage process, direct electromagnetic emission is a very efficient process. The conversion of the unstable electrostatic waves to electromagnetic waves depends strongly on the details of the wave-wave interaction or wave-wave-particle interaction that are well understood only in specific limits (e.g., using the weak turbulence approximation). Direct electromagnetic emission on the other hand depends only on the shape of the velocity distribution; mildly relativistic or relativistic electrons can transfer their energy directly to electromagnetic radiation. Direct electromagnetic emission processes are split into two groups: (a) spontaneous emission and (b) stimulated emission. Gyrosynchrotron and bremsstrahlung emission are spontaneous processes (see Melrose, 1985, and references therein). Stimulated emission is a collective process and requires an unstable velocity distribution that can excite the high frequency electromagnetic modes (e.g., the electron cyclotron mode and its harmonics).

In this review we will discuss briefly the characteristics of the linear and nonlinear evolution of the loss-cone and ring velocity distributions. Wu (1985) reviewed the theory

* Proceedings of the Second CESRA Workshop on Particle Acceleration and Trapping in Solar Flares, held at Aubigny-sur-Nère (France), 23–26 June, 1986.

Solar Physics **111** (1987) 155–166.

of the ECM and its application to the Auroral Kilometric Radiation (AKR). Our discussion will focus on the applications of ECM to solar flares.

The interest in direct electromagnetic emission processes was renewed in solar physics recently (Holman *et al.*, 1980; Melrose and Dulk, 1982a; Sharma *et al.*, 1982) following a series of high time and spectral resolution microwave observations (Slottje, 1978; Zhao and Jin, 1982; Benz, 1985). Microwave bursts with duration less than 100 ms have been observed by Dröge earllier (Dröge, 1967, 1977) and named 'spikes' by de Groot. In this review our discussion will be focused exclusively on the very fast 'spikes', the millisecond spikes as they now called. Dulk (1985) and Benz (1986) reviewed the characteristics of the millisecond spikes and concluded that (a) the duration of each spike can be as short as 1 ms. (b) They are intense, with brightness temperature as high as 10^{15} K. (c) Their polarization can be as high as 100%. (d) Their bandwidth is extremely narrow ($\delta f/f_0 < 0.1\%$) and they appear only in the centimetric-decimetric frequency band.

It is easy to show that the above characteristics of the millisecond spikes rule out the 'conventional' plasma emission processes (see Melrose, 1987). By the term 'conventional' we mean here the interaction of broadband wave packets in which the 'random phase approximation' is valid. Vlahos *et al.* (1983) pointed out that if the interacting plasma waves are almost monochromatic or 'narrow' band wave-packets the character of the nonlinear wave-wave interaction is different. In fact, nonlinear interaction of narrow band electrostatic waves can also be a candidate for the interpretation of millisecond spikes (Vlahos *et al.*, 1983), but the coherent electrostatic wave-wave interaction is not a direct emission process and will not be reviewed here.

The plan of the review is as follows:In Section 2 we review the characteristics of the linear growth for two well-known velocity distributions (loss cone and ring distribution), that can excite the ECM. In Section 3 we review the saturation and efficiency of the ECM instability, and in Section 4 we discuss the application of the ECM to the millisecond spikes in solar flares. The escape of ECM radiation from the solar surface and the interaction of the strong electromagnetic radiation with the plasma outside the radiation is discussed in Section 5. Finally in Section 6 we summarize the successes, the shortcomings and the outstanding questions on the theory of the ECM and we compare the theoretical predictions with the existing observations.

2. The Linear Growth Rate of the ECM Instability

The starting point for the understanding of the ECM is the study of the linear growth rate. The general expression for the magnetoionic mode is given by Melrose (1980), but for our discussion it is more instructive to study the simplified expression for the growth of waves travelling perpendicular to the magnetic field (see Wu, 1985). The dispersion relation for the extraordinary (x) mode, $|\omega_r|^2 \simeq \Omega_e^2 \simeq k^2c^2 \gg \omega_e^2$, is

$$1 - \frac{c^2k^2}{\omega^2} + \frac{\omega_e^2}{\omega^2} \int \mathrm{d}^3v \left(\Omega_e \frac{\partial F}{\partial v_\perp} + k_\parallel v_\perp \frac{\partial F}{\partial v_\parallel} \right) \frac{v_\perp (J_1'(b))^2}{(\omega - \Omega_e/\gamma - k_\parallel v_\parallel)} = 0 \qquad (1)$$

and for the ordinary (o) mode

$$1 - \frac{c^2k_\perp^2}{\omega^2} + \frac{\omega_e^2}{\omega^2} \int \mathrm{d}^3 v \left(\Omega_e \frac{\partial F}{\partial v_\perp} + k_\| v_\perp \frac{\partial F}{\partial v_\|} \right) \frac{v_\|^2 J_1^2(b)}{v_\perp(\omega - \Omega_e/\gamma - k_\| v_\|)} = 0\,, \tag{2}$$

where $\Omega_e = |e|B/m_e c$, $J'(b) = \mathrm{d}J(b)/\mathrm{d}b$, $b = k_\perp v_\perp/\Omega_e$, $\gamma = 1/(1 - v^2/c^2)^{1/2}$, $k_\perp/k_\| \ll 1$, ω_e is the plasma frequency, F is the velocity distribution, v is the velocity of the particles and we use the indices $\|$, $\perp$ to denote the component parallel and perpendicular to the magnetic field. Assuming that $\omega = \omega_r + i\gamma$ and $\gamma \ll \omega_r$ we estimate the linear growth rate for the x-mode,

$$\gamma_x = \frac{\pi^2\omega_e^2}{4\omega_r} \int_{-\infty}^{\infty} \mathrm{d}v_\| \int_0^\infty \mathrm{d}v_\perp v_\perp^2\, \delta\left[\omega_r - \Omega_e\left(1 - \frac{v^2}{2c^2}\right) - k_\| v_\|\right] \Omega_e \frac{\partial F}{\partial v_\perp}\,, \tag{3}$$

and for the o-mode,

$$\gamma_0 = \frac{\pi^2\omega_e^2}{4\omega_r} \int_{-\infty}^{\infty} \mathrm{d}v_\| \int_0^\infty \mathrm{d}v_\perp (v_\perp^2/c^2) v_\|^2\, \delta\left[\omega_r - \Omega_e\left(1 - \frac{v^2}{2c^2}\right) - k_\| v_\|\right] \Omega_e \frac{\partial F}{\partial v_\perp}\,, \tag{4}$$

where ω_r is the real frequency. In deriving the growth rates γ_x, γ_0 from the dispersion relation (Equations (1) and (2)) we made several assumptions: (a) We assumed that the particles are midly relativistic ($(v/c) \ll 1$) and the relativistic factor was approximated by $\gamma \simeq (1 - v^2/2c^2)$, (b) we studied only the first harmonic $\omega_r \simeq \Omega_e \simeq kc$, (c) $b = v_\perp/c$ is small and we approximated the Bessel functions $J_1^2(b)$ with $v_\perp^2/4c^2$ and $(J_1'(b))^2$ with $\frac{1}{4}$, (d) the term $\partial F/\partial v_\|$ is neglected when $\partial F/\partial v_\perp \simeq \partial F/\partial v_\|$ since $k_\| v_\perp(\partial F/\partial v_\|)$ is much smaller than $\Omega_e(\partial F/\partial v_\perp)$, (e) finally we ignored the ambient plasma.

Several intresting conclusions can be drawn from a study of the argument of the δ-function in Equations (3) and (4): (1) If $\gamma = 1$, the relativistic correction is ignored, then the growth rates are always negative. (2) The perpendicular velocity is limited from the resonance condition ($\omega - \Omega_e(1 - v^2/2c^2) - k_\| v_\| = 0$). (3) The growth rate for the x-mode is higher than the growth rate for the o-mode.

With the use of graphic representation of the resonance condition in the $v_\perp$, $v_\|$ space we can study the characteristics of the growth rate for a given distribution (Wu, 1985).

The linear growth rate as a function of the angle $\theta = \cos^{-1}(\bar{k} \cdot \bar{\mathrm{B}})$ for a loss cone velocity distribution chosen as a Maxwellian with a hole,

$$F(v_\perp, v_\|) = n g_N(a)\,(2\pi m K_B T)^{-3/2} \exp(-mv^2/2K_B T) \tag{5}$$

with

$$g_N(a) = \begin{cases} a_N\,, & a \le a_0\,, \\ a_N[\sin\{(\tfrac{1}{2})\pi(\pi - a)/(\pi - a_0)\}]^N, & a > a_0\,, \end{cases} \tag{6}$$

is plotted in Figure 1 using the complete dispersion relation (Equations (9) and (10) in Hewitt *et al.*, 1982)]. The radiation is confined inside a narrow cone around 90°. In

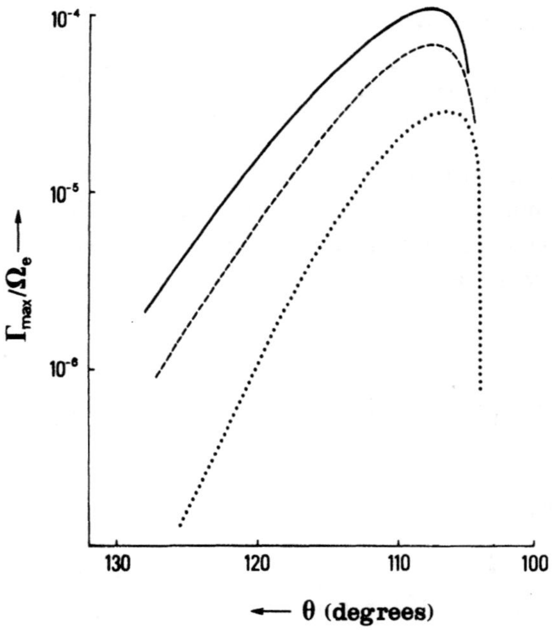

Fig. 1. The maximum growth rate as a function of angle $\theta = \sin^{-1}(\mathbf{k} \cdot \mathbf{B})$ for $N = 1$ (dotted curve), $N = 3$ (dashed curve), and $N = 6$ (solid curve) (Hewitt *et al.*, 1982).

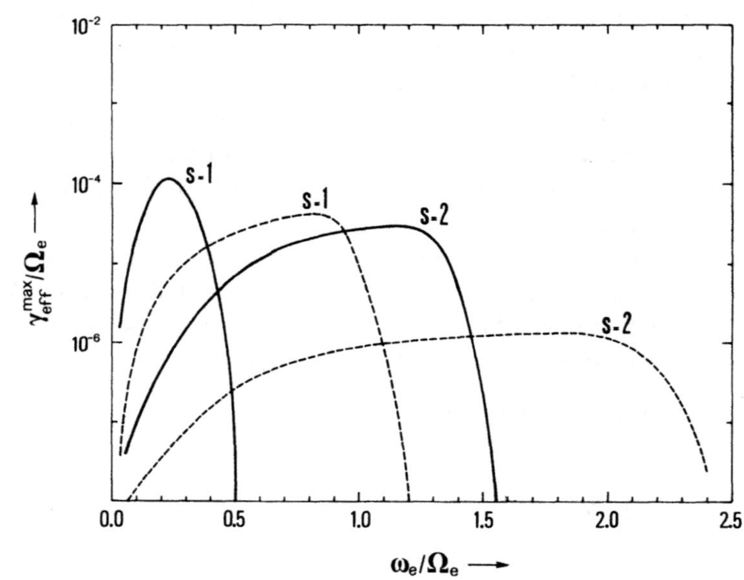

Fig. 2. The maximum growth rate as a function of ω_e/Ω_e. The first and second harmonics for the *x*-mode (solid line) and *o*-mode (dashed line) are shown (Sharma and Vlahos, 1984).

Figure 1 the ratio ω_e/Ω_e is small. As the ambient plasma density increases the linear growth rate of the first and higher harmonics of the extraordinary and ordinary mode changes dramatically (Figure 2). In fact for $\omega_e/\Omega_e < 0.3$, the fundamental x-mode has the highest growth rate; for $0.3 < \omega_e/\Omega_e < 1$ the fundamental o-mode and the second harmonic x-mode dominate (Sharma and Vlahos, 1984; Winglee, 1985a). As the ratio ω_e/Ω_e increases above one, no direct emission is expected (the Z-mode and the electrostatic modes have the largest growth rates). It is easy to show that the mode with the largest growth rate will be the dominant mode and will saturate the instability before the other modes grow to any appreciable level (Sharma *et al.*, 1982). The effects of finite plasma temperature and superthermal tails on the ECM have been studied in detail by many authors (Sharma and Vlahos, 1984; Winglee, 1985b; Vlahos and Sharma, 1985). These results are important for the application of the ECM in solar flares since the ECM will co-exist with 'hot', dense plasma in the base of the corona.

The results obtained above are similar for the hollow beam or ring velocity distribution (Wu and Freund, 1984),

$$F(u_\parallel, u_\perp) = (\pi a_\perp^2)^{-1}(\pi a_\parallel^2)^{-1/2} A \exp\left(-\frac{(u_\perp - u_{\perp 0})^2}{a_\perp^2} - \frac{(u_\parallel - u_{\parallel 0})^2}{a_\parallel^2}\right), \tag{7}$$

where $u = p/m$, p is the momentum of the particle, $a_\parallel$ and $a_\perp$ denote the characteristic momentum spread parallel and perpendicular to the ambient magnetic field, $u_{\parallel 0}$ is the beam velocity and $u_{\perp 0}$ describes a 'population inversion' in the perpendicular direction, $A = 2\exp(x_0^2)(1 - ix_0 Z(-ix_0))^{-1}$, $x_0 = u_\perp / a_\perp$ and Z is the plasma dispersion function. In Figure 3 we plot the linear growth rate for the fundamental as a function of ω_e/Ω_e assuming that $u_{\parallel 0} < u_{\perp 0}$ (Freund *et al.*, 1983). Higher harmonics will dominate the emission when ω_e/Ω_e increases and when $\omega_e/\Omega_e > 1$ no direct electromagnetic emission is possible. For $u_{\parallel 0} \gg u_{\perp 0}$ the ring velocity distribution can excite three modes along the ambient magnetic field ($k_\parallel \gg k_\perp$); (a) The Whistler mode, (b) the ECM with real frequency $\omega_r \gtrsim \Omega_e$, (c) the high-frequency ECM with real frequency $\omega_r \simeq \gamma\Omega_e$. Sprangle and Vlahos (1986) and Vlahos and Sprangle (1987) analysed the linear and nonlinear evolution of the high-frequency ECM. The high-frequency ECM emission does not grow when $\omega_e/\Omega_e > 1$. An important characteristic of the high-frequency ECM is that the real frequency ω_r is much higher than the ambient cyclotron frequency ($\omega_r \simeq \gamma\Omega_e$) when the electrons are relativistic.

3. Saturation and Efficiency of the ECM

The linear theory discussed in Section 2 only suggests the frequency range and the wave numbers that will be excited by the unstable velocity distribution. The exponential growth lasts only 50–100 growth times, $t_g \simeq \gamma_j^{-1}$ (where γ_j is the growth rate of the mode j). The saturation time for typical solar parameters is $\simeq 10^{-6}\ \mathrm{s}^{-1}$ and the nonlinear effects which are discussed in this section will eventually saturate the ECM.

The amplification of the ECM radiation is due to the positive slope in the direction perpendicular to the magnetic field. One possible way to saturate the instability is for

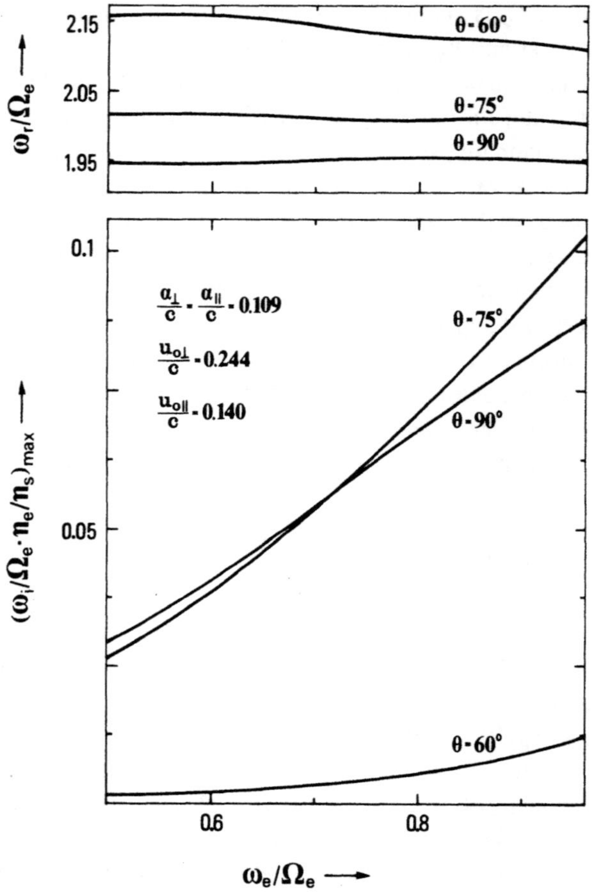

Fig. 3. The maximum growth rate for the *x*-mode of a hollow beam as a function of ω_e/Ω_e (Freund *et al.*, 1983).

the growing waves to flatten the slope of the velocity distribution in the direction perpendicular to the magnetic field. Wu *et al.* (1981) used the quasi-linear approximation to estimate the saturation level for the growing waves. Using the following initial distribution function for the electrons,

$$F = \frac{n_e}{\pi^{3/2} a (a_1^2 - a_2^2)} \exp\left(-\frac{v_\parallel^2}{a^2}\right)\left[\exp\left(-\frac{v_\perp^2}{a_1^2}\right) - \exp\left(-\frac{v_\perp^2}{a_2^2}\right)\right], \tag{8}$$

they found that the saturated wave energy $W_s = \int \mathrm{d}^3 k E_k(t)$ is

$$\frac{W_s}{n_e m_e a_1^2} = \frac{2 a_2 \Omega_e}{\pi a_1 \omega_e}\left(\frac{k_{0\parallel} a}{\Omega_e}\right)\left(\frac{\gamma_{j,\,\max}}{\Omega_e}\right)\left(\frac{n_0}{n_e}\right), \tag{9}$$

where a_1, a_2 are the velocity spreads in the direction perpendicular to the magnetic field

$(a_1^2 > a_2^2)$, a is the velocity spread in the parallel direction, $\gamma_{j,\,\max}$ is the maximum growth rate, n_0 is the density of the ambient plasma, n_e is the density of the energetic electrons and $k_{0\parallel}$ is the wave number of the fastest grown waves. Sharma *et al.* (1982), using solar parameters, showed that the efficiency (W_s/W_F) is 1–10%, where W_F is the total energy of the energetic electrons in the loss-cone. Numerical simulations by Wagner *et al.* (1984), Pritchett (1984), Lin *et al.* (1984), and Vlahos and Sprangle (1987) showed that the efficiency is $\simeq$ 1–5% (see Figure 4). Li (1987) studied the role of inhomogeneity of magnetic field along the ray path of the emitting wave and found that it limits the spatial region of ECM growth and the saturation level of the emitting wave.

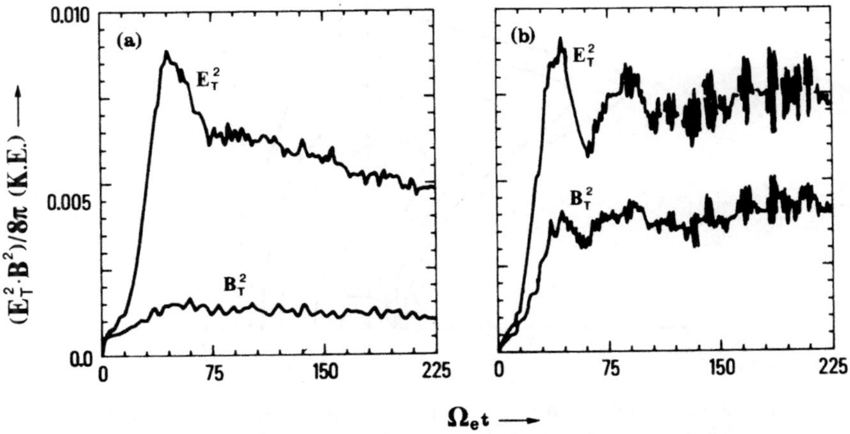

Fig. 4. The time histories of the transverse electric (E_T^2) and magnetic (B^2) energy normalized to the initial electron kinetic energy (Pritchett, 1984).

Melrose and Dulk (1982b) pointed out that the results presented above have assumed an infinite uniform velocity distribution. Electrons trapped inside a magnetic trap will continue losing energy as they bounce inside the loop. Melrose and Dulk (1982b) argued (qualitatively) that the energy transferred to the ECM by the trapped electrons could be much higher than the saturation values estimated above.

4. Application of the ECM to the Millisecond Spikes in Solar Flares

Holman *et al.* (1982) proposed that the millisecond spikes are the signature of the ECM radiation driven by the trapped energetic electrons inside a flaring loop. A study on the formation of a loss cone velocity distribution inside a magnetic arch was presented by Kovalev and Korolev (1981) and White *et al.* (1986). They analysed Liouville's equation for the free-streaming evolution of the velocity distribution $F(\mathbf{r}, \mathbf{v}, t)$:

$$\frac{\partial F}{\partial t} + \dot{\mathbf{r}}\frac{\partial F}{\partial \mathbf{r}} + \dot{\mathbf{p}}\frac{\partial F}{\partial \mathbf{p}} = Q(\mathbf{r}, \mathbf{v}, t)\,, \tag{10}$$

where $Q(\mathbf{r}, \mathbf{v}, t)$ is a source function term describing the rate with which energetic particles are 'created' at the point $\mathbf{r}$. The magnetic field geometry used is $B(z) = B_0(1 + z^2/d^2)$, where d is the scale length of the variation of the magnetic field. Solving Equation (10) we can find the slope of the velocity distribution of the electrons streaming towards the chromosphere or reflected, by the converging magnetic field, back to the solar corona. In Figure 5 the velocity distribution is plotted in the $v_\perp$, $v_\parallel$ plane.

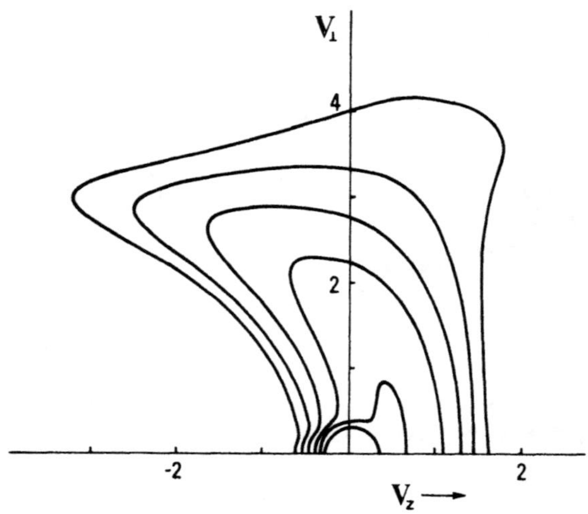

Fig. 5. The free-streaming velocity distribution in $v_\perp, v_\parallel$ space away from the injection region. The distribution starts to resembes a ring distribution (White *et al.*, 1986).

We observe the formation of the loss-cone on the particles streaming away from the chromosphere (e.g., $v_\parallel < 0$). The solution of the free-streaming equation is not valid when the velocity distribution is unstable. The shape of the velocity distribution changes dynamically once the waves start to grow and the wave-particle interaction should be incorporated in Equation (10). The evolution of an unstable velocity distribution inside a magnetized plasma is described by the two-dimensional $[(r(z, r), v(v_\parallel, v_\perp)]$ quasilinear equations

$$\frac{\partial F}{\partial t} + \dot{\mathbf{r}} \frac{\partial F}{\partial \mathbf{r}} = L_w F + Q(r, v, t) , \tag{11a}$$

where

$$L_w = \left[\frac{\partial}{\partial v_\parallel} D_\parallel \frac{\partial}{\partial v_\parallel}\right] + \left[-\frac{\Omega_e}{k_\parallel v_\perp}\frac{\partial}{\partial v_\perp} + \frac{\partial}{\partial v_\parallel}\right] D_\perp \left[-\frac{\Omega_e}{k_\parallel v_\perp}\frac{\partial}{\partial v_\perp} + \frac{\partial}{\partial v_\parallel}\right], \tag{11b}$$

$D_\parallel(W)$, $D_\perp(W)$ are the diffusion coefficients, and W is the wave amplitude. The wave amplitude is a function of the linear growth rate $(\partial W/\partial t + v\,(\partial W/\partial r) = \gamma_j W)$, which in turn is a function of F (see Equation (3)). Equation (11) is not easy to solve. In the analysis presented in Section 3, we assumed an initial shape for the velocity distribution (e.g., loss-cone) and estimated the linear growth and then solving the quasilinear equations (ignoring the terms $\partial F/\partial r$, D, $\partial W/\partial r$, and Q), we estimated the relaxation of the velocity distribution. The evolution of an inverted population (loss cone or ring distribution) inside a magnetic trap has not been studied in detail yet.

Melrose and Dulk (1982a) and Sharma *et al.* (1982) analysed the ECM instability for a given loss-cone velocity distribution in the solar corona and concluded that the millisecond bursts can be interpreted as the signature of the ECM only if $0.3 < \omega_e/\Omega_e < 1$ since for these parameters the dominant mode is the second harmonic x-mode which could be possible to escape the gyroresonance absorption from the solar corona.

5. Absorption of the ECM Generated Radiation from the Solar Corona

Gyromagnetic absorption of electromagnetic waves propagating out of the solar corona in a magnetic field topology whose strength decreases away from the solar surface has been discussed extensively by Zheleznyakov (1970). His results are correct for obliquely propagating electromagnetic waves in a homogeneous stratified atmosphere. Maser emission at the sth harmonic occurs at $\omega > s\Omega_e$, and propagates upwards towards lower B, and, hence, lower Ω_e; it does not pass through the sth absorbing layer, but it must passes through the layers $s+1$, $s+2$, etc. The optical depth of the sth layer ($s \geq 2$) for the extraordinary mode is (Zheleznyakov, 1970),

$$\tau_s = \frac{\omega_e^2}{\omega^2} \frac{s^{2s}}{2^{s-1}} \frac{\sqrt{2}}{s!} \frac{\omega L_B}{c} \left(\frac{T_e}{mc^2}\right)^{s-1}, \tag{12}$$

with $L_B = B/|\text{grad}\, B|$. For $\omega/2\pi \cong 3$ GHz, $L_B \cong 10^4$ km, $T_e \cong 10^7$ K, and $\omega_e/\omega \cong 0.1$ we conclude that the fundamental cannot escape through the second harmonic absorbing layer under normal circumstances. Holman *et al.* (1980) suggested that maser emission at $s = 1$ might escape through a window at small θ where the absorption is much weaker than Equation (12) implies. However, it is hard to see how the radiation which is generated at large θ can reach the $s = 2$ absorption layer at very small θ. Stepanov (1978), repeated the calculations of gyroresonance absorption for waves propagating perpendicular to the magnetic field ($k_{\parallel} = 0$) and found that the gyroresonance absorption at $(s+1)\Omega_e$ layer is reduced considerably. We believe that these calculations should be further extended to include the escape of electromagnetic waves from an active region where the magnetic field is concentrated in loop-like magnetic structures and falls off quickly away from the loop. Another important point if that if the electromagnetic radiation approaches the $(s+1)\Omega_e$ layer inside a flaring loop is likely to get amplified if the plasma is unstable. It is clear from our discussion so far that the ray tracing of ECM radiation away from the active region is an unresolved, but fundamental problem.

Vlahos and Sprangle (1987) studied the excitation of the high-frequency maser. They assumed that the interaction of a flare energized plasma by the passage of a quasi-perpendicular shock can accelerate electrons to relativistic energies along the magnetic field. Perpendicular to the magnetic field the distribution in velocity space is a ring, since electrons with $v_\perp \cong 0$ are transmitted through the shock. The accelerated particles have perpendicular energy much smaller than their axial energy. The electromagnetic radiation emitted along the ambient magnetic field can be intence, 100% polarized and emitted at a frequency $\omega \cong \gamma\Omega_e$. For $\gamma \gg 1$ the emitted electromagnetic radiation will be far above the fundamental electron-cyclotron harmonic.

Absorption of electromagnetic emission by the solar atmosphere will mean local heating or particle acceleration (Melrose and Dulk, 1982b, 1984; Sprangle and Vlahos, 1983; Karimabadi *et al.*, 1987).

Melrose and Dulk (1982b, 1984) proposed that the maser may operate in localised regions (called elementary bursts, EB) for short duration, leading to radiation in the form of many elementary bursts. They assumed that a large fraction of the available free energy in the source region is radiated away. The energy in each elementary burst is absorbed in a second-harmonic layer, i.e., where the cyclotron frequency is half its value in the source region. In the absorption volume for a given elementary burst the temperature is increased only slightly but the overall heated region is the envelope of the absorption volumes for many elementary bursts. Melrose and Dulk proposed that the temperature in the absorption volume may reach as high as a few $\times 10^7$ K, which may be observable as soft X-ray enhancements.

Sprangle and Vlahos (1983) and Karimabadi *et al.* (1987) studied the interaction of intense, narrow bandwidth electromagnetic radiation, generated by masering with the tail of a Maxwellian velocity distribution in the absorption source. They found, using flare parameters, that the tail electrons, in the absorption volume, can be accelerated up to 100 keV. These results suggest that there is a link between the energy released inside a flaring trap and the electrons accelerated in open field lines (e.g., type III bursts, see Benz, 1986). Sprangle and Vlahos assumed that only 1% of the energy available in the reflected electrons was transferred to electromagnetic waves, absorption of this energy by the bulk distribution will not effect the temperature of the absorbing source.

The models discussed above are the first attempts to use the ECM for fast cross field transfer.

6. Summary and Discussion

The linear growth and the nonlinear evolution of the loss cone and ring velocity distribution driven ECM are now well understood. The main conclusions reached so far are the following:

(1) The loss cone or ring distributions will excite the ECM. Depending on the ratio of plasma to gyrofrequency (ω_e/Ω_e) different harmonics will dominate the emission. The polarization of the emitted radiation depends on the ratio ω_e/Ω_e but it can reach as high as 100% for $\omega_e/\Omega_e < 0.3$. For $\omega_e/\Omega_e > 1$ the ECM will be weak and other plasma instabilities will control the evolution of the loss cone or ring velocity distributions.

(2) The nonlinear evolution of the loss cone or ring distributions suggests that the instability will saturate in 10^{-6} s in the solar atmosphere and only 1–5 % of the available energy will be transferred to the electromagnetic waves.

(3) The relativistic ring distribution, with large axial velocity $v_{\parallel 0} \gg v_{\perp 0}$, will excite a high frequency mode ($\omega = \gamma \Omega_e$).

Let us now assume that electrons accelerated inside a flaring loop form a loss cone velocity distribution in the footpoints of the loop. The velocity distribution is unstable to the ECM and will saturate quickly, radiating away a small fraction of the available energy. The emitted radiation is narrow bandwidth, it can reach brightness temperatures as high as 10^{15}–10^{20} K and it can be highly polarised. The frequency of the emitted radiation will be around the first or second harmonic of the cyclotron frequency (for solar parameters around 2–3 GHz).

The results reported above seem to explain most of the known characteristics of the millisecond spikes, unfortunately there are several difficulties as well: (a) We mentioned above that the ECM is sensitive to the ω_e/Ω_e. This parameter changes dramatically during a solar flare (especially around the transition region). (b) The gyroresonance absorption from the upper atmosphere will absorbe the first and sometimes the second harmonic of the ECM radiation. Higher harmonics will not grow fast enough to reach appreciable levels in the solar atmosphere. The high-frequency maser can overcome some of the problems reported above, since it will be excited in frequencies much above the first and second harmonic if the electrons are relativistic ($\omega = \gamma \Omega_e$). It propagates along the ambient magnetic field and is formed by outward moving disturbances (thus the ring distribution is formed away from the transition region).

The ECM radiation can transfere energy across the ambient magnetic field lines. This energy may heat the ambient plasma or accelerate the tail of the local Maxwellian distribution. We should then expect a correlation of the millisecond spikes with soft X-ray kernels or energetic electrons screaming away from the solar atmosphere (e.g., type III bursts).

There are still several open questions on the application of the ECM to solar flares, we will mention here only a few.

(1) Is the efficiency of the ECM from a trapped population of energetic electrons much higher than the efficiency of the homogeneous velocity distribution reported above?

(2) What is the time evolution of the ECM?

Answering these questions will require a detailed understanding of the electron motion inside the flaring loop and estimating the formation time and saturation time of the loss cone velocity distribution. Finally:

(3) How does the ECM escape from the flaring loop and interact with the flare environment? Ray tracing calculations performed in the Earth's aurora (Omidi *et al.*, 1984) are difficult to apply in the solar atmosphere since the magnetic topology of the active region is not well known and it changes from event to event.

We believe that our theoretical understanding of the millisecond spikes has improved dramatically the last few years but we still have a long way to go.

References

Benz, A. O.: 1985, *Solar Phys.* **96**, 357.
Benz, A. O.: 1986, *Solar Phys.* **104**, 99.
Dröge, F.: 1967, *Z. Astrophys.* **66**, 176.
Dröge, F.: 1977, *Astron. Astrophys.* **57**, 285.
Dulk, G.: 1985, *Ann. Rev. Astron. Astrophys.* **23**, 169.
Freund, H. P., Wong, H. K., Wu, C. S., and Xu, M. J.: 1983, *Phys. Fluids* **26**, 2263.
Fung, P. C. W. and Yip, W. K.: 1966, *Australian J. Phys.* **19**, 759.
Hewitt, R. G., Melrose, D. B., and Ronnmark, K. G.: 1982, *Australian J. Phys.* **35**, 447.
Holman, G. D., Eichler, D., and Kundu, M. R.: 1980, in M. R. Kundu and T. Gergely (eds.), 'Radio Physics of the Sun', *IAU Symp.* **86**.
Karimabadi, Menyuk, C. R., Sprangle, P., and Vlahos, L.: 1987, *Astrophys. J.* **316**, 462.
Kovalev, V. A. and Korolev, O. S.: 1981, *Soviet Astron.* **25**, 215.
Li, H. W.: 1987, *Solar Phys.* **111**, 167 (this issue).
Lin, A. T., Chang, W. W., and Lin, C. C.: 1984, *Phys. Fluids* **27**, 1054.
Mangeney, A. and Veltri, P.: 1976, *Astron. Astrophys.* **47**, 165.
Melrose, D. B.: 1980, *Plasma Astrophysics*, Gordon and Breach, New York.
Melrose, D. B.: 1985, *Solar Radiophysics*, Cambridge Univ. Press, Cambridge.
Melrose, D. B.: 1987, *Solar Phys.* **111**, 89 (this issue).
Melrose, D. B. and Dulk, G. A.: 1982a, *Astrophys. J.* **259**, 844.
Melrose, D. B. and Dulk, G. A.: 1982b, *Astrophys. J.* **259**, L41.
Melrose, D. B. and Dulk, G. A.: 1984, *Astrophys. J.* **282**, 308.
Omidi, N., Wu, C. S., and Gurnett, D. A.: 1984, *J. Geoph. Res.* **89**, 883.
Pritchett, P. L.: 1984, *Phys. Fluids* **27**, 2393.
Ramaty, R.: 1969, *Astrophys. J.* **178**, 241.
Sharma, R. R. and Vlahos, L.: 1984, *Astrophys. J.* **280**, 405.
Sharma, R. R., Vlahos, L., and Papadopoulos, K.: 1982, *Astron. Astrophys.* **112**, 377.
Slottje, C.: 1978, *Nature* **275**, 520.
Sprangle, P. and Vlahos, L.: 1983, *Astrophys. J.* **273**, L95.
Sprangle, P. and Vlahos, L.: 1986, *Phys. Rev.* **A33**, 1261.
Stepanov: 1978, *Soviet. Astrophys. Letters* **4**, 103.
Twiss, R. Q. and Roberts, J. A.: 1958, *Australian J. Phys.* **11**, 424.
Vlahos, L. and Sharma, R. R.: 1985, *Astrophys. J.* **290**, 347.
Vlahos, L. and Sprangle, P.: 1987, *Astrophys. J.* (in press).
Vlahos, L., Sharma, R. R., and Papadopoulos, K.: 1983, *Astrophys. J.* **275**, 374.
Wagner, J. S., Lee, L. C., Wu, C. S., and Tajima: 1984, *Radio Sci.* **19**, 509.
White, S. M., Melrose, D. B., and Dulk, G. A.: 1986, *Astrophys. J.* **308**, 424.
Winglee, R. M.: 1985a, *J. Geophys. Res.* **90**, 9663.
Winglee, R. M.: 1985b, *Astrophys. J.* **291**, 160.
Wu, C. S.: 1985, *Space Sci. Rev.* **41**, 215.
Wu, C. S. and Freund, H. P.: 1984, *Radio Sci.* **19**, 519.
Wu, C. S., Tsai, S. T., Xu, M. J., and Shen, J. W.: 1981, *Astrophys. J.* **248**, 384.
Zao, R. and Jin, S.: 1982, *Sci. Sinica*, Ser. A, **25**, 422.
Zheleznyakov, V. V.: 1970, *Radio Emission of the Sun and Planets*, Pergamon Press, Oxford.

THE SATURATION OF ELECTRON-CYCLOTRON MASER AND THE TIME PROFILE OF EMITTED SPIKES*

HONG-WEI LI**

Osservatorio Astrofisico di Arcetri, Firenze, Italy

Abstract. In this paper the evolution of the hollow beam distribution of energetic electrons giving rise to ECM instability is investigated and the spatial dispersion term is included in the equation of wave energy. The instability causes the growth of wave energy, while the propagation of waves evacuates the electromagnetic energy from the source region. By analysing these two effects spike-like time profiles of waves are obtained. It is found that the saturation time t_s of ECM emission and the duration of spikes increase with the decrease of the frequency of solar radio spike emission. The approximate expressions of t_s and of the peak wave energy density are derived.

1. Introduction

A hollow beam distribution of energetic electrons with energy of a few tens of keV has been proved to be able to excite electron cyclotron maser (ECM) instability and was applied for explaining solar radio spike emissions (Freund *et al.*, 1983; Li *et al.*, 1985; Li, 1986).

In this paper the variation of the magnetic field along the ray path, i.e., the spatial dispersion, and the propagation of waves are taken into account. The emitted wave energy increases owing to the instability growth, which abates due to the diffusion of energetic electrons in velocity space. On the other hand, the propagation of waves evacuates the electromagnetic energy from the source region. By analysing these two effects we obtain time profiles of spike emission. The improved expressions of the saturation time of ECM emission and of the peak wave energy density are also obtained.

2. Evolution of the Distribution Function and the Saturation of ECM

The ECM instability can be excited by a hollow beam distribution of the form

$$f(p_\perp, p_\parallel) = \frac{1}{2\pi^2 p_{0\perp}^2 p_{0\parallel} \varepsilon_\perp \varepsilon_\parallel} \exp\left\{ -\left(\frac{P_\perp - P_{0\perp}}{\varepsilon_\perp P_{0\perp}}\right)^2 - \left(\frac{P_\parallel - P_{0\parallel}}{\varepsilon_\parallel P_{0\parallel}}\right)^2 \right\}, \quad (1)$$

where $p_\perp$ and $p_\parallel$ are the electron momentum components transverse and parallel to the magnetic field $\mathbf{B}$, respectively, $\varepsilon_\perp$ and $\varepsilon_\parallel$ are the width of the beam distribution. The growth rate of ECM at the sth harmonic is (Melrose and Dulk, 1982)

$$\omega_i(\mathbf{K}) = \int \mathrm{d}\mathbf{P}\, A_s(\mathbf{P}, \mathbf{K})\, \delta(\omega - s\Omega_e/\gamma - K_\parallel v_\parallel) \left(\frac{s\Omega_e}{\gamma v_\perp} \frac{\partial}{\partial p_\perp} + K_\parallel \frac{\partial}{\partial p_\parallel} \right) f(p_\perp, p_\parallel), \quad (2)$$

* Proceedings of the Second CESRA Workshop on Particle Acceleration and Trapping in Solar Flares, held at Aubigny-sur-Nère (France), 23–26 June, 1986.

** On leave from the Department of Astronomy, Nanjing University, Nanjing, People's Republic of China.

Solar Physics **111** (1987) 167–173.

where $A_s(\mathbf{P}, \mathbf{K})$ is a complicated function (see Appendix B in Melrose and Dulk, 1982) and γ the Lorentz factor.

The ECM emission causes the decrease of $p_\perp$, which in turn reduces and eventually quenches the maser instability. Assuming that the quasilinear process prevails over other nonlinear processes, we have (Wu *et al.*, 1981)

$$\frac{\partial}{\partial t} f(t, v_\perp, v_\parallel) = D(t, v_\perp, v_\parallel) \frac{1}{v_\perp} \frac{\partial}{\partial v_\perp} v_\perp \frac{\partial}{\partial v_\perp} f(t, v_\perp, v_\parallel), \tag{3}$$

where

$$D(t, v_\perp, v_\parallel) = 4\pi^2 e^2 m_e^{-2} \int \mathrm{d}\mathbf{K}\, \mathscr{E}_\mathbf{K}(t)\, \delta(\omega - s\Omega/\gamma - K_\parallel v_\parallel) \tag{4}$$

is the diffusion coefficient and $\mathscr{E}_\mathbf{K}(t)$ the wave spectral density. Denoting by $W(t) = \int \mathrm{d}\mathbf{K}\, \mathscr{E}_\mathbf{K}(t)$ the wave energy density and noticing the property of ECM emission that $\mathscr{E}_\mathbf{K}$ is narrow and peaks at a wavenumber $K_{\parallel *}$ where $\omega_i(\mathbf{K}, t)$ has its maximum $\omega_i(t)$, we have approximately

$$\frac{\partial}{\partial t} W(t) = 2\omega_i(t) W(t) \tag{5}$$

in a homogeneous background. If the magnetic field varies along the ray path, the spatial region of the growth of ECM is limited by the narrow width of ECM emission (Winglee, 1985) and a spatial dispersion term $V_g\, \partial W(t)/\partial l$ should be added to the left-hand side of Equation (5), where V_g is the group velocity of waves. The length scale of this region can be evaluated by $l = L_B\, |\Delta B/B| = L_B\, |\Delta\omega/\omega|$, where $L_B = B/|\partial B/\partial l|$ is the length scale of magnetic field along the ray path (approximately perpendicular to the magnetic field). Those waves originating in this region will not grow any more once they propagate out of this region (Figure 1). Accordingly, with an approximation $\partial W(t)/\partial l \sim W(t)/l$, Equation (5) becomes

$$\frac{\partial}{\partial t} W(t) + V_g W(t)/l = 2\omega_i(t) W(t)\,. \tag{6}$$

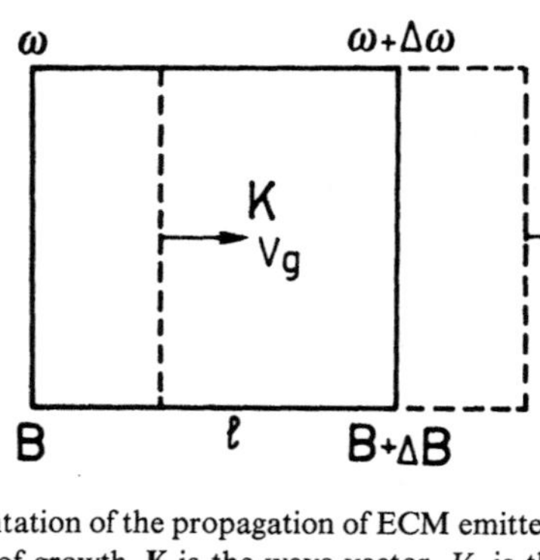

Fig. 1. Schematic representation of the propagation of ECM emitted waves originating in a limited spatial region of growth. **K** is the wave vector, V_g is the group velocity.

On the other hand, as $l \ll L$, we can neglect the inhomogeneity of the magnetic field in the region l and use the dispersion relation for homogeneous plasma in this study. In the following we will investigate ECM emission with the set of Equations (1), (2), (3), (4), and (6).

Introducing

$$\tau(t) = \int_0^t \mathrm{d}t' \, D(t', v_\perp, v_\parallel) \tag{7}$$

in (3), we get the distribution of energetic electrons for $\tau > 0$ (i.e., $t > 0$)

$$f(\tau, v_\perp, v_\parallel) = \frac{1}{2\tau} \int_0^\infty \mathrm{d}x \, x f(0, x, v_\parallel) I_0\left(\frac{x v_\perp}{2\tau}\right) \exp\left\{-\frac{x^2 + v_\perp^2}{4\tau}\right\}, \tag{8}$$

where $f(0, v_\perp, v_\parallel)$ is the initial distribution (1) in terms of velocity components $v_\perp$ and $v_\parallel$, $I_0(z)$ is the zero-order modified Bessel function.

Substituting (8) into (2), we calculated the variations of growth rates of ECM at the second, third and fourth harmonics in X-mode versus τ (Figure 2) and versus t (Figure 3), and the variations of the wave energy density versus time, i.e., the time profiles of the flux density (Figure 4).

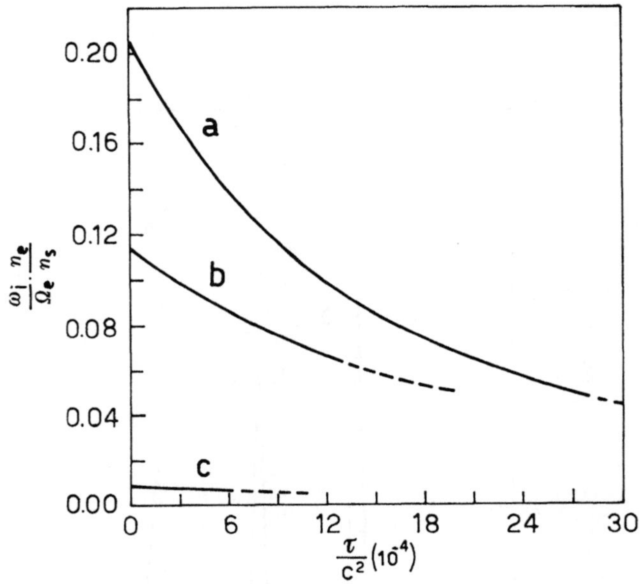

Fig. 2. Variations of the growth rates $(\omega_i/\Omega_e)(n_e/n_s)$ at the second (a), third (b), and fourth (c) harmonics of ECM in X-mode vs τ/c^2, where c is the speed of light, τ is given by (7), the energy of the energetic electrons is 50 keV, $n_s/n_e = 4 \times 10^{-5}$, $\Delta\omega/\omega = 0.01$, and other parameters are (a): $B = 400$ G, $L_B = 500$ km, $A = L_{Bt}/L_B = 10$, $\omega_p/\Omega_e = 0.4$, $T_e = 10^6$ K; (b): $B = 100$ G, $L_B = 3000$ km, $A = 5$, $\omega_p/\Omega_e = 1.4$, $T_e = 5 \times 10^6$ K; (c): $B = 50$ G, $L_B = 4 \times 10^4$ km, $A = 2$, $\omega_p/\Omega_e = 2.2$, $T_e = 10^7$ K.

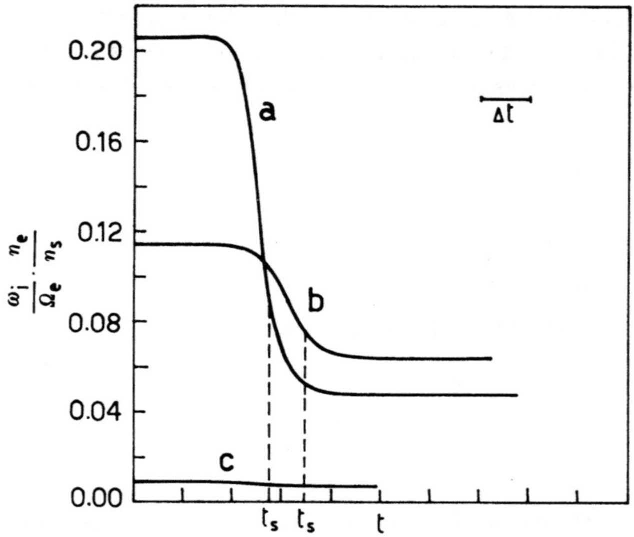

Fig. 3. Variations of the growth rates at the second (a), third (b), and fourth (c) harmonics of ECM in X-mode vs time t. The time-scale $\Delta t = 0.1$ ms for (a), 1 ms for (b), and 50 ms for (c). The parameters are the same as in Figure 2.

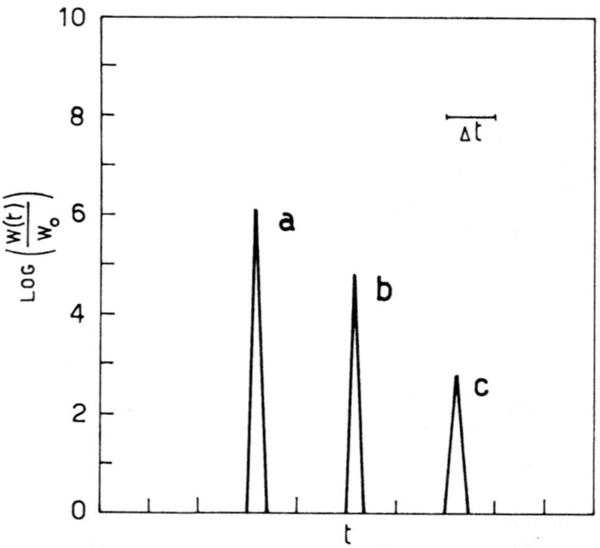

Fig. 4. The time profiles of emitted spikes at the second (a), third (b), and fourth (c) harmonics of ECM. $W(t)$ and W_0 are the wave energy density at time t and the initial level, respectively, the time-scale $\Delta t = 2$ ms for (a), 20 ms for (b), and 500 ms for (c). The parameters are the same as in Figure 2.

Defining the time corresponding to the peak wave energy density W_s as the saturation time t_s of ECM emission, we find out that growth rates almost do not change until time $t = at_s$, where $a \approx 0.8$ for all these harmonics, then fall down to a low-value steeply and almost linearly, and the maser instability is quenched. Hence, the time profiles of the wave energy density or of the flux density do appear like spikes (Figure 4).

Bearing the above properties of ECM emission in mind we derive approximate expressions for t_s, W_s and for the peak flux density S as follows. From (4) and (7) we have

$$\tau \approx 4\pi^2 e^2 (m_e^2 K_{\parallel *} v_{0\parallel})^{-1} \int_0^t \mathrm{d}t' \; W(t') \,. \tag{9}$$

Then Equations (6), (9), and the definitions of W_s and t_s yield

$$W_s = W_0 \exp\left\{2 \int_0^{t_s} \mathrm{d}t \; \omega_i(t) - V_g t_s / l\right\}, \tag{10a}$$

$$2\omega_i(t_s) = V_g / l \,, \tag{10b}$$

$$\tau_s = 4\pi^2 e^2 (m_e^2 K_{\parallel *} v_{0\parallel})^{-1} \int_0^{t_s} \mathrm{d}t \exp\left\{2 \int_0^{t} \mathrm{d}t' \; \omega_i(t') - V_g t / l\right\}. \tag{10c}$$

In Equation (10) W_0 is the initial wave energy density assumed to be the level of thermal fluctuation. In order that strong spike emission can be produced it is necessary that $(2\omega_{i0} - V_g/l)t_s \gg 1$, where ω_{i0} is the initial growth rate. In this case, and taking the properties of growth rates mentioned above into account, we obtain

$$t_s \approx (1 + a)^{-1} (\omega_{i0} - V_g/2l)^{-1} \, \mathrm{Ln}(W_s/W_0) \,, \tag{11}$$

$$W_s \approx (4\pi^2 e^2)^{-1} \, m_e^2 \tau_s v_{0\parallel} K_{\parallel *} (\omega_i - V_g/2l) \,, \tag{12}$$

which are the improved ones of that given by Li *et al.* (1984) and Li (1986) in that the spatial dispersion is included. Then the peak flux density of emitted spikes received at the Earth is

$$S = 4\pi c W_s A \, e^{-\tau_x} / (\Delta\omega \Delta\Omega_r R^2) \,, \tag{13}$$

where R is the distance of the Earth from the Sun, $\Delta\Omega_r$ the emission solid angle of value about 1 (Li *et al.*, 1984), τ_x the optical depth of the gyroresonance absorption, $A \sim L^2$ the surface area of the radio source and $L \approx \min(V_g t_s, L_{Bt} \, |\Delta\omega/\omega|)$, in which L_{Bt} is the length scale of the magnetic field along the flux tube and is taken to be several times of L_B.

In Equation (12) τ_s can be evaluated by analysing the condition for maser instability

$\partial f/\partial v_\perp > 0$ (Wu *et al.*, 1981; Li *et al.*, 1984). From the distribution function (8) we have

$$\frac{\partial f}{\partial v_\perp} = (8\pi^{5/2} m_e^3 v_{0\perp}^2 v_{0\parallel} \varepsilon_\perp \varepsilon_\parallel v_\perp^{1/2} \tau^{1/2})^{-1} \int_0^\infty \mathrm{d}x\, x^{1/2} \left(\frac{x - v_\perp}{\tau} - \frac{1}{v_\perp} \right) \times$$

$$\times \exp\left\{ - \left(\frac{x - v_{0\perp}}{\varepsilon_\perp v_{0\perp}} \right)^2 - \left(\frac{x - v_\perp}{4\tau} \right)^2 - \left(\frac{v_\parallel - v_{0\parallel}}{\varepsilon_\parallel v_{0\parallel}} \right)^2 \right\}. \tag{14}$$

As $(\partial f/\partial v_\perp)_{t=0}$ has its maximum at $v_\perp \approx (1 - \varepsilon_\perp/\sqrt{2}) v_{0\perp}$, we have $\partial f/\partial v_\perp < 0$ and, hence, $\omega_i < 0$ when $\tau > 0.5\varepsilon_\perp v_{0\perp}$. If we take the spatial dispersion into account, $W(t)$ will decrease before the growth rate reduces to zero. So we can take approximately

$$\tau_s \approx \varepsilon_\perp v_{0\perp} (\omega_{i0} - V_g/2l)/(8\omega_{i0}). \tag{15}$$

For comparison, we use the same parameters as in Figure 2 to calculate τ_s, t_s, W_s, and S with Equations (11)–(15). These results are in agreement with the numerical ones (Table I).

TABLE I

Comparison between numerical and analytical results of ECM emission in X-mode

s	τ_s/c^2		t_s (ms)		W_s/W_0		S (s.f.u.)	
	a	*b*	*a*	*b*	*a*	*b*	*a*	*b*
2	0.0011	0.0018	0.28	0.27	1.2×10^6	1.9×10^6	1.4×10^3	2.3×10^3
3	0.00074	0.00086	3.5	3.2	5.9×10^4	8.2×10^4	2.2×10^2	3.0×10^2
4	0.00035	0.00037	140	110	5.6×10^2	6.5×10^2	0.9×10^2	1.1×10^2

The parameters are the same as in Figure 2. *a*: numerical results; *b*: analytical results.

It can be seen from (12) that a strong spike emission can be produced only for the harmonics having higher growth rates. Under the conditions in the solar corona we have $V_g \approx 0.8c$, $|\Delta\omega/\omega| \approx 0.01$, and $L_B \gtrsim 10^4$ km at meter wavelengths, hence, $V_g/(2l) \approx 10^3\ \mathrm{s}^{-1}$. On the other hand, Li (1986) obtained

$$(\omega_{i0}/\Omega_e)(n_e/n_s) \approx 10^{2-s}, \tag{17}$$

for the sth harmonic in X-mode. For meter wavelength emissions we can take $B < 20$ G if $s > 4$, so these harmonics will not produce strong spike emission unless $n_s/n_e > 10^{-2}$, which is too large to be acceptable. Therefore, we conclude that in general the harmonics higher than the 4th of ECM in X-mode can not considerably contribute to solar radio spike emissions.

As the sth harmonic of ECM in X-mode will be cut-off if $\omega_p/\Omega_e > (s(s-1))^{1/2}$, higher harmonics dominate at larger values of ω_p/Ω_e which correspond to a higher altitude in the solar corona and, hence, to a lower emission frequency. Then, as can be seen from

Figures 2–4 and Table I, the saturation time t_s and the duration of spikes which is about twice t_s will increase with the decrease of the emission frequency.

3. Summary and Conclusions

The inhomogeneity of magnetic field along the ray path and the propagation of emitted waves are studied in this paper for the ECM emission mechanism. The wave energy increases owing to the ECM instability, while the propagation of waves evacuates the electromagnetic energy from the source region. By including these effects in the quasi-linear theory the variations of ECM growth rates with time are calculated and spike-like time profiles of waves are obtained. The improved expressions of the saturation time t_s of ECM emission, of the peak wave energy density W_s and of the peak flux density S are derived. Main conclusions of this paper are as follows.

(1) The inhomogeneity of magnetic field limits the spatial region of ECM growth and the growth of the emitted wave energy.

(2) The saturation time t_s and the duration of spikes increase with the increase of the harmonic number of ECM and, hence, with the decrease of the emission frequency for solar radio spike emission.

(3) The growth rates of ECM at the second, third and fourth harmonics almost do not change until $t \approx 0.8 t_s$ and then fall down rapidly, so the time profiles of the energy density of emitted waves appear like spikes.

(4) Harmonics higher than the 4th of ECM in X-mode can not considerably contribute to solar radio spike emission.

The assumption that the quasi-linear process dominates the phenomena has been imposed in this paper. It is worthy to study ECM emission including possible nonlinear processes. Probably it will reveal some new interesting features.

Acknowledgements

The author acknowledges the referee for his valuable comments and is grateful to the hospitality of Arcetri Astrophysical Observatory, where this work was carried out.

References

Freund, H. P., Wong, H. K., Wu, C. S., and Xu, M. J.: 1983, *Phys. Fluids* **26**, 2263.
Hewitt, R. G. and Melrose, D. B.: 1983, *Australian J. Phys.* **36**, 725.
Li, H.-W.: 1986, *Solar Phys.* **104**, 131.
Li, H.-W., Li, C.-S., and Fu, Q.-J.: 1985, in C. de Jager and Chen Biao (eds.), *Proceedings of Kunming Workshop on Solar Physics and Interplanetary Travelling Phenomena*, Vol. 2, p. 579.
Melrose, D. B. and Dulk, G. A.: 1982, *Astrophys. J.* **259**, 844.
Winglee, R. M.: 1985, *J. Geophys. Rev.* **90**, 9663.
Wu, C. S., Tsai, S. T., Xu, M. J., and Shen, J. W.: 1981, *Astrophys. J.* **248**, 384.

HARMONIC EMISSION AND POLARIZATION OF MILLISECOND RADIO SPIKES*

A. O. BENZ and M. GÜDEL

Institute of Astronomy, ETH, Zürich, Switzerland

Abstract. The spectral distribution of millisecond radio spikes observed by the Zürich spectrometers in the 200–1100 MHz range has been studied. In one event out of a total of 36 we have found clearly developed harmonic structure. The ratio between the two bands of emission was $1:1.39 \pm 0.01$. We have also determined the sense of circular polarization of the spike events and compared it to the magnetic polarity of the leading spot of the flaring active region. According to the 'Leading Spot Rule' the majority of the events (10 out of 13) were emitted in the ordinary mode.

1. Introduction

Millisecond radio spikes have brightness temperatures up to 10^{15} K (cf. review by Benz, 1986). For this reason the very efficient electron cyclotron instability (often referred to as 'maser') has been evoked as the cause of emission. The first quantitative theories (Holman *et al.*, 1980; Melrose and Dulk, 1982) predict emission at the lowest harmonics of the electron-cyclotron preferentially in the extraordinary mode. This model and predictions are similar to auroral kilometric radiation, where usually (but not always) x-mode is observed and there is evidence for harmonics (Benson, 1984). More recent calculations, however, yield ordinary mode emission at plasma densities expected in solar flares. Although there is no alternative emission process proposed, the maser model for solar spikes needs to be observationally confirmed. The goal of this study was to search for harmonics predicted by the above theory and to study the mode.

2. Observations

Spike events can be readily identified in the film recordings of the Zürich analog spectrograph (DAEDALUS) and analyzed from digital recordings of the Zürich frequency-agile spectro-polarimeter (IKARUS). Both instruments have operated in the 100–1100 MHz range during the past solar cycle (Perrenoud, 1981). The resolution of the digital instrument varied in frequency from 1–3 MHz and in time from 25–100 ms. The analog instrument observes the full spectrum is a continuous sweep, whereas the digital spectrometer only observes in certain bands selected for their absence of interference.

We have selected 36 spike events for further analysis, excluding fine structure of type IV events. All of these events were associated with type III activity at lower frequency. An example of a digital registration is shown in Figure 1. The selection contains both small groups of spikes as investigated by Benz *et al.* (1981) as well as large

* Proceedings of the Second CESRA Workshop on Particle Acceleration and Trapping in Solar Flares, held at Aubigny-sur-Nère (France), 23–26 June, 1986.

Solar Physics **111** (1987) 175–180.

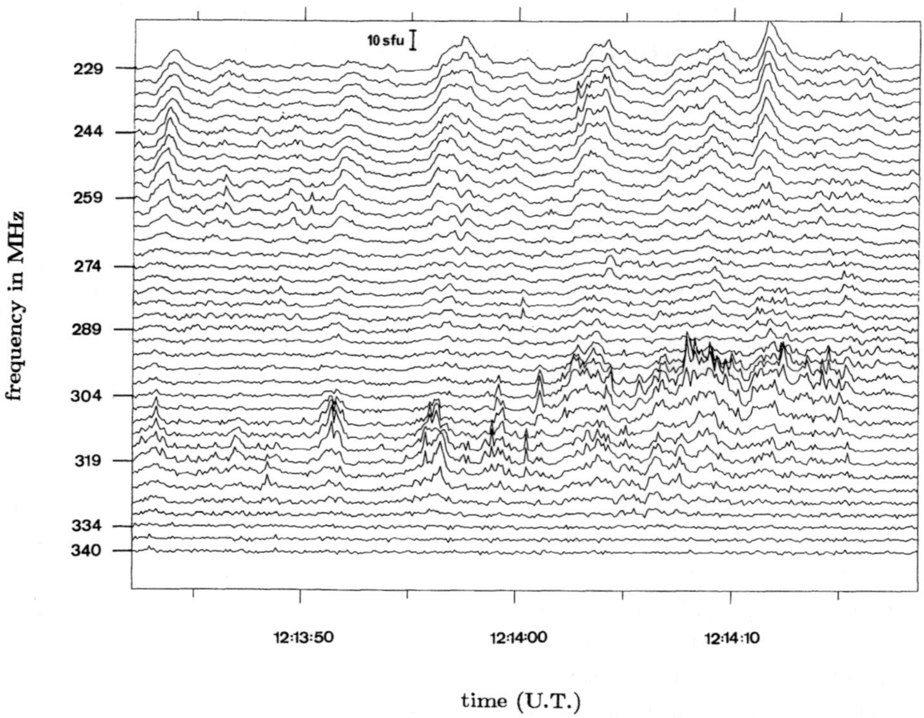

Fig. 1. Example of spike event observed by the IKARUS digital spectro-polarimeter on 1980, December 10. The time profile of the flux in 38 channels is shown. Spikes are visible in the lower part of the figure, type III bursts in the upper part.

groups containing some ten thousand spikes. As much as we could confirm from hard X-ray and Hα observations, they all occurred during the impulsive phase of flares.

3. Harmonics

We have inspected the 36 selected events for harmonic structures both in the analog and the digital data. We frequently noticed the spikes to form clusters in frequency and time. The clusters usually do not seem to be related to each other and do not occur simultaneously. Only one case, presented in Figure 2, showed a clear splitting in frequency. Terrestrial interference at constant frequency has been removed for clarity. Weak type III activity is visible at frequencies below 450 MHz. The spikes occurred during the rise of the impulsive phase of a major flare which later produced type IV emission. The two bands of spikes have also been digitally recorded and are displayed in Figure 3. There seems to be a correspondence between the global structures in the two bands indicating that they are related to each other. A frequency ratio can be determined between corresponding clusters of spikes in the two bands of Figures 2 and 3. It is $1:1.39 \pm 0.01$ in frequency. The scatter of this value is surprisingly small and excludes any chance coincidences.

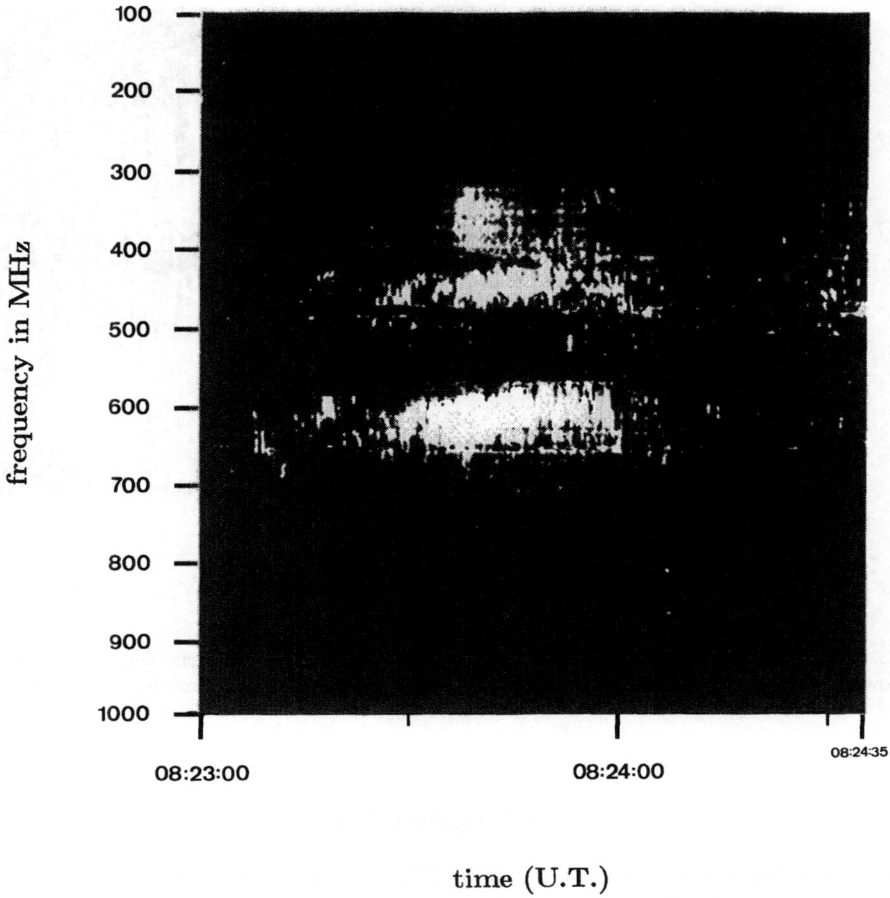

Fig. 2. Analog recording of harmonic structure in a spike event by the DAEDALUS spectrograph on 1981, May 16. The radio emission (bright) is shown in the frequency-time plane.

We have also noted a correlation between single spikes in the two bands. Their frequency ratio, however, seems to be less stable.

The spikes of the 16 May, 1981 event were weakly circularly polarized. As in all weakly polarized spike events the polarization at peak flux scatters considerably from spike to spike. The average polarization of the spikes in the low-frequency band was significantly higher ($16.4 \pm 3\%$ left) than in the high-frequency band ($7.6 \pm 5\%$ left). The sense of polarization (left) was independently confirmed by observations of the Trieste Observatory (courtesy Dr P. Zlobec). It corresponds to ordinary mode according to the Leading Spot Rule (cf. next section).

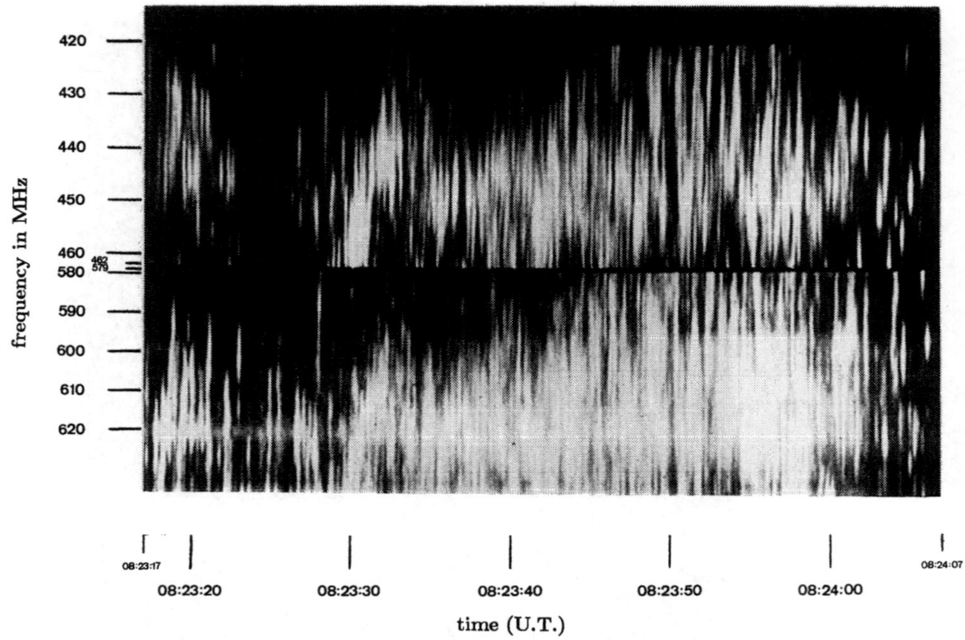

Fig. 3. The same event as in Figure 2 observed by the digital IKARUS instrument. Only the two bands showing spikes are shown. The low-frequency band is enlarged in frequency scale by a factor of 1.4.

4. Leading Spot Rule

In 13 out of the 36 events we were able to determine a well defined sense of circular polarization as well as the position of the associated Hα flare. For the remaining bursts we either did not have polarization data, the polarization was close to zero, or no Hα flare was reported. The magnetic field polarity of the leading spot was looked up in *Solar-Geophysical Data* using mainly Kitt Peak observations. The Leading Spot Rule suggests that the leading polarity of the active region determines the magnetic field direction of the source. This rule would predict that radio sources emitting extraordinary mode are left circularly polarized for negative (S) leading polarity and vice versa. Applying the rule to these spike data, however, would indicate that 10 out of the 13 events were polarized in the ordinary mode.

In this set of observations the Leading Spot Rule suggests ordinary mode also for the associated type III bursts in 10 cases out of 12 which could be measured. The associated type III bursts generally have the same polarization as the spikes (Benz *et al.*, 1982).

5. Discussion

These observations of spikes contradict early theories of maser emission. However, they do not exclude maser emission in general for the following reasons:

The observed ratio of the harmonic structure in a spike event cannot be the result of two integer multiples of the electron gyrofrequency unless unreasonably high harmonics are assumed. However, maser emission does not occur exactly at the harmonics of the gyrofrequency. Particularly the fundamental is shifted to higher frequency due to refraction. Fundamental and second harmonic ($n = 2$) of the X-mode have a frequency ratio of 1 : 1.39 for a ratio of plasma frequency to gyrofrequency of 0.52 (Aschwanden and Benz, 1987). This possibility does not seem to exist for O-mode emission, which is suggested by the Leading Spot Rule. There is an alternative: the discrepancy would disappear if one of the bands is assumed to be the result of electrostatic waves at the upper hybrid frequency ω_{uh} (sometimes referred to as Z-mode) which are also well-known under similar conditions. They can emit radio waves at their second harmonic by coalescence. The observed ratio then is given by

$$\frac{\omega}{2\omega_{uh}} = \frac{n\Omega_e}{2\Omega_e(1 + \omega_p^2/\Omega_e^2)^{1/2}} = \begin{cases} 1.39 \\ 1/1.39 \end{cases},$$

where n is the harmonic of the maser, Ω_e and ω_p are the electron cyclotron and the plasma frequency, respectively. The two cases refer to the Z-mode emission being at the lower-frequency band (top) or at the higher frequency band (bottom), respectively. E.g., for $n = 2$ the ratio ω_p/Ω_e is 0.967 for the second case, the first case having no real solution.

The Leading Spot Rule is not a compelling proof of the mode of emission. It has produced consistent results for type I bursts (90% ordinary mode, cf. Elgarøy, 1977, and references therein) and decimetric pulsations (extraordinary mode, Aschwanden, 1986). The results for type III (predominantly ordinary mode) are less clear (cf. Krüger, 1979, and references therein). A detailed analysis of U-bursts by Benz *et al.* (1979) has given further evidence for ordinary mode emission of electron beams. The emission mode of spikes determined by the leading Spot Rule may thus be taken as an indication requiring further corroboration. Most importantly, however, preferential emission in the ordinary mode is predicted for the maser instability at $\omega_p/\Omega_e \gtrsim 0.4$ for $n = 1$ and $\omega_p/\Omega_e \gtrsim 1.4$ for $n = 2$ by recent calculations (Sharma and Vlahos, 1984; Melrose *et al.*, 1984). The predominant maser mode in sources at wave saturation is a topic of ongoing theoretical research and of great importance for any quantitative model.

Acknowledgements

This work has profited from various discussions at the 1986 CESRA workshop in Aubigny. In particular we thank M. J. Aschwanden and one of the referees for constructive criticism and suggestions, and P. Zlobec for supplementary data. The construction of the Zürich spectrometers is being financially supported by the Swiss National Science Foundation (Grant No. 2.211–0.84).

References

Aschwanden, M. J.: 1986, *Solar Phys.* **194**, 57.
Aschwanden, M. J. and Benz, A. O.: 1987: in preparation.
Benson, R. F.: 1984, *Radio Sc.* **19**, 397.
Benz, A. O.: 1986, *Solar Phys.* **104**, 99.
Benz, A. O., Urbarz, H. W., and Zlobec, P.: 1979, *Astron. Astrophys.* **79**, 216.
Benz, A. O., Zlobec, P., and Jaeggi, M.: 1982, *Astron. Astrophys.* **109**, 305.
Elgarøy, Ø.: 1977, *Solar Noise Storms*, Pergamon Press, Oxford.
Holman, G. D., Eichler, D., and Kundu, M. R.: 1980, in M. R. Kundu and T. Gergeley (eds.), 'Radio Physics of the Sun', *IAU Symp.* **86**, 465.
Krüger, A.: 1979, *Introduction to Solar Radio Astronomy and Radio Physics*, D. Reidel Publ. Co., Dordrecht, Holland.
Melrose, D. B. and Dulk, G. A.: 1982, *Astrophys. J.* **259**, 844.
Melrose, D. B., Hewitt, R. G., and Dulk, G. A.: 1984, *J. Geophys. Res.* **89**, 897.
Perrenoud, M. R.: 1981, *Solar Phys.* **81**, 197.
Sharma, R. R. and Vlahos, L.: 1984, *Astrophys. J.* **280**, 405.

EVIDENCE OF HARMONIC MICROWAVE RADIATION DURING SOLAR FLARES*

M. STÄHLI, A. MAGUN, and E. SCHANDA

Institute of Applied Physics, University of Bern, CH-3012 Bern, Switzerland

Abstract. First observational evidence of harmonic radiation at microwave frequencies during solar bursts is presented for the event of April 28, 1983. The recordings between 3.1 and 19.6 GHz show a typical continuum with a spectral maximum near 5.2 GHz. Superimposed fine structures with durations in the order of some seconds exhibit a very unusual spectral behaviour. Narrow-banded intensity peaks appeared at 5.2 and 11.8 GHz which were barely visible at 3.1, 8.4 and 19.6 GHz. These structures can be interpreted as harmonic emission. Harmonic radio emission can be generated either by plasma radiation, gyroradiation, electron-cyclotron maser or by nonlinear conversion processes. However, all of those mechanisms require extreme assumptions on the source and the ambient plasma in order to account for the observations.

1. Introduction

Harmonic radiation from the sun during flares is well known at metre and decametre wavelengths. In the frequency range 1–200 MHz type II and type III bursts sometimes show simultaneous emission at two different frequencies with a ratio of about 1 : 2 (e.g., Dulk, 1985). At decimetric wavelengths up to a frequency of about 1000 MHz the harmonic branch of the radiation is the one usually observed. Both fundamental and harmonic radiation are often characterized by narrow-banded and short-lived emission features. It is believed that harmonic radiation below 1 GHz is due to Langmuir turbulences at the local plasma frequency, which are partly converted into radio waves at the plasma frequency and its second harmonic.

Harmonic radiation at microwave frequencies might occur during the emission of millisecond spikes which occur up to $\simeq 5$ GHz. They are believed to be emitted by the electron cyclotron maser mechanism at the low harmonics of the gyrofrequency. Simultaneous observations at the second and third harmonic (3.47 and 5.2 GHz) however failed to detect harmonic emission (Stähli and Magun, 1986).

The detailed analysis of one event with millisecond spikes reported here shows, for the first time, harmonic radiation with a frequency ratio of about 1 : 2. The observed harmonic structures are not millisecond spikes, but intensity peaks with durations in the order of seconds. We now present the observational results and discuss possible emission processes for harmonic radiation.

2. Observations

2.1. Instruments

At the University of Bern solar microwave bursts are observed regularly between 3.1 and 92.5 GHz (Magun *et al.*, 1981). This frequency range is covered by a total of eight

* Proceedings of the Second CESRA Workshop on Particle Acceleration and Trapping in Solar Flares, held at Aubigny-sur-Nère (France), 23–26 June, 1986.

Solar Physics **111** (1987) 181–188.

TABLE I

Microwave instruments for the observation of solar bursts at the University of Bern

Instrument	Frequency (GHz)	Wavelength (cm)	Bandwidth (MHz)	Time resolution (ms)
Polarimeter	3.1	9.7	120	5 (100)
Spectrometer[a]	3.47	8.6	200	5
Polarimeter	5.2	5.8	200	5 (100)
Polarimeter	8.4	3.6	6	100
Polarimeter	11.8	2.5	250	100
Polarimeter	19.6	1.5	600	100
Polarimeter	35.0	0.86	1000	100
Polarimeter	50.0	0.60	1000	100
Polarimeter	92.5	0.32	1000	100

[a] No recording of circular polarization.

polarimeters and a spectrometer (Table I). The total flux and the circular polarization are recorded digitally with a temporal resolution of 5 to 100 ms, depending on the instrument.

2.2. Event of April 28, 1983

Until now we did not detect any harmonic structures in our microwave data. This is partly because our instruments are not optimized for such observations. The observing frequencies are not related by integer ratios and the frequency range is not sufficiently covered. In addition we did not analyze our data in detail for this unexpected phenomenon.

The event of April 28, 1983 (Figure 1), investigated for the spectral behaviour of millisecond spikes, shows a typical microwave continuum with the spectral maximum near 5.2 GHz (Figure 2). Below 5.2 GHz there are spiky parts at the beginning (10 : 48 : 30–10 : 49 : 00 UT) and at the end (10 : 51 : 00–10 : 52 : 00 UT) of the burst which are not temporally resolved in the plot. These structures will not be discussed here. However, we now look at the structured part (10 : 49 : 00–10 : 51 : 00 UT) of the event in more detail.

2.3. Harmonic peaks

A first look at the expanded intensity plot shows that 5.2 and 11.8 GHz have simultaneous structures whereas the 8.4 GHz-intensity is almost smooth (Figure 3). A detailed analysis of all the structures with durations between 0.5 and 5 s shows that at least 5 out of 18 structures have a very unusual spectrum. They appear only at two frequencies with an approximate ratio of 1 : 2 and are not visible at frequencies in between. There are two different groups of such peaks: The first group has simultaneous maxima at 3.47 and 8.4 GHz and only small fluxes at 5.2 GHz. The second group has maxima at 5.2 and 11.8 GHz with small fluxes at 8.4 GHz. Three evident examples of these structures are marked in Figure 3 by arrows. No measurable time delay occurs

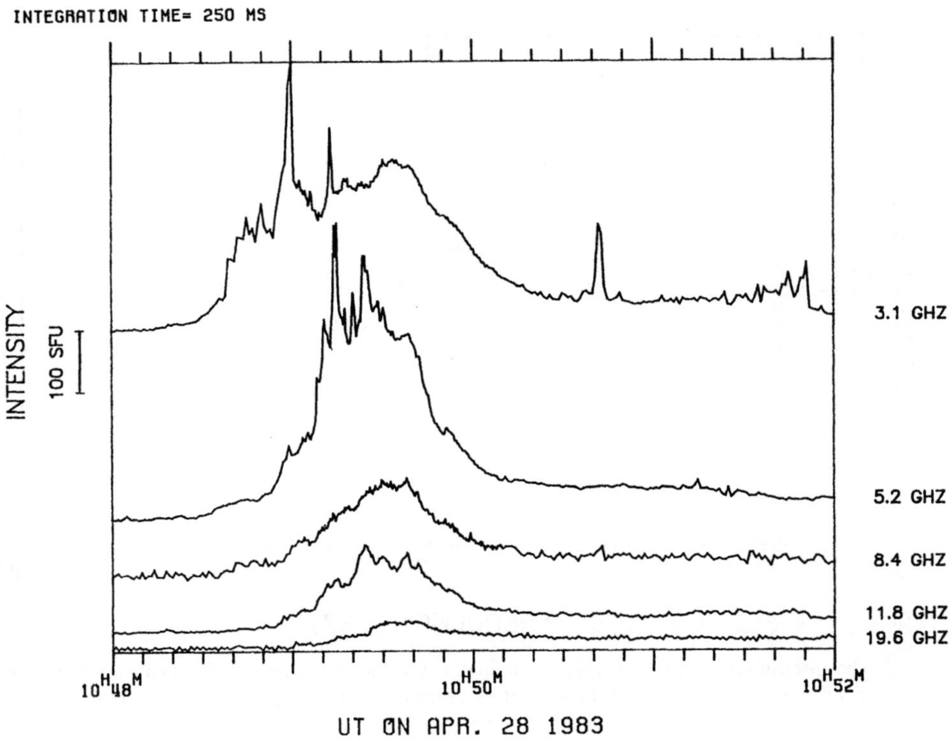

Fig. 1. The event of April 28, 1983: The intensities at five frequencies in the range between 3.1 and 19.6 GHz are plotted as a function of the time with an integration time of 250 ms.

between the correlated peaks at the two different frequencies. In order to determine their spectrum the underlying slowly varying continuum was subtracted and the spectrum of the peaks plotted (Figure 4). The observing frequencies at which the intensity-peaks were recorded do not have integer ratios. Therefore the data points were fitted with cubic polynomials in order to illustrate the possible spectra and the peak frequencies. These coarse fits lead to the suggestion that the ratio between the two frequencies is roughly 1 : 2 (Figure 4). With other ratios as, e.g., 2 : 3 it is not possible to fit the data points with the same accuracy. The features of three prominent harmonic peaks are summarized in Table II.

The maxima of the spectra are at $\simeq 4$ GHz and $\simeq 8$ GHz and at $\simeq 6$ GHz and $\simeq 12$ GHz, respectively. There are several possibilities to explain this ratio: the observed radiation could be the emission at the fundamental and the second harmonic, or at higher harmonics such as 2 : 4. The latter, however, can be excluded as the emission at the intermediate observing frequencies (e.g., near the third harmonic) should also have been observed. Therefore, the observed peaks seem to be generated by a process with fundamental radiation in the range 4 to 6 GHz and the second harmonic between 8 and 12 GHz. As one might expect for harmonic emission, the intensity of the peaks at the

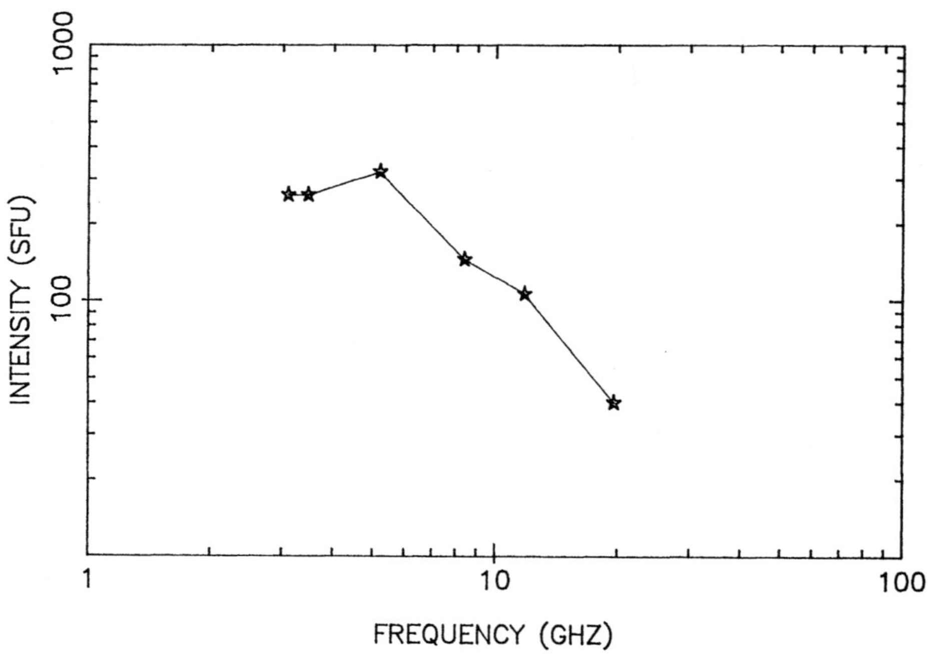

Fig. 2. The spectrum of the slowly varying continuum of the microwave burst. The intensities at the time of the maximum (10 : 49 : 30 UT) are plotted against the frequency on logarithmic scales.

lower frequency is higher than at the upper frequency. The fits in Figure 4 allow us to estimate the bandwidth of the structures. It is about 1.5 to 3 GHz for the fundamental and 4 to 7 GHz for the harmonic components, which correspond to relative bandwidths of approximately 50%.

For the identification of the emission mechanism the circular polarization of the structures is important. After the subtraction of the underlying continuum the fundamental peaks are slightly polarized (5–15%). The sense of polarization is opposite for the two components. As the microwave continuum is expected to be polarized in the extraordinary mode, the fundamental peaks seem to be emitted in the ordinary mode. The polarization of the harmonic components is only weak ($<10\%$) and cannot be determined accurately because of the small fluxes relative to the receiver noise.

The observations presented show a phenomenon with solar origin. Instrumental effects can be excluded because harmonic pairs of structures were observed at different frequencies. Furthermore, structures with non-harmonic spectra were recorded during this event, and the spectrum of the continuum (Figure 2) appears to be very broad-banded. Local atmospheric effects or interferences can also be excluded as frequencies from 3.1 to 5.2 GHz and frequencies between 8.4 and 92.5 GHz are observed from locations 20 km apart. However, with the present observations at only a few frequencies it cannot be excluded, that the spectrum exhibited many peaks of non-harmonic origin. Some of them could occasionally be close to harmonic ratios.

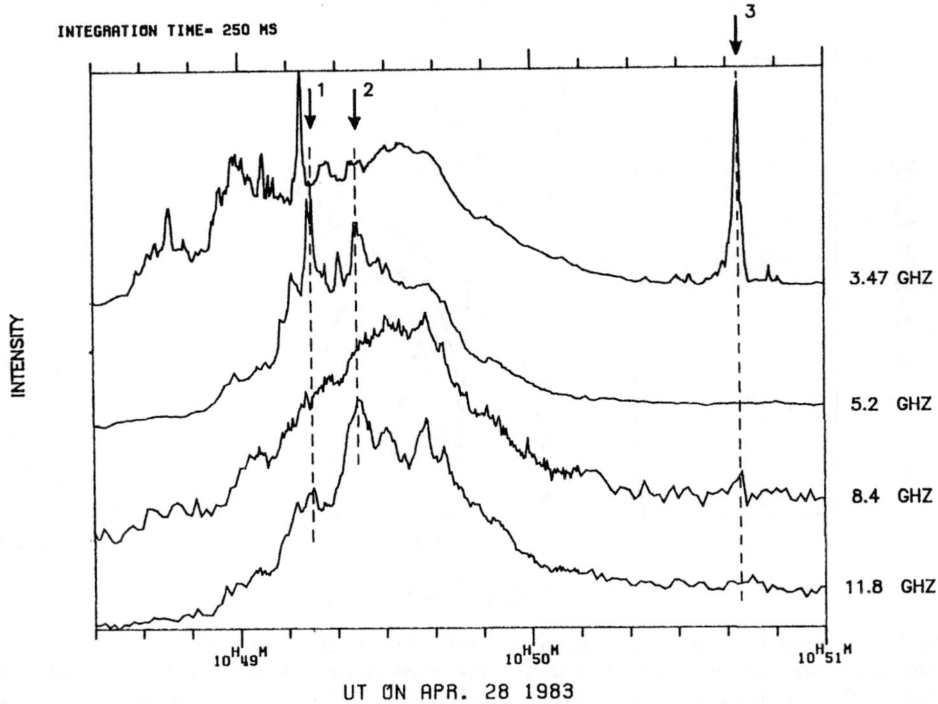

Fig. 3. The structured part of the event. For easier comparison of the fine structures the intensities at the four different frequencies are scaled to equal amplitudes. Three evident examples of harmonic peaks are marked by arrows.

2.4. Other events with harmonic peaks

A quick look through our data of the period from June 1982 to April 1986 shows that harmonic structures are rare. Out of several hundred events probably less than 5 gave evidence of harmonic radiation. However, we have to take into account that our instruments are not suitable for the detection of harmonic structures, especially if their relative bandwidth is less than $\simeq 50\%$.

3. Theoretical Considerations

Assuming it to be real, what is the emission process for the observed phenomenon? Harmonic structures can be explained by several mechanisms, e.g., plasma radiation, gyroradiation, electron cyclotron maser or by other nonlinear conversion processes. In the following we shall briefly discuss these different possibilities in connection with the observational results presented here.

3.1. Plasma radiation

It is well known that the harmonic emission of metric type II and type III bursts can be explained by plasma radiation. Plasma radiation at higher frequencies, however, is

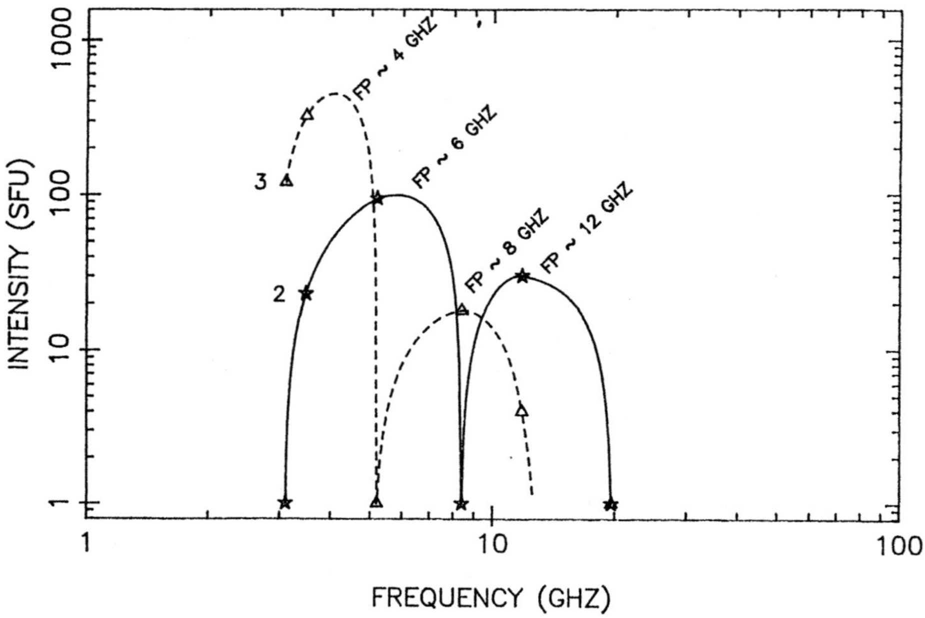

Fig. 4. The spectra of two selected peaks (peaks 2 and 3 of Figure 3) on logarithmic scales. The intensity of the underlying continuum has been subtracted. The measured data points are marked by asterisks or triangles. The solid curves are cubic polynomials that pass through the data points, demonstrating possible harmonic structure with peaks at 6 and 12 GHz, respectively. The dashed curves demonstrate a case with peaks at about 4 and 8 GHz, respectively.

TABLE II

Data of three prominent harmonic peaks

Peak No.	Time (UT)	Peak-frequencies (GHz)	Peak-fluxes (SFU)	Flux-ratio S_1/S_2	Bandwidth Abs. (GHz)	Bandwidth Rel. (%)	Circular polarization (%)
1	10:49:14	$f_1 = 6$	170		3	50	10*R*
		$f_2 = 12$	13	13	7	60	?
2	10:49:24	$f_1 = 6$	100		3	50	5*R*
		$f_2 = 12$	30	3	5	40	?
3	10:50:42	$f_1 = 4$	350		1.5	40	15*R*
		$f_2 = 8$	18	20	4.5	60	?

strongly absorbed during the propagation to the observer. The important absorption process is either gyroresonance absorption or free-free absorption, depending on the temperature of the plasma and the strength of the magnetic field. An estimate of the free-free absorption by the ambient electrons with a temperature of 5×10^5 K shows that fundamental plasma radiation is observable up to approximately 1 GHz assuming density scale heights down to 100 km. Perpendicular to the magnetic field in loops, such strong electron density decreases may be possible. However, for fundamental plasma radiation at 6 GHz, density scale heights of only a few kilometers are necessary. Probably even ducting of the radio waves (Duncan, 1979) is not able to overcome this strong absorption. The required electron density for fundamental plasma radiation at 6 GHz would be $4.4 \times 10^{11}\ \mathrm{cm}^{-3}$.

3.2. Gyroradiation

Incoherent gyrosynchrotron radiation from electrons spiraling in a magnetic field loop can explain most of the solar microwave observations. For the explanation of the observed harmonic structures gyroradiation at low harmonics, i.e., the emission of non-relativistic particles must be considered. However, a very high magnetic field of more then 2 kG is necessary to explain fundamental cyclotron radiation up to 6 GHz. A simple estimate shows that further difficulties exist. On the assumption of cyclotron emission from thermal electrons and a homogeneous optically thick source with a homogeneous magnetic field, an upper limit of the brightness temperature of 10^8 K can be estimated from the bandwidth (Lang, 1974). A source with a diameter of $\simeq 10^4$ km is necessary to account for the observed fluxes in the order of one hundred SFU. This is a reasonable value. However, strong gyroresonance absorption at the second and third harmonic of the local gyrofrequency is expected (Zheleznyakov, 1970).

3.3. Electron cyclotron maser

The electron cyclotron maser generates intense radiation with short rise times near the local gyrofrequency and its low harmonics (e.g., Melrose and Dulk, 1982). Therefore, it can be used to explain the millisecond spikes which have been observed up to $\simeq 5$ GHz (Stähli and Magun, 1986). There are at least two major problems with this emission process. Fundamental radiation of the electron cyclotron maser is believed to be reabsorbed totally by gyroresonance absorption. Furthermore again a high gyrofrequency of up to 6 GHz is necessary. The generation of the harmonic structures by maser emission at the second and fourth harmonic can be excluded because no flux near the third harmonic was observed.

3.4. Other nonlinear conversion processes

A nonlinear wave-wave coupling mechanism seems to be able to produce conversion between fundamental and harmonic radiation with sufficient efficiency to overcome absorption processes (Schanda, 1972). This mechanism depends strongly on the ratio of the plasma frequency ω_p to the electron cyclotron frequency Ω_e. It can be shown that in a very narrow range of this ratio ($1.3 < \omega_p/\Omega_e < 1.5$) an efficient wave coupling can be achieved over a reasonable bandwidth of several per cent. The conversion takes place within typically a hundred to a thousand wavelengths. Within this distance the momentum condition is fulfilled and the parameter ω_p/Ω_e remains in the narrow range mentioned above. First calculations yield values of the wave energy needed for this process which are compatible with the observed flux densities.

This approach seems promising and will now be studied in more detail. In particular the influences of possible absorption mechanisms have to be investigated.

4. Conclusions

Our microwave observations show that some solar bursts exhibit evidence of harmonic structures at frequencies up to $\simeq 12$ GHz. Several possible emission mechanisms for this phenomenon have been briefly discussed. However, all of them require extreme assumptions on the magnetic field and the ambient plasma to account for the observations. This is probably the reason for the rarity of harmonic structures in the microwave range. In order to investigate the harmonic peaks in more detail, further observations with better spectral resolution and coverage including circular polarization are necessary. A detailed analysis of our digital recordings of the last few years could be interesting too.

We are in the process of extending our observations in the range between 3 and 12 GHz. Two frequency ranges at the fundamental (3–4 GHz) and the second harmonic (5–9 GHz) will be covered by two different spectrometers with high spectral and temporal resolution. Additionally the existing fixed frequency polarimeters (Table I) will extend the frequency range.

Acknowledgements

This work was supported by the Swiss National Science Foundation under grant No. 2.420–0.84.

References

Dulk, G. A.: 1985, *Ann. Rev. Astron. Astrophys.* **23**, 169.
Duncan, R. A.: 1979, *Solar Phys.* **63**, 389.
Lang, K. R.: 1974, *Astrophysical Formulae*, Springer-Verlag, Berlin.
Magun, A., Fuhrer, M., Kämpfer, N., Schöchlin, W., Stähli, M., and Wülser, J. P.: 1981, *Bern Solar Obs.* **46**.
Melrose, D. B. and Dulk, G. A.: 1982, *Astrophys. J.* **259**, 844.
Schanda, E.: 1972, *Proc. Summer School on Plasma Physics and Solar Radio Astronomy*, Ille de Re, p. 291.
Stähli, M. and Magun, A.: 1986, *Solar Phys.* **104**, 117.
Zheleznyakov, V. V.: 1970, *Radio Emission of the Sun and Planets*, Pergamon Press, Oxford.

LOW-LEVEL DECIMETRIC (1.6 GHz) SOLAR BURST ACTIVITY*

H. S. SAWANT, J. E. R. COSTA, R. H. TREVISAN, C. J. B. LATTARI, and P. KAUFMANN

INPE; Instituto de Pesquisas Espaciais, C.P. 515, 12200 S. José, Campos, SP, Brasil

Abstract. Observations of solar bursts at 1.6 GHz were carried out in the month of July 1985 for about two weeks. Five intervals of solar burst activity, each one lasting for a couple of minutes, were observed. Predominantly, two classes of fast bursts were observed: viz: 'spike' and 'blips'. However, some of these bursts were two orders of magnitude less intense than those reported earlier.

Low-level blips have typical duration ~ 350 ms, excitation time ~ 200 ± 25 ms, decay time ~ 130 ± 25 ms and a low degree of circular polarization of about 15%. Detailed investigations of decay times of the blips have been carried out in terms of collisional damping and Landau damping. Observed decay times of the blips seem to favour the hypothesis of collisional damping. This investigation suggests that blips probably originate at second harmonic by beam plasma interaction as that of metric type III bursts. Also, low-level ms-spikes with the half power duration in the range of 5 to 20 ms suggest that source sizes be smaller than 50 km if the process of emission is electron-cyclotron maser.

1. Introduction

Decimetric solar radio burst observations have been carried out for more than the last two decades. Earlier work has been well summarized by Tarnstrom and Philip (1971). Recently, decimetric bursts have been classified by Wiehl *et al.* (1985) depending upon their time constants, as fast 'transients' bursts with duration less than one second and 'trap' type with duration ~ 5 s. Majority of the decimetric bursts are of fast transient types, i.e., mainly ms-spikes, with duration ≤ 100 ms and blips, with duration ~ 200 ms. Intense ms-spikes in cm wavelength region were first reported by Slottje (1978) and recently up to 4000 MHz by Stähli and Magun (1978). Weak (~ 20 s.f.u.) ms-spike-like structures were reported earlier by Dröge (1977) at 1415 MHz and their VLBI observations by Tapping *et al.* (1983) at 1663 MHz. Estimated brightness temperatures of the intense spikes are in the range of 10^{11} to 10^{15} K. The most accepted process of emission is electron-cyclotron maser, first proposed by Holman *et al.* (1980) and Melrose and Dulk (1982). A summary of last 25 years research on spike work has been made recently by Benz (1986a).

Blip is another predominant short-duration narrow-bandwidth burst discovered by Benz *et al.* (1981) and well investigated in the decimeter range by Fürst *et al.* (1982) and Benz *et al.* (1983). These authors have found that blips resemble in properties the type III bursts except for their narrow bandwidth and hence they have suggested that beam plasma instabilities similar to those producing type III bursts may also produce blips. Also it has been realized that ms-spikes and blips originate near the acceleration region or in the acceleration region where electron densities are around 10^9 to 10^{11} cm^{-3}

* Proceedings of the Second CESRA Workshop on Particle Acceleration and Trapping in Solar Flares, held at Aubigny-sur-Nère (France), 23–26 June, 1986.

Solar Physics **111** (1987) 189–199.

(Wiehl *et al.*, 1985). Hence, a further investigation may provide some information on processes of primary energy release. In this note we are reporting weaker, one order of magnitude less intense, ms-spikes than reported so far and their properties are discussed in terms of electron-cyclotron maser process. We also report blips two-order of magnitude less intense than reported earlier. Detailed investigations of their time profile have been carried out and their implications have been discussed in terms of beam-plasma interactions.

2. Instrumentation

Solar radio observations at 1663 MHz were carried out by using the 13.7 m Itapetinga antenna in R and L circular polarization with time-resolution of 3 ms and sensitivity of 0.08 s.f.u. Half-power beamwidth of the antenna (51′) and efficiency of antenna (~50%) at 1663 MHz were estimated by using Taurus A as a standard radio source. The total power receiver had a dynamic range ~35 dB. Isolation between R and L polarization was 22 dB. Observations were recorded on seven channel Ampex analog tape recorder, and on seven channel thermic recorder, HP-7418A. Later on, data were digitized by using a digital oscilloscope Tektronix 5223, with time resolution of 3 ms.

3. Observations

Solar observations were carried from July 1–13, 1985. Sun was tracked daily from 11 to 20 UT. The total number of hours of observations were ~120 hours. From July 7–10,

TABLE I

1.6 GHz radio observations obtained at Itapetinga (Brasil) and other associated solar activity

Date	Interval of the 1.6 GHz activity (UT)	Type of 1.6 HGz activity	H-α		Flare		
			Start	Max.	End	Imp.	Location station
07/07/85	16:32:00–16:33:31	Fast transients mostly ms-spikes				SF	S21 W03 Pale
08/07/85	14:57:30–14:58:30	Blips	14:37	14:40	14:46	–	–
09/07/85	15:30:05–15:30:57	Bursts with duration 1 s	–	–	–	–	–
	16:48:17–16:54:03	Bursts with duration of 0.3–5 s	16:46	16:51	17:07	SB	S16 W35 Pale
10/07/85	14:43:48–14:44:01	Blips	–	–	–	–	–

five intervals of low-level solar burst activity were observed. Mostly, activity consisted of ms-spikes and blips. The predominant types of decimetric activities observed during these intervals and the associated solar activities are summarized in Table I. During solar activity days, solar flux densities of the quiet Sun at 2695 and 1415 MHz were on an average 95 and 75 s.f.u., respectively (S.G.D. 1985) and, hence, the estimated quiet-sun flux density at 1663 MHz was $\sim$82 s.f.u. Antenna temperature contribution due to Sun was estimated as 1.2×10^4 K by using the calibration process of the antenna, where the size of source (Sun) is taken into consideration as suggested by Giudice and Castelli (1971). The system noise temperature was negligible compared to these values. We present here two sets of observations and discuss them in terms of possible emission mechanism.

3.1. Low-level millisecond spikes

Millisecond fine structures at 1663 MHz were recorded on July 7, 1985 in *R* circular polarization only, around 16:32:00 UT for about 100 s. There was no detectable enhancement in the background continuum level prior to or during the interval of millisecond spike activity. No Hα flare has been reported around this time. The occurrence of ms-spike nearly coincides in time with the start time of 2800 and 2695 MHz bursts and an intense spike at 1663 MHz of $\sim$4.5 s.f.u. was observed around the time of the maxima of the above bursts (see Table I, and Figure 1(a)).

Peak flux values of ms-spikes mostly ranged between 4.5 and 0.4 s.f.u. Minimum detected peak flux value of ms-spikes was well above the sensitivity (0.1 s.f.u.) of the system. Thus the lowest observed flux values were at least one order of magnitude less than reported by Dröge (1972) at 1420 MHz and recently by Tapping *et al.* (1983) at 1663 MHz. A total of 35 events of ms-spikes, isolated or in groups, were observed

Radio bursts			Noise storm 169 MHz	Blen decimetric band/type
Ottawa 2800 MHz UT/s.f.u	AFGL 2695 MHz UT/s.f.u.	AFGL 606 MHz UT/s.f.u.		
16:32:12–16:32:36 3.8 Max	16:32:35–16:32:55 5	–	Started	–
13:30–15:35 GRF 2.6	–	–	Maximum	–
13:40–17:00 GRF 3.0	–	–	Continued	–
–	–	16:46:25–16:54:00 50	–	16:46:42–16:51:18 Type III GG RS
–	–	14:43:10–14:44:15 20	On west limb 9 s.f.u.	

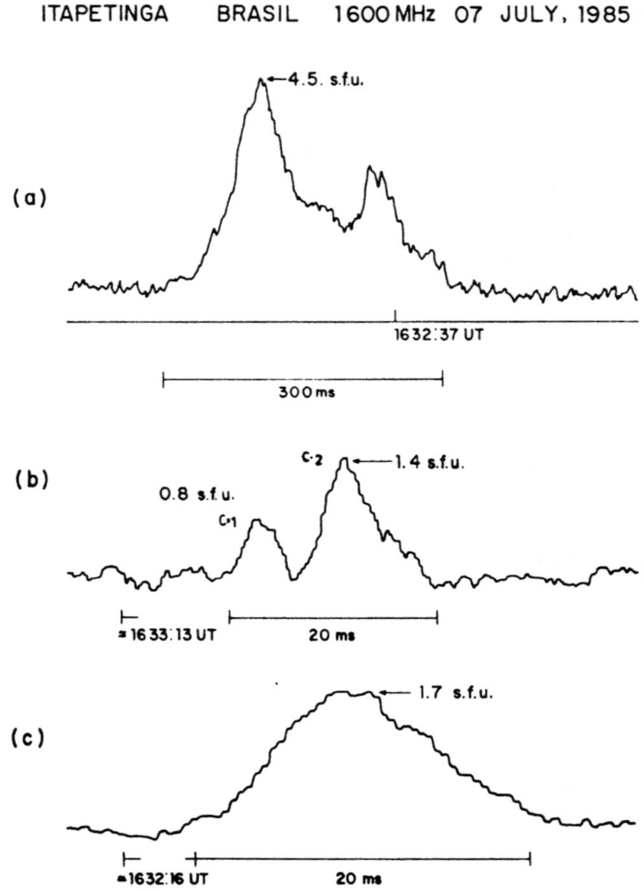

Fig. 1a–c. Examples of intensity time profiles of the ms-spike bursts observed at 1.6 GHz. (a) One of the intense ms-spike bursts. (b) Spike bursts C.1 and C.2. (c) An isolated ms-spike burst. The ms-spike burst C.1 exhibit rise time and half power duration nearly equal to instrumental time resolution.

during this interval. Beside this, 20 gradual rise and fall (GRF) type bursts with duration ~250 ms were also observed. Occasionally ms-spikes were found to be superimposed on them. No periodicity was observed either in the occurrence of ms-spike or GRF type of bursts. Examples of weak and intense spikes are given in Figure 1(a–c). Half-power durations of ms-spikes are plotted against peak flux values in Figure 2. The half-power duration rise and fall times were between 3 and 60 ms. Sixty percent of the observed spikes have half-power duration less than 20 ms. The observed cutt-off in half-power duration around 3 ms is due to instrumental time resolution. Typical durations of ms-spikes reported at 1415 MHz by Dröge (1967) are around 3 to 7 ms.

As described above, ms-spikes with extremely low-peak flux (~0.4 s.f.u.) values have been observed, thus suggesting that ms-spikes activity continues even at extremely low level of intensities. If we assume that the spike duration is governed by the excitation

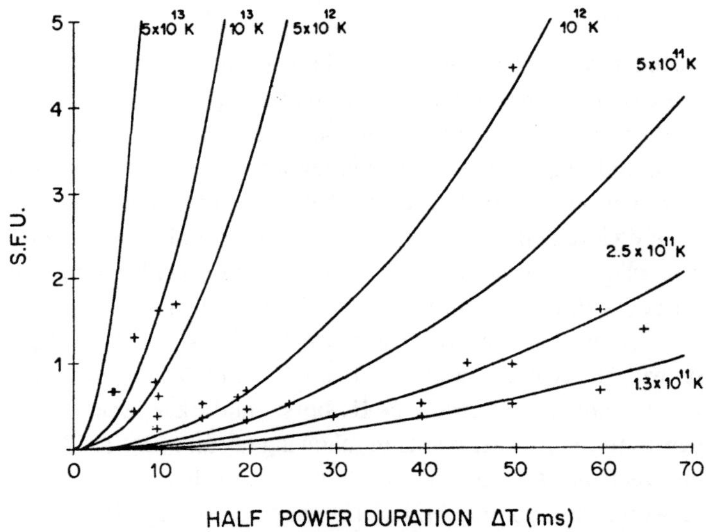

Fig. 2. The peak flux values of the ms-spike are plotted against their half-power duration. Assuming Alfvén speed times the half power duration as a dimension of the spike source, curves of a constant temperature are drawn. Cut-off below 3 ms is due to instrumental time resolution.

period, and that the m.h.d. instability (Spicer, 1977) propagating with Alfvén speed is the source of generation, then Alfvén speed times the half power duration will yield the source dimensions. Further, by using Rayleigh–Jeans equation, one can estimate the brightness temperature. Assuming Alfvén velocity $\sim 3000 \text{ km s}^{-1}$ for a reasonable magnetic field of 200 G and electron densities $\sim 2 \times 10^{10} \text{ cm}^{-3}$, source sizes and brightness temperature have been deduced which are in the range of 10–200 km and 2×10^{13} K–1.5×10^{11} K, respectively. Hence, constant brightness temperature curves are plotted on the flux density versus half-power duration plot of Figure 2. From the same figure one can see that 60% of the spikes have half-power duration less than 20 ms, or estimated source sizes less than 60 km. Solar VLBI observations at 1663 MHz (Tapping *et al.*, 1982) of decimetric spikes burst have also suggested the possibility of existence of such small (≤ 50 km) source sizes.

Alternatively, we may assume that the amplification is due to mildly relativistic electrons accelerated at the top of the loop, some of which are lost at the foot points of the loop by collisions to form a loss cone and some are reflected back and drive the maser instability as suggested by Melrose and Dulk (1982). In this case, electron-cyclotron maser saturates fast compared to the time-scales of ms-spikes. Hence, upper limit of dimensions of source sizes of ms-spikes can be obtained by rise time times the velocity of light. If we were to estimate and use the dimensions of source sizes, as stated above, to determine the brightness temperature, then only in seven cases, estimated brightness temperatures reached up to 1.2×10^{10} K. These lower temperatures may be partially due to observed low values of the flux densities and (i) overestimation of the source sizes or (ii) rise times independent of source, as suggested by Stähli and Magun (1986).

3.2. Low-level blips

A total of nine bursts were observed with *e*-folding duration in the range of 0.2 to 0.5 s. Out of these, three bursts were isolated and the remaining were in groups of two or three. Flux values of these bursts ranged between 9 and 0.5 s.f.u. For two cases, polarization can be measured and degree of circular polarization of these bursts was $\sim 15\%$ right-handed circular. Benz *et al.* (1983) have reported for the first time such a short duration, 0.2 s, narrow bandwidth and at least two orders of magnitude more intense bursts than reported here as 'blips'. As discussed below, circumstantial evidence suggests that seven out of nine bursts, with typical duration of 0.35 s are 'blips'.

The type of the bursts that concern us here, are 'blips', fast drift type $\mathrm{III}_{\mathrm{dm}}$ and normal type III bursts with the typical duration of 0.2, 0.7, and 0.8 s (Wiehl *et al.*, 1985). Fast drift type $\mathrm{III}_{\mathrm{dm}}$ are similar to metric type III bursts and have large bandwidths. Blips differ in properties to the type III bursts in following (Benz *et al.*, 1983), that blips have higher starting frequencies (~ 500 MHz), have narrow bandwidth (~ 50 MHz) and have broad maxima around 600 MHz. The number of type III bursts at a given frequency increase with decreasing frequency. Also blips have higher rate of association with hard X-ray compared to type III bursts (Benz *et al.*, 1983). High-frequency/time

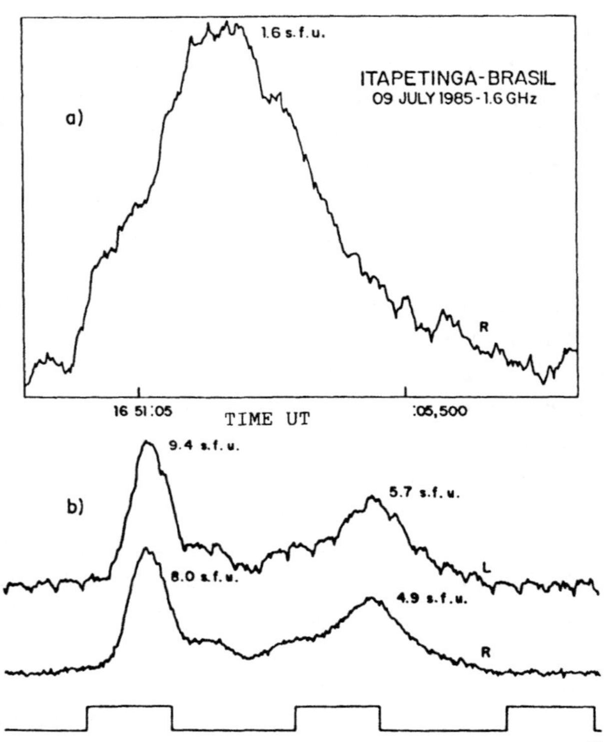

Fig. 3. Examples of intensity time profiles of the blips observed at 1.6 GHz. (a) Isolated blip in *R* polarization. (b) Group of blips in *R* and *L* polarization.

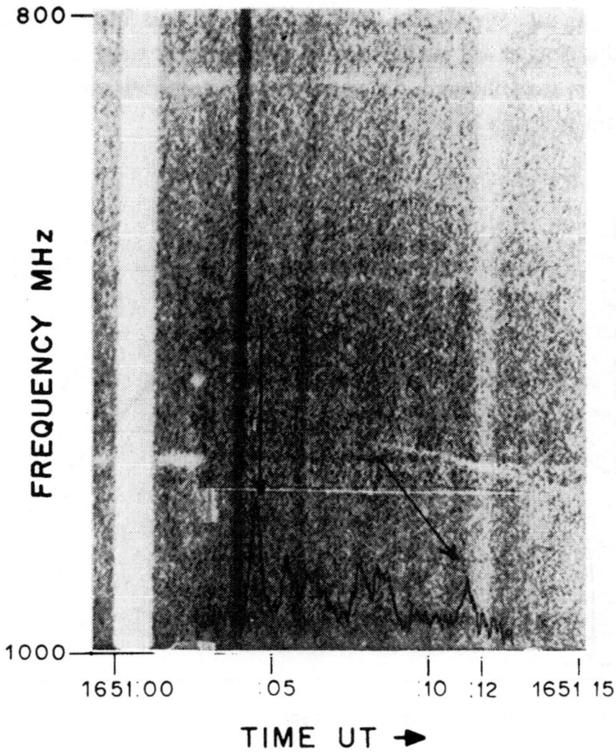

Fig. 4. Time profile of 1.63 GHz bursts observed on 9 July, 1985 at Itapetinga is superimposed nearly at the bottom of the dynamic spectra obtained from Zürich (courtesy of Dr A. O. Benz). Vertical arrow indicates 'blips' observed at 1.63 GHz (illustrated in Figure 3(a)) and has no counterpart below 1000 MHz, where as oblique arrow indicates the bursts observed at 1.63 GHz and in the decimetric range around 16:51:12 UT.

resolution dynamic spectra in the frequency range of 1000–100 MHz were obtained from Zürich Radio Astronomy Group (Benz, 1986b) for the time intervals mentioned in Table I. Comparison of the observed short duration bursts at 1663 MHz with the dynamic spectra suggested that seven out of nine short duration (~ 0.35 s) bursts were not associated with type $\mathrm{III}_{\mathrm{dm}}$ or metric type III bursts. In addition to that higher (5%) probability of occurrence of blips compared to negligible probability of occurrence of type III bursts at 1000 MHz (Benz *et al.*, 1983) suggests that observed short duration (0.35 s) bursts are very likely to be blips. Examples of blips are shown in Figure 3(a–b) and comparison with dynamic spectra in Figure (4), respectively.

Time-profiles of these low level blips are less affected by the dynamic range of the receiver and hence are most suitable for accurate determination of excitation and decay time. Damping of the radiation can be collisional (i.e., plasma wave energy is lost by electron-ion collisions) or Landau damping (i.e., plasma wave energy is absorbed by a beam of particles having velocity nearly equal to the phase velocity). In both cases, the decay is exponential. In that case, if one plots logarithm of the burst intensity versus time when decay is an exponential, the log-linear plot is linear, which enables us to determine the duration of the exciter and the decay time more accurately (Aubier and Boischot, 1972). In the following, we examine in detail both the damping mechanisms and discuss them in terms of observations.

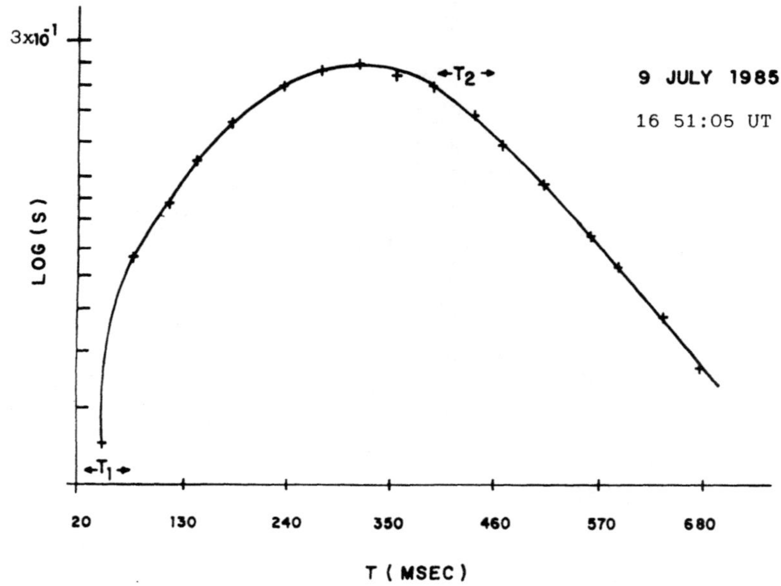

Fig. 5. Log-linear plot of the isolated blip which has been illustrated in Figure 3(a). Error bars around T_1 and T_2 are uncertainties in the measurements of excitation 'D_e' and decay 'T_D' times.

Figure 5 is the log-linear plot of the isolated blips observed at 16:51:05 UT on 9 September, 1985, and illustrated in Figure 3(a). The linear part of the curve corresponds to decay time, T_D. From the start time, T_1, to the beginning of the linear parts, T_2, is the excitation time D_e (i.e., exciter duration). Adopting the same method for the other blips, we have found an average excitation time $D_e = (200 \pm 25)$ ms and decay time $T_D = (130 \pm 25)$ ms.

The average values of the decay times versus frequency were plotted by Benz *et al.* (1983) in their Figure 5. We have plotted average value of decay time at 1663 MHz on their figure and reproduced here as Figure 6. Observed decay time at 1663 is slightly higher than expected from the equation of fit $\tau f^{+0.95} = 10^{7.71}$ as given by Alvarez and Haddock (1973). From the observed average decay time, the temperature of the corona can be obtained assuming emission of the blips at second harmonic and damping of the plasma wave as due to electron-ion collisions. The observed low degree of polarization at 1663 MHz favours the hypothesis that emission of the blips should be at second harmonic. The temperature of the corona is estimated by the equation

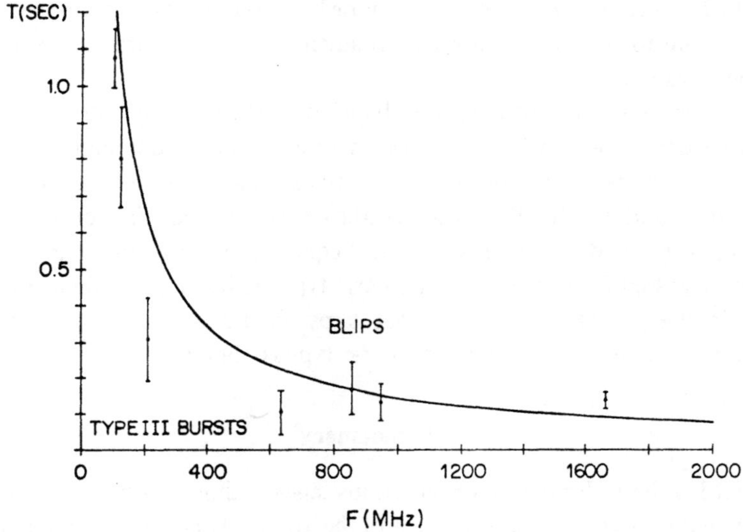

Fig. 6. Decay time versus frequency of the type III bursts (≤ 500 MHz) in comparison with blips (≥ 500 MHz). The curve is the fit to the data of the type III bursts as given by Alvarez and Haddock (1973).

$T = 0.65 \times 10^{-4} f^{4/3} \tau^{2/3}$ (Kundu, 1965) and is $= 1.19 \times 10^7$ K, which is one order of magnitude higher than expected, implying that observed decay times are larger than expected. Observed larger decay times are conventionally attributed to generation of radiation by slow beam particles during the main decay phase, thus effectively increasing the observed decay time.

If the decay of the Langmuir waves, is due to Landau damping, then assuming an exciter to be uniform in density and velocity, Subramanian *et al.* (1981) have given the following equations for excitation and decay time:

$$T_D = \frac{1}{6\pi^{3/2}} \frac{v_1^2}{v_2^2} \frac{v_1^0}{v_1} \left(\frac{n_0}{n_1}\right)^2 \omega \exp\left\{\left[\frac{v_1 - v_2}{v_{01}}\right]^2\right\}, \tag{1}$$

$$D_e \simeq \left[\frac{3N/z}{4\pi n}\right]^{1/3} \left[\frac{1}{v_1} + \frac{1}{v_2}\right], \tag{2}$$

where (n_1, v_1) and (n_2, v_2) are electron density and velocity of faster and slower particles of the beam, respectively. $V_{01} = v_{th}$ thermal velocity of the electrons and N is the total number of particles, where $V_1 = V_2$, $10V_{01} = 0.3c$, c is the velocity of light and $n_0 = 7.7 \times 10^9$ cm^{-3} electron density corresponding to 800 MHz.

By substituting the average values of T_D and D_e in Equations (4) and (5), one obtains the number density of energetic particles as 3.34×10^4 cm^{-3} or $n_1/n_0 = 10^{-4}$ and total number of particles in the beam $= 10^{32}$. These numbers are almost the same as numbers obtained by Fürst *et al.* (1982), for exciter of the blips, on the assumption that decimetric radiation is due to plasma emission mechanism at second harmonic similar to the type III burst radiation.

Thus, as seen from above discussions, 'blips' at 1663 MHz originate at lower corona, i.e., at higher density level where collisions should be more dominant and decay time of the 'blips' should have been a few millisecond (for source parameters of 7.7×10^9 cm^{-3} and $T = 10^6$ K). However, the observed decay time of the blips and its good fit to Alvarez and Haddock's empirical equation of decay time versus frequency (originally suggested for metric and kilometric type III bursts) and their observed low degree of circular polarization, suggest that 'blips' should originate at second harmonic by beam plasma interaction similar to metric type III bursts.

4. Summary

The present low-level decimetric observations suggest that (i) low-level ms-spikes put constraint on the source size. Source sizes of the low level ms-spikes should be less than 50 km, if electron-cyclotron maser were to be the emission process for generation of ms-spikes.

(ii) Detailed investigations of the time profile of the low-level blips, discovered first time above 1000 MHz, suggest that blips are produced due to beam plasma interaction, like type III burst at second harmonic. The beams which produce blips, decay in short time. The observed decay time of the blips favours the hypothesis that decay is due to collisional damping like metric type III bursts.

Acknowledgements

The 1.63 GHz receiver was developed by the Observatoire de Paris, Meudon. Its use, at Itapetinga, in the present solar study was made possible by the kind permission of Dr A. Boischot, F. Biraud, and C. Rosolen. The new sub-reflector assembly at Itapetinga, to operate at 1.63 GHz was developed by Drs E. Scalise Jr., R. E. Schall, and Y. Bakor. Thanks are due to Drs A. O. Benz for providing solar dynamic spectra and G. Dulk and R. P. Kane for useful discussions. INPE operates CRAAM and Itapetinga Radio Observatory. Suggestions of anonymous referee considerably improved the paper.

References

Alvarez, H. and Haddock, F. T.: 1973, *Solar Phys.* **30**, 175.
Aubier and Boischot, A.: 1972, *Astron. Astrophys.* **19**, 343.
Benz, A. O., Fürst, E., Hirth, W., and Perrenoud, M. R.: 1981, Nature **291**, 239.
Benz, A. O.: 1986a, *Solar Phys.* **104**, 99.
Benz, A. O.: 1986b, Personal communication.
Benz, A. O., Bernold, T., and Dennis, B.: 1983, *Astrophys. J.* **271**, 355.
Dröge, F.: 1967, *Z. Astrophys.* **66**, 176.
Dröge, F.: 1977, *Astron. Astrophys.* **57**, 285.

Fürst, E., Benz, A. O., and Hirth, W.: 1982, *Astron. Astrophys.* **107**, 178.

Giudice, D. A. and Castelli, I. P.: 1971, *IEEE Trans. on Aerospace and Electronic Sys.*, AES 7, No. 2, 226.

Holman, G. D., Eichler, D., and Kundu, M. R.: 1980, in M. R. Kundu and T. E. Gergely (eds.), 'Radio Physics of the Sun', *IAU Symp.* **86**, 457.

Kundu, M. R.: 1965, *Solar Radio Astronomy*, John Wiley and Sons, New York, p. 329.

Melrose, D. B. and Dulk, G. A.: 1982, *Astrophys. J.* **259**, 844.

Slottje, C.: 1978, *Nature* **275**, 520.

Solar Geophysical Data: 1985, No. 492, part I.

Spicer, D. S.: 1977, *Solar Phys.* **53**, 305.

Stähli, M. and Magun, A.: 1986, *Solar Phys.* **104**, 117.

Subramanian, K. R., Krishan, V., and Sastry, Ch. V.: 1981, *Solar Phys.* **70**, 375.

Tapping, K. F., Kuijpers, J., Kastra, J. S., Van Nieuwkoop, J., Graham, D., and Slottje, C.: 1983, *Astron. Astrophys.* **122**, 177.

Tarnstrom, G. L. and Philip, K. W.: 1971, *Scientific Report*, No. UAG R-217, University of Alaska.

Wiehl, H. J., Benz, A. O., and Aschwanden, M. J.: 1985, *Solar Phys.* **95**, 167.

DIRECT GENERATION OF SOLAR AND STELLAR RADIO BURSTS BY ENERGETIC ELECTRON MASER*

P. LOUARN, D. LE QUÉAU, and A. ROUX

Centre de Recherches en Physique de l'Environnement Terrestre et Planétaire, CNET, 92131 Issy-les-Moulineaux, France

Abstract. A fully relativistic electron maser is proposed for the explanation of certain non-thermal solar and stellar radio bursts. This mechanism (maser synchrotron) is based on a gyroresonant interaction between waves and electrons of high energies and uses the free energy contained in an electronic distribution function that peaks for energies around 1 MeV.

By a calculation of the growth rates of the three electromagnetic modes, we show that the X-mode prevails for values of ω_p/ω_c up to 2 or 3. This result is very different from the standard cyclotron maser case where such values of ω_p/ω_c lead to quench the X-mode growth. Hence, the synchrotron maser instability appears to be a direct and efficient amplification process for considerably larger physical conditions than the cyclotron maser. In addition, the radiation, emitted over the second gyroharmonic, freely propagates without a strong reabsorbtion at the $2\omega_c$ layer. All these points can constitute major advantages of this mechanism in an astrophysical context.

1. Introduction

It is presently generally admitted that the cyclotron maser instability (Wu and Lee, 1979) is the most likely process for explaining the non-thermal redio emissions emerging from planetary auroral zones: the terrestrial and saturnian kilometric radiations and the Jovian decametric radiations. This instability has also been invoked for explaining radio emissions related to solar flares (Mangeney and Veltri, 1976; Sharma *et al.*, 1982) and stellar flares (Melrose and Dulk, 1982; Dulk, 1985). In these different studies, the instable distribution function was characterized by a relatively low mean energy ($E < 1000$ keV). Growth rate calculations showed that the different modes are only amplified near gyroharmonics and that the predominant mode or harmonic are determined by the ratio ω_p/ω_c (where ω_p is the plasma frequency and ω_c the gyrofrequency). For low values of this ratio ($\omega_p/\omega_c < 0.3$) the X-mode is efficiently amplified near its cut-off and dominant. For higher values of ω_p/ω_c, apart a predominance of the fundamental O-mode or second harmonic X-mode under special conditions (Winglee, 1985), the Z-mode near the upper hybrid frequency (Melrose *et al.*, 1984) and the whistler mode (Sharma and Vlahos, 1984) are generally the most amplified. As far as planetary emissions are concerned, these characteristics justify the interest for the CMI. Indeed, non-thermal planetary emissions are produced within magnetospheric regions where ω_p/ω_c is low (typically $\omega_p/\omega_c < 0.2$) and consequently, the CMI can work efficiently as a direct amplyfying process. Conversely, a direct amplification of stellar or solar radio bursts by the CMI seems less convincing. In stellar environments, the

* Proceedings of the Second CESRA Workshop on Particle Acceleration and Trapping in Solar Flares, held at Aubigny-sur-Nère (France), 23–26 June, 1986.

Solar Physics **111** (1987) 201–206.

relatively hot background plasma ($T > 10^6$ K) strongly damps radiation at the first gyroharmonics. Therefore, fundamental X- and O-modes cannot escape out of the stellar magnetosphere (Sharma and Vlahos, 1984). Moreover, radiation at higher harmonics in the X- or O-mode is unexpected since the predominance of the fundamental X-mode (if $\omega_p/\omega_c < 0.3$) or the upper hybrid Z-mode (if $\omega_p/\omega_c > 0.3$) would preclude the emergence of harmonics before the maser saturates. A possible solution would be a conversion process: the mother wave (Z-mode) is generated by a masering process and then converted into X- or O-mode at frequencies lying above the 2nd gyroharmonic (Melrose *et al.*, 1984; Dulk, 1985). This conversion mechanism is, however, expected to greatly reduce the efficiency of the emission process. For all these reasons, cyclotron masers appear less attractive in the context of astrophysical radio emissions than in the context of planetary radio emissions.

We develop here a complementary approach of the cyclotron maser instability. Our calculation is based upon the hypothesis that electronic population having fully relativistic energies ($E \sim 1$ MeV) and exhibiting various non-thermal features could well be produced during flaring events. In a recent article (Louarn *et al.*, 1986), we have investigated some of the possibilities offered by a fully relativistic electron maser (Synchrotron Maser, SMI hereafter), but we only considered the amplification of the X-mode in the special case of a strictly perpendicular propagation. In the present paper, we calculate the growth rates of various electromagnetic modes: X, O, and Z (thus including the electrostatic upper hybrid branch), for any angle of propagation and for various modelized distribution functions.

2. Calculation of the Growth Rates

2.1. Basic calculations

Our calculation follows the usual perturbation method outlined by Wu (1985) in a recent review paper ans also used in Louarn *et al.* (1986).

We assumed that the plasma is infinite and homogeneous and that its electronic population can be splitted into two components: (i) a cold population (density n_c, plasma frequency ω_{pc}) and (ii) a hot population (n_h, ω_{ph}). The assumption: $n_c \gg n_h$ allows to neglect the effects of the hot electrons in the dispersion equation of the wave. Thus, the non thermal energetic electrons only contribute to wave growth or damping.

The growth rate is obtained by a summation over the contributions of successive resonant poles (Louarn *et al.*, 1986). These poles defined ellipses in the velocity space with equations $\omega - k_\parallel v_\parallel - m(\omega_c/\Gamma) = 0$ ($m = 1, 2, 3, \ldots$, and Γ is the Lorentz factor: $(1 - v^2/c^2)^{-1/2}$). The study of this summation over the various resonant poles reveals an important difference between the CMI (mildly relativistic electrons; $E < 100$ keV) and the SMI (fully relativistic electrons; $E \sim 1$ MeV). The resonant ellipses corresponding to a given supraluminous wave (characterised by its frequency ω and wave vector $\mathbf{K}$) have a width which rapidly increases with the order m of the contributing pole. This corresponds to a rapid increase of the energy of the resonant electrons. In the CMI case, only a few electrons are present in the high energy tail of the distribution function and thus practically only one pole significantly contributes to the wave growth or damping. This is linked to the fact that instability of the various supraluminous modes is only possible over a very small bandwidth, in the immediate vicinity of gyroharmonics. Conversely, in the SMI case, many resonant poles simultaneously contribute to the wave growth or damping. This leads to significant changes in the frequency and bandwidth of the emission. For example, we obtain a quasi continuous spectrum instead of a line spectrum. These modifications are examined in the next section.

3. Numerical Results

We shall first consider the case where the instable distribution function is modelled by a DGH:

$$F(u_\perp, u_\parallel) = \frac{1}{(2\pi u_T)^{3/2}} \frac{1}{n!}\left(\frac{u_\perp}{\sqrt{2}\, u_T}\right)^{2n} \exp[-(u_\perp^2 + u_\parallel^2)/2u_T^2]\,,$$

where $u_{\perp,\parallel} = p_{\perp,\parallel}/mc$ and p, m are the electron momentum and mass, respectively.

We have chosen: $u = 0.85$ and $n = 2$. This function is peaked at energies of the order of 950 keV but efficient resonant contributions can arise from electrons with energies up to 2 MeV.

We calculate the spatial growth rates of the three X-, O-, and Z-modes ($\gamma_N N_i$ where N_i is the imaginary part of the refractive index). This study includes the upper hybrid electrostatic branch which is nothing but the continuation of the Z-mode at small wavelengths, at least in the hypothesized context of a cold plasma propagation.

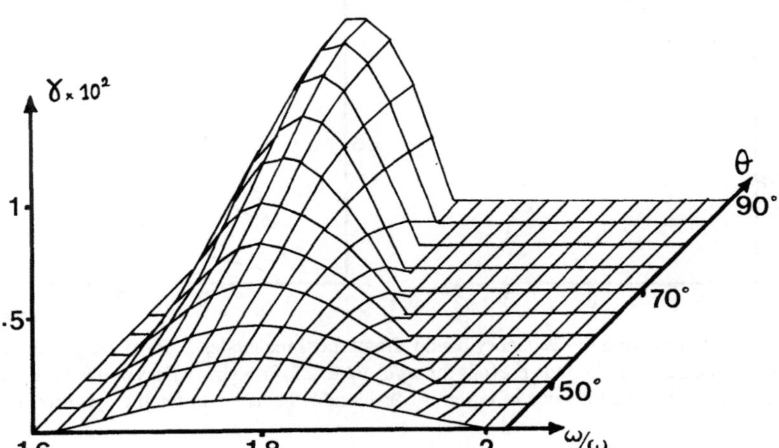

Fig. 1. The graph represents the normalised growth rate ($\gamma = (N_i/N_r)/(n_c/n_h)$ of the X-mode when it is destablished by a DGH distribution function peaked at 1 MeV. The variables are the angle of propagation θ and the normalised frequency ω/ω_c. The maximum occurs near the cut-off, for a perpendicular propagation and it is relatively large (see text). Both the frequency bandwidth and the angular range for amplification (10% and 20°, respectively) are very large when compared to those obtained in the cyclotron maser case.

On Figure 1, we have plotted the growth rate of the X-mode when $\omega_p/\omega_c = 1$. The two variables are the angle of propagation θ and the normalised frequency ω/ω_c.

We find the largest growth rate for a nearly perpendicular propagation and the growth rate seems to be relatively insensitive to variation of θ between $\theta = 90°$ and $\theta = 70°$. These results justify the hypothesis made by Louarn *et al.* (1986) who only considered perpendicular propagation. The emission is concentrated near the X-mode cut-off, the maximum of the growth rate occurs for $\omega = 1.7\,\omega_c$ and the frequency bandwidth is of the order of $\Delta\omega/\omega_c \sim 0.2$ at $\theta = 90°$. When the angle of propagation diminishes, the frequency bandwidth increases and the growth rate decreases (Figure 1).

These conditions for amplification are not too restrictive: $\Delta\omega/\omega \sim 10\%$ and $\Delta\theta \sim 30°$ ($\Delta\theta$ is the angular range which corresponds to a division of the growth rate by 2). This has to be compared with the amplification by the CMI which is only possible for a restricted range of frequencies ($\Delta\omega/\omega < 1\%$) and over a very narrow conical sheet ($\Delta\theta \sim 1°$).

We have also calculated the growth rates of the O- and the Z-modes. For $\omega_p/\omega_c = 1$, we find that the Z-mode is not amplified and that the O-mode has a growth rate smaller than the X-mode (for $\theta = 90°$: $N_i = 0.8 \times 10^{-2}$ for the X-mode and $N_i = 0.6 \times 10^{-3}$ for the O-mode).

Finally, we have investigated how the growth rate and the frequency bandwidth of the three modes: X, O, Z vary as a function of the ratio ω_p/ω_c. Our results are plotted on Figure 2.

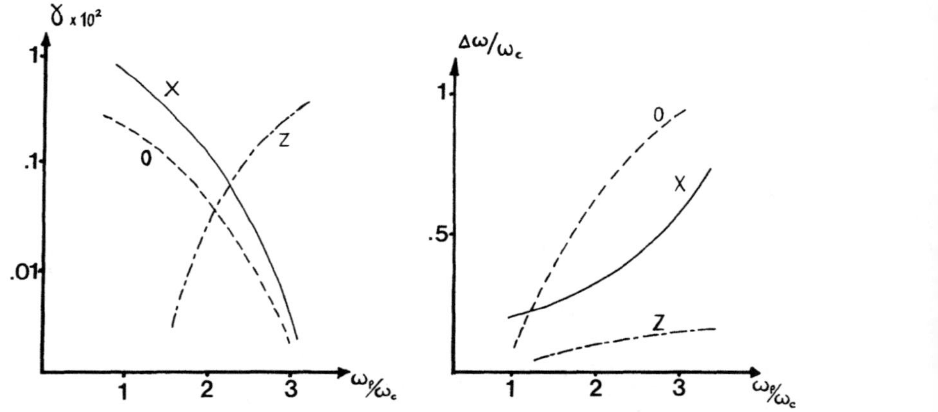

Fig. 2. The growth rates and the frequency bandwidth of the three modes X, Z, O are plotted for various values of the ratio ω_p/ω_c. The X-mode prevails for ω_p/ω_c up to 2.3, fir higher values of this ratio the Z-mode predominates. This critical value of ω_p/ω_c is considerably higher than for the cyclotron maser.

For the X-mode, we find that the maximum of the growth rate decreases when ω_p/ω_c is increazed, in opposite, for the Z-mode, this maximum increases. The growth rate of the O-mode is always smaller than those of the X- or Z-mode. The Z-mode has a roughly constant frequency bandwidth ($0.2\,\omega_c$). The X-mode has larger frequency bandwidths which increase with ω_p/ω_c.

Because of larger growth rates and frequency bandwidth, the X-mode is dominant when ω_p/ω_c is smaller than 2. For higher values of this ratio, the predominant mode is certainly the Z-mode. We have also made this ascertainment for other kinds of distribution functions.

For example, in the case of a ring-like function that reaches its maximum for electronic energy of the order of 800 keV, we find that the X-mode could predominate for ω_p/ω_c as high as 4.

Thus, in both cases (SMI and CMI), the Z-mode is the most instable mode for large values of ω_p/ω_c. But the critical value at which the Z-mode becomes dominant is far larger in the SMI case than in the CMI one. This effect is likely to be enhanced if a finite temperature is taken into account for the background plasma. As the index of refraction of Z-mode (on the electrostatic upper-hybrid branch) is large, this mode will be more sensitive to cyclotronic absorbtion by the thermal background than the X-mode. Thus, it will be more damped than the X-mode for non perpendicular propagation, so the angular range over which it can be more unstable than the X-mode is likely to be considerably reduced. Consequently, the critical value at which the Z-mode grows faster than the X-mode is very sensitive to the shape of the distribution function. The relevance of these considerations to the radio emissions of flaring loops will be examined in the following discussion.

3. Discussion and Conclusion

Because of the predominance of the X-mode for ω_p/ω_c as high as 4, the SMI is a direct amplification mechanism for a larger range of physical conditions than the CMI. Moreover, in a source where ω_p/ω_c is high ($\omega_p/\omega_c > 1.4$), the X-mode is emitted above the second gyroharmonic or even the third gyroharmonic (if $\omega_p/\omega_c > 2.4$). Thus, the radiation is not strongly reabsorbed by gyroabsorbtion at the $2\omega_c$ or $3\omega_c$ layers during its propagation towards the observer. This possibility of a direct amplification and the easy access of the destabilized waves to the free space are two important advantages of the SMI process. Furthermore, the SMI is able to amplify the waves far from gyroharmonics, then conversely to the CMI (Sharma and Vlahos, 1984), this amplification is largely independant from the temperature of the background plasma in the source region. This argument is only valid for supraluminous modes (X- or O-modes). Indeed, infraluminous modes (electrostatic Z-mode or whistler mode) easily interact with thermal electrons, then, they can be damped when supraluminous modes are not.

We have not included whistler waves in the present analysis, this will be done in future works. Nevertheless, the relaxation process due to these waves (infraluminous waves) is certainly very different from those due to supraluminous ones. Therefore, the saturation of the maser by infraluminous modes is still a matter of hypothesis.

The peak in the distribution function at energy around 1 MeV is a crucial ingredient of our mechanism. This peak has never been observed but, due to the relaxation of the maser, it would be erased in less of 10^{-3} s. Obviously, this very short time-life of the non-thermal distribution makes it practically invisible.

Some numerical applications can help us to appreciate the efficiency of the process. Taking an average value of $N_i/N_r \sim 5 \times 10^{-3}$ and n_h/n_c as small as 10^{-4} or 10^{-5}, we find that the wave is amplified by a factor 10^{10} after a travel approximately equal to 10^7 wavelengths. For centimetric and decimetric waves, this corresponds to a distance of about 10^2 or 10^3 km. The number of energetic electrons is difficult to appreciate, we have chosen $n_h/n_c \sim 10^{-4}$ to 10^{-5} which is consistent with an overall power law distribution function ($f(E) = E^{-\gamma}$) with $\gamma \sim 3$–4 and n_c being the density of low energy electrons ($E \leq 20$ keV). The source size is quite similar to the transverse dimension of a flaring loop and consistent with the very spiky temporal variations of the radio bursts (duration of the order of 10^{-3} s). The needed number of energetic electrons within the source region is about 10^{22}–10^{24} if one supposes that $n_c/n_h \sim 10^{-4}$, $n_c \sim 10^{11}$ cm^{-3} and a source volume of the order of 10^6 km^3. This number of electrons is relatively small as compared to the estimated number of energetic electrons involved in a flare phenomena (10^{35} electrons with an energy greater than 20 keV).

The effects of the shape of the distribution function will be discussed in a forthcoming paper. Future works will be oriented towards the evaluation of the effective gain, in a realistic geometry, which implies the inclusion of the role played by various plasma inhomogeneities.

References

Dulk, G. A.: 1985, *Ann. Rev. Astron. Astrophys.* **23**, 169.
Hewitt, R. G., Melrose, D. B., and Dulk, G. A.: 1983, *J. Geophys. Res.* **88**, 10065.
Louarn, P., Lequeau, D., and Roux, A.: 1986, *Astron. Astrophys.* **165**, 211.
Mangeney, A. and Veltri, P.: 1976, *Astron. Astrophys.* **47**, 165.
Melrose, D. B. and Dulk, G. A.: 1982, *Astrophys. J.* **259**, 844.
Melrose, D. B., Hewitt, R. G., and Dulk, G. A.: 1984, *J. Geophys. Res.* **89**, 897.
Sharma, R. R. and Vlahos, L.: 1984, *Astrophys. J.* **280**, 405.
Sharma, R. R., Vlahos, L., and Papadopoulos, K.: 1982, *Astron. Astrophys.* **112**, 317.
Winglee, R. M.: 1985, *J. Geophys. Res.* **90**, 9663.
Wu, C. S.: 1985, *Space Sci. Rev.* **41**, 215.
Wu, C. S. and Lee, L. C.: 1979, *Astrophys. J.* **230**, 621.

HARD X-RAY EMISSION PROCESSES IN SOLAR FLARES*

(Invited Review)

N. VILMER

Observatoire de Paris, Section d'Astrophysique de Meudon, 92190 Meudon, France

Abstract. Solar hard X-ray emission is one of the most direct diagnostics of accelerated particles during solar flares. In this review, the current understanding of hard X-ray emission processes is discussed: first the different emission mechanisms (in particular inverse Compton radiation, energetic ion or electron bremsstrahlung) are presented and the plausibility of each of these mechanisms is discussed. Then, different types of hard X-ray models (thermal or non-thermal, homogeneous or inhomogeneous emission regions) are presented together with the comparison of their predictions with X-ray observations (spectral, spatial and temporal informations – directivity and polarization).

1. Introduction

A large fraction of the energy released during solar flares goes to energetic particles (electrons or ions) which then lose their energy through radiation (X-ray, γ-ray, radio wavelengths) and heating of the ambient medium, when interacting with the solar atmosphere. The understanding of the particles' energization processes is then crucial to the understanding of the solar flare development.

It is now well accepted that X-ray and radio emissions produced during the different phases of solar flares come from a common injection of energetic electrons (see Pick, 1986; Trottet, 1986, and references therein). In order to get a quantitative interpretation of both emissions, it is then necessary to know the energetic electron population giving rise to these different emissions. Although X-ray (and γ-ray) emissions are the most direct diagnostics of accelerated particles in the solar atmosphere, models are nevertheless necessary to infer from X-ray (and γ-ray) observations the energetic electron (or ion) population as well as information on the X-ray source.

In this review, the plausible hard X-ray emission processes as well as the different models of the X-ray emitting sources proposed so far in the literature will be discussed (Sections 2 and 3) and the predictions of the models will be compared with the observations. Although, strictly speaking, emission processes by energetic electrons cannot be studied apart from the energy release and acceleration mechanisms, these last mechanisms will not be discussed here (see the reviews on acceleration processes by Mangeney and Benz in this issue).

2. Hard X-Ray Emission Mechanisms

Solar hard X-ray emission (photon energy greater than a few dekakeV) is a continuum emission which can be produced either through the interaction of energetic particles

* Proceedings of the Second CESRA Workshop on Particle Acceleration and Trapping in Solar Flares, held at Aubigny-sur-Nère (France), 23–26 June, 1986.

Solar Physics **111** (1987) 207–223.

(electrons or ions) with a cold ambient medium (see below) or through bremsstrahlung of a heated plasma ($T \geq 10^8$ K).

2.1. Energetic electrons interactions

Up to now, most work on the subject has been concerned with energetic electrons (10 keV to a few MeV) since for particles with the same energy, radiation losses are largest for electrons than for protons. X-ray emission can in principle be then produced either through gyrosynchrotron emission, inverse Compton radiation or bremsstrahlung with the ambient medium. As concerns gyrosynchrotron emission, Korchak (1971) argued that this mechanism is very improbable for the production of X-ray since, to be dominant in the 10–100 keV range, it would require a low density of the emitting region, a strong magnetic field ($\sim 10^4$–10^5 G) and a large number of very energetic electrons (energies up to 1–10 GeV) (Švestka, 1976) which do not seem to be produced in most flares. This mechanism can then be completely ruled out for the production of X-rays.

Korchak (1971) argued that inverse Compton scattering by relativistic electrons on a thermal radiation field can dominate bremsstrahlung radiation from the same electron population only if the emission region has a low density. More recently, Kaufmann *et al.* (1986) and McClements and Brown (1986) have discussed for one event the possible interpretation of hard X-ray emission through inverse Compton scattering on a highly relativistic electron population either of ambient photons or of the synchrotron photons emitted by the electron population itself (synchrotron self-Compton mechanism). McClements and Brown (1986) show that both X-ray and microwave spectral characteristics – including the exceptionally high (90–100 GHz) microwave spectrum turnover frequency could be reproduced through both inverse Compton-scattering mechanisms. However, such interpretations suffer for many problems. First, to ensure that bremsstrahlung radiation does not dominate, limits the ambient plasma density, so that the energetic particles density required to produce the observed X-ray and microwave radiations is high compared to the ambient plasma density itself. Moreover, in both cases, the electron population responsible for both emissions has a low-energy cutoff above 1 MeV and such a population may be difficult to produce. On the other hand, Klein (1987) showed that the spectral characteristics of the event in the X-ray and microwave domains can be explained in terms of conventional models with X-ray emission produced through electron-ion bremsstrahlung.

It may then be concluded that inverse Compton radiation is probably not the *dominant* X-ray emission mechanism for *usual* solar flares so that the dominant interaction of energetic electrons for X-ray emission is bremsstrahlung with the ambient medium.

2.2. Energetic protons interactions

Although it had been mentioned in the sixties by several authors (e.g., Elliot, 1964) that fast protons could generate hard X-ray bremsstrahlung, most of the later work on the subject has been devoted to X-ray emission produced by energetic electrons (10 keV to a few MeV). Indeed, X-ray emission produced by protons would require either a large number of protons in the same energy range which is not observed in the interplanetary medium (Peterson and Winckler, 1959) or higher energy protons. However, the recent discovery through simultaneous observations in both hard X-rays and γ-rays that fast protons (energy of a few MeV to a few GeV) are present in the solar atmosphere in the impulsive phase of a flare (see, e.g., Chupp, 1984) led to new insights into the production of X-ray emission by protons.

Bremsstrahlung of fast protons (10–100 MeV) with ambient electrons ('inverse bremsstrahlung') has been recently reconsidered by Emslie and Brown (1985). In a thick target, the bremsstrahlung yield of a proton is $\sim 2 \times 10^3$ times greater than that of an electron with the same velocity. The production of X-ray emission through inverse bremsstrahlung may then present several advantages compared to the case of electron bremsstrahlung. First, the required beam current and then the return current induced to maintain plasma neutralization are reduced, thus solving the problem of the stability of the reverse current (see next section). Second, the required acceleration efficiency is reduced and the particle energy content does not exceed the total flare energy. The model fails however for several reasons. In particular, the number of energetic protons required to produce the observed X-rays is about 10^3 times the number required to produced the associated γ-ray emission (Emslie and Brown, 1985). Another difficulty arises when column densities deduced from UV observations and from X-ray observations using 'inverse bremsstrahlung' computations are compared. These discrepancies could perhaps be reduced if some of the X-ray emitting protons are partially trapped in a region not dense enough to produce γ-rays, thus requiring a complete treatment of the dynamics of the X-ray emitting particles. Moreover, a more detailed analysis of the proton-electron bremsstrahlung ('inverse bremsstrahlung') in the electron rest frame (Heristchi, 1986) indicates that protons can produce in fact higher energy X-rays than electrons with equivalent velocities. This effect could possibly affect the number of energetic protons required to produce the observed X-ray flux and could be looked into more details in the future.

Energetic proton beams can also generate indirectly X-ray emission either through acceleration of energetic secondary electrons by near head – on collisions with the beam protons (Elliot, 1964) or through rapid (time-scale of a few seconds) intense localized heating of plasma electrons to temperatures suitable for thermal hard X-ray bremsstrahlung (Simnett, 1986). The former process is about 10 times less efficient than 'inverse bremsstrahlung' at least at low energies (Emslie and Brown, 1985; Heristchi, 1986). Anyhow, in such a context, X-ray emission is a secondary effect of the primary energy release and conventional non-thermal models are required to predict the produced X-ray flux. Although the second process may seen attractive to explain recent observations suggesting energy transfer from the release site to the emitting regions by protons beams (100 keV–1 MeV) (Simnett and Strong, 1984; Orwig and Woodgate, 1985), it is not obvious that, because of different plasma boundary conditions, the heating mechanism proposed by Haines and Marsh (1982) for laboratory plasmas may operate efficiently in solar flares (Emslie and Brown, 1985). Moreover, a quantitative estimate of the consequent X-ray flux has not yet been established.

3. Thermal and Non-Thermal Hard X-Ray Models

As shown in the previous section, the most probable hard X-ray emission mechanism is energetic electron bremsstrahlung produced with the ambient medium. The X-ray flux depends then on the target and on the energetic electron population. In the literature, two classes of models have been distinguished depending on the nature of the energetic electron population whether the high energy electrons are part of a thermally relaxed distribution – at temperatures greater than 10^8 K (thermal models) (see, e.g., Crannell *et al.*, 1978; Mätzler *et al.*, 1978; Brown *et al.*, 1979) or consist of a distribution of suprathermal electrons incident on a cold target (non-thermal models) (see, e.g., Brown, 1971, 1972). In the first case, the energy released during flares will go preferentially to heating of the ambient medium, while in the other case, it will go to particle acceleration. In relation to this question, there has been in the past a long controversy between non-thermal and thermal interpretations of hard X-ray solar flares (see, e.g., the review paper by Brown and Smith, 1980). We shall discuss here only some characteristics and predictions of both models.

3.1. Thermal Models

In most of these models, the plasma is heated in the coronal part of a magnetic loop ($T \cong 10^8$ K). In the case of adiabatic models, the hot plasma is confined by a strong time varying magnetic field and a correlation between X-ray emission measures and plasma temperatures is predicted, correlation which has indeed been observed for some bursts (Crannel *et al.*, 1978; Mätzler *et al.*, 1978). The main problem of such models is the short conductive cooling time of the hot source which may be much shorter than the duration of the observed burst. This can be overcome if the energy release site is extremely localized and the energy preferentially deposited on the electrons. Indeed, the development of an ion acoustic turbulent front which is associated with the hot free streaming electron beam and its return current leads to a higher plasma confinement (see, e.g., Brown *et al.*, 1979; Smith and Lilliequist, 1979; Vlahos and Papadopoulos, 1979). This type of thermal model is then characterized by the appearance of two collisionless conduction fronts propagating symetrically away from the energy release site at a speed close to the local ion sound speed. Estimates of the X-ray yield of thermal models demands knowledge of how well the plasma may be confined. This point is still controversial. Indeed, Batchelor *et al.* (1985) argued that a 3-dimensional treatment of the interactions between the hot electron population and the turbulent ion sound waves leads to a better confinement of the hot plasma electrons than previously assumed, because of the effect of the thermoelectric field or of magnetic mirroring on high pitch angle electrons. This also implies that the heated region may produce microwave emission since the high energy tail of the maxwellian distribution remains populated. Their model predicts then a correlation between the X-ray rise time (assumed to be approximately the propagation time of the conduction front from the top of the loop to the chromosphere) and a characteristic time deduced from the observed microwave flux in the optically thick part of the spectrum. Surprisingly, despite the simplicity of the

X-ray and microwave computations and of the treatment of the plasma confinement, such a correlation is indeed observed for some events (Figure 1). However, it is not obvious that the X-ray rise time may be interpreted in terms of the propagation time of the conduction front and this time could be in fact characteristic of the rate of energy input in the source. In these conditions, the observed correlation does no longer constitute evidence for the present thermal model (MacKinnon, 1985).

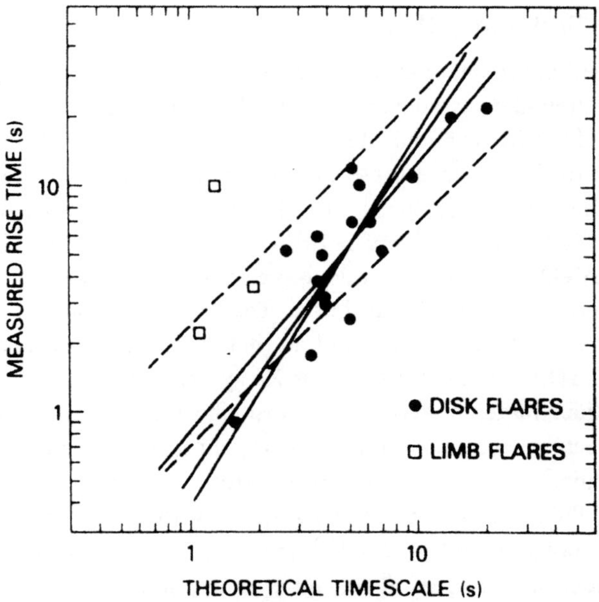

Fig. 1. Correlation diagram for 26 flares of the measured X-ray rise time and the characteristic time scale $\tau_0 = 9.4 \times 10^{17} S_2^{1/2}/f_2 T_e$ where S_2 is the microwave flux density at Earth in s.f.u. at a frequency f_2 (Hz) below the observed turnover frequency and T_e is the source temperature in K deduced from X-ray observations. Solid lines indicate best fits found by linear least-squares fitting. Dashed lines are boundaries of the expected positions of disk flares, if the sources are arches from 2 to 4 times as long as they are thick (from Batchelor *et al.*, 1985).

On the other hand, Smith (1986) showed that the presence of fast electrons in the heated plasma enhances the heat flux to the cold region. Although his treatment has severe limitations due to the use of a fluid code to study the plasma evolution, it shows nevertheless that this effect has important implications for thermal models. Indeed, it increases the conduction front speed thus decreasing the temperature obtainable with a given energy input in the loop and thus the efficiency of thermal models for the production of high energy X-rays. High-energy X-ray emission must then be produced by non-thermal energetic electrons, being possibly runaway electrons from the hot source (see, e.g., Brown *et al.*, 1979; Emslie and Vlahos, 1980) and is then difficult to interprete in terms of purely thermal models. Moreover, the good observed association between X-ray and microwave emission and the fact that the two sources may not be cospatial (Kundu, 1984) are better understood if microwave and hard X-ray emissions are produced by a common injection of non-thermal electrons (see, e.g., Trottet, 1986).

3.2. Non-thermal models

Several classes are distinguished depending on the relative magnitudes of the electron injection timescale in the emitting source, the escape time from the source and the energy loss time of the electrons in the ambient medium.

3.2.1. *Thin and Thick Target Models*

In the 'thin target' approximation, the electron escape time is much smaller than the energy loss time (low-density source) and the electrons lose only a small part of their energy in the medium (Brown and Melrose, 1977; Kane *et al.*, 1980b). The X-ray flux depends then on the instantaneous electron number and the X-ray yield is low. On the other hand, in the 'thick target' beam model, energetic electron beams are injected in high density media (electron density $> 10^{11}$ cm^{-3}) the energy loss time is very short (< 1 s for 100 keV electrons) and a continuous electron injection is required to reproduce the observed duration of the X-ray burst. The instantaneous electron spectrum is different from the injection spectrum and, in that case, the X-ray flux depends on the injection spectrum (Brown, 1971, 1973). The X-ray yield is higher than in the 'thin target' case. Although thick target models are the most efficient non-thermal ones, the yield is still low since electrons lose most of their energy through non-radiative collisions with the ambient medium. A large number of electrons is then required to produce the observed X-ray fluxes ($\geq 5 \times 10^{34}$ electrons s^{-1} with energy > 20 keV or electron energy content above 20 keV of 2×10^{27} erg s^{-1}). This is the main problem of all non-thermal models and the reason why other emission processes (see Section 2) have been investigated. Indeed, the electron energy content is comparable to the total energy released during the flare and the large electron beam currents induced (10^{16}–10^{21} Å) demand that a neutralizing reverse current be rapidly established among the ambient electrons (Brown and Melrose, 1977; Heyvaerts, 1981). Of course to be rigorous, the formation of the electron beam and of its associated return current should be treated simultaneously with the electron acceleration process.

The major effect of the reverse current is to enhance the beam energy losses as compared to collisional energy ones, because of ohmic losses due to the induced electric field driving the return current (see, e.g., Heyvaerts, 1981). Following an earlier work (Knight and Sturrock, 1977), Emslie (1980), and more recently Emslie (1985) showed that, assuming a steady state for the beam-return current system, ohmic energy losses become important for electron fluxes greater than $10^{17} e^{-}$ cm^{-2} s^{-1} and overcome collisional losses for ambient densities smaller than 10^9 cm^{-3}. There has been a controversy in the literature concerning the generation of the electric field driving the return current. Indeed, it can be established electrostatically by a space charge layer at the beam head (and as a consequence the beam-return current system can be treated in a steady state) (see, e.g., Brown and Bingham, 1984) or inductively because of an incomplete current neutralization in the beam-return current system (Spicer and Sudan,

1984). However, in any case, the effect of this electric field is to decelerate the beam electrons. This deceleration can still be enhanced if the return current becomes unstable to the generation of ion acoustic waves. In all cases, since the beam lifetime is greatly reduced when compared to the collisional one, the non thermal hard X-ray yield is reduced and the beam energy is converted into enhanced plasma heating, thus affecting the observed thermal radiation (Brown and Bingham, 1984; Spicer and Sudan, 1984; Cromwell *et al.*, 1985). The beam becomes then an efficient way of producing hard X-ray emission through thermal bremsstrahlung. It can be noticed, however, that the effects of the return current may be reduced if it consists of a small number of high energy electrons rather than on the bulk of ambient electrons (Rowland and Vlahos, 1985).

The electron beam itself can also become unstable to the growth of Langmuir waves due to the positive slope the beam develops through Coulomb collisions during its propagation (Emslie and Smith, 1984). In the case of a quasi-linear relaxation through the generation of plasma waves, this effect is however important neither on the hard X-ray yield or spectrum nor on the beam energy deposition rate, since the bremsstrahlung cross section acts as a filter on the exact form of the electron distribution (McClements *et al.*, 1985). However, such a beam-plasma instability may have a microwave signature through electromagnetic conversion of the high level of the produced Langmuir waves (Emslie and Smith, 1984; McClements, 1987).

In conclusion, beam plasma instabilities may play a great part in the dynamics of the electron beams responsible for hard X-ray emission and these effects, studied only recently in the X-ray field, should be investigated further.

3.2.2. *Trap Plus Precipitation Model*

A more realistic situation for the production of hard X-rays through non thermal bremsstrahlung is to inject and trap particles in a magnetic loop where they will lose all their energy. Following earlier works (Takakura and Kai, 1966; Brown, 1974; Melrose and Brown, 1976; Bai and Ramaty, 1979). Vilmer *et al.* (1982) developed a model where hard X-ray emission is produced by electrons continuously injected in a region of uniform density (n_0) where they are partially trapped. The electron injection function is supposed to have a temporal behavior independent of the energy, peaking at a certain time t_{max} and lasting for t_0. Its energy dependance is a power law with a spectral index γ constant during the whole injection (Vilmer *et al.*, 1982; MacKinnon *et al.*, 1983; Trottet and Vilmer, 1983). Energy losses are supposed to be entirely due to electron-electron collisions, other energy losses (e.g., due to collective phenomena) being neglected as compared to collisional ones. Scattering of the electrons into the loss cone and subsequent precipitation in denser layers are considered in the weak (electron diffusion due to collisions) and strong limits (electron diffusion due to plasma waves, such as whistlers, resonating with the electron loss-cone distribution in the trap) (Kennel, 1969; Melrose and Brown, 1976; Kawamura *et al.*, 1981). This gives rise to a chromospheric thick target component, whose relative importance is a function of scattering mechanism (Melrose and Brown, 1976; Trottet and Vilmer, 1983). The total X-ray flux depends then on the source parameters (trap density, magnetic field gradient in the loop) and on the energetic electron population (injection hardness and duration). As concerns plasma instabilities, it may be noticed that depending on the characteristics (density, magnetic field) of the trap region, other instabilities than the whistlers mentioned above may develop, either with the trapped population (e.g. electron cyclotron masers (Melrose and Dulk, 1982)) or with the beams of precipitating electrons (e.g., electrostatic waves in the upper hybrid branch (Vlahos *et al.*, 1983)). These instabilities may produce either directly or through electromagnetic conversion solar microwave spikes with high brightness temperatures and high polarization (see, e.g., the review by Vlahos in this issue).

In a trap-plus-precipitation model, delays in the maxima of high-energy X-ray emission as compared with low-energy ones are predicted. They are due to the increase with energy of the electron lifetime in the trap. Such delays are indeed observed in some events (Bai and Ramaty, 1979; Vilmer *et al.*, 1982; Bai and Dennis, 1985). For a given injection spectrum, delays increase with decreasing density (see, e.g., Figure 2), with

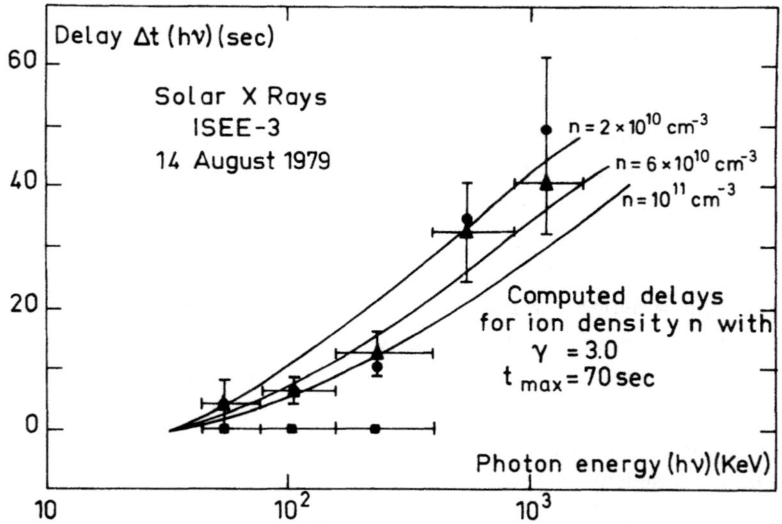

Fig. 2. Comparison of computed (solid lines) and observed delays (4 s av) $\Delta t(h\nu)$ as a function of the photon energy $h\nu$. ■ refers to the impulsive maximum, and ▲, ● to the first and second gradual studied maxima respectively (from Vilmer *et al.*, 1982).

photon energy and with the injection duration. The delay magnitudes as well as the hardness and the temporal evolution of the X-ray flux depend also on the precipitation regime (Figure 3), the delays being very small when electron scattering and precipitation are in the strong limit (e.g., wave particle interactions). In the case, of a homogeneous trap density, MacKinnon (1986) showed moreover that trap density temporal variations (e.g., due to chromospheric evaporation) are also important when looking at hard X-ray delays. Thus, depending on the characteristics of the emission region and of the electron

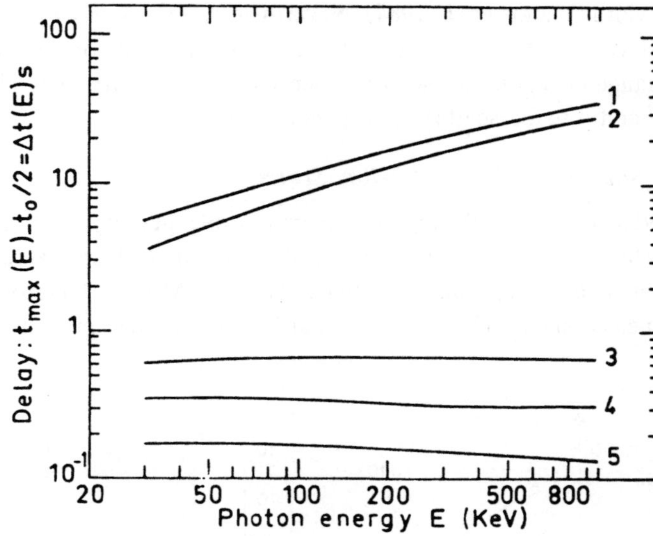

Fig. 3. Delays $\Delta t_{(E)} = t_{max(E)} - t_0/2$ of the X-ray maximum as a function of the photon energy E for $\gamma = 3$, $t_{max} = t_0/2 = 60$ s (see text), $n = 10^{10}$ cm^{-3} (ambient density) in the case of (1) perfect trap (2) leaking trap in the weak diffusion limit (3), (4), (5) leaking trap in the strong diffusion limit with $\alpha_0^2/L = 10^{-5}$, 2×10^{-5}, 5×10^{-5} (α_0 is the loss cone angle given by the ratio of the magnetic field at the top of the loop and at the mirror point and L is the characteristic length of the loop) (from Vlahos *et al.*, 1984).

injection, various delays and various spectral and temporal evolutions of the X-ray events (impulsive and gradual with or without delays) are reproduced with such a model. This allows one to infer from the observations, characteristics of the electrons and of the emitting region. Impulsive and complex bursts are produced by several successive electron injections in high density media (thick target approximation) (Trottet and Vilmer, 1984; Hernandez *et al.*, 1986). On the other hand, large gradual hard X-ray events, which exhibit large delays in the times of maxima at different X-ray energies may be reproduced with an energetic electron population continuously injected in a medium density region ($n \sim 6 \times 10^{10}$ cm^{-3}) where they are trapped (Figure 2). In these conditions, hard X-ray time delays do not necessarily reflect a 'second step' acceleration of high-energy electrons, since particle interactions with the medium may also produce a similar phenomenon. The 'second step' acceleration is a delayed (by a few seconds) acceleration of electrons with energies greater than a few hundred keV as compared with the acceleration of low-energy electrons and is generally invoked to explain delays between the times of maxima of high- and low-energy X-rays (Bai and Ramaty, 1979; Bai *et al.*, 1983). For some events, the observed X-ray delays increase continuously with energy (as on Figure 2), while for other events, there is a sudden increase of the delays above a given photon energy which may vary from one flare to the other (Schwartz, 1984). This sudden increase in delay times has been interpreted as an evidence in favor of a 'second step' acceleration (Vlahos *et al.*, 1984). It must be reminded, however, that this effect may be purely instrumental, being due to the increase with energy of the detector bandwidth (Vlahos *et al.*, 1984). Nevertheless, if the sudden increase is real, it may also possibly be due to a change in the electron precipitation rate above a given energy, this change being due to the finite energy range of electron energies which can resonate with, and thus be scattered by, plasma waves.

3.2.3. *Spatial Structure of the Hard X-Ray Source*

For all the models presented above, the only observable quantity which is predicted is the total X-ray flux. With the advent of X-ray observations with spatial resolution such as the ones obtained through direct imaging (HXIS–SMM or Hinotori) or through stereoscopic measurements (Kane, 1983), hard X-ray models predicting the spatial

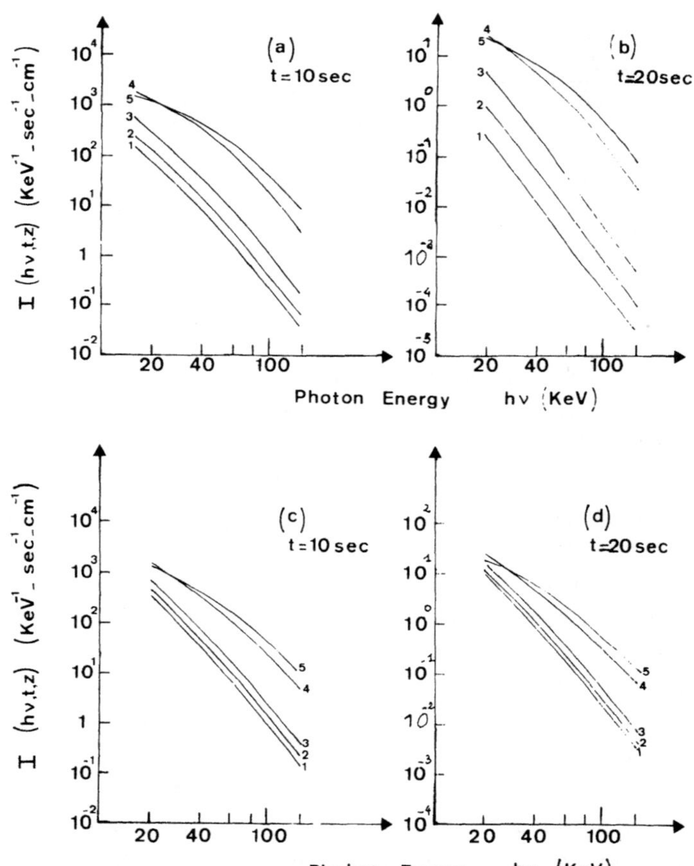

Fig. 4. Evolution with depth and time of the X-ray flux $I(h\nu, t, z)$ produced by energetic electrons injected at the top ($z = 0$) of a converging magnetic structure (mirror ratio = 2). Figures 4(a) and 4(b) correspond to a beamed injection (injection pitch angle cosine 0.5) while Figures 4(c) and 4(d) to an isotropic one. Curves 1, 2, 3, 4, 5 correspond, respectively, to $z = 5 \times 10^7$ cm, 10^8 cm, 2×10^8 cm, 5×10^8 cm, and 6×10^8 cm. The other parameters are $\gamma = 3$ (spectral index at injection), $t_{\max} = 10$ s (injection maximum and half duration), $n_0 = 10^{10}$ cm^{-3} (density at $z = 0$), density scale height $H = 10^8$ cm^{-3}. Scales of Figures (b) and (d) have been corrected of typographic errors as compared to Vilmer *et al.* (1986b).

structure of the source are demanded. Such predictions have been made in a steady state situation by Emslie (1981) for different empirical model atmospheres in the case of thermal or non-thermal bremsstrahlung. However, his treatment cannot give high-temporal resolution and spatial evolutions of the X-ray flux which requires the knowledge of the evolutions of the energetic electrons. This evolution is properly described by a Fokker–Planck equation which relates the spatial, temporal and spectral evolution of a non-thermal electron population continuously injected in an inhomogeneous medium and evolving collisionally and under the action of non uniform magnetic field structures. This equation has been solved in the past for steady state (Leach and Petrosian, 1981; Kovalev and Korolev, 1981) and impulsive injections (Kovalev and Korolev, 1981). Leach and Petrosian (1983) used this method in a steady-state situation to evaluate the height distribution, spectrum, polarization and directivity of the X-ray emission produced by a non-thermal electron population injected with a Gaussian distribution in pitch angle at the top of a magnetic structure. The X-ray

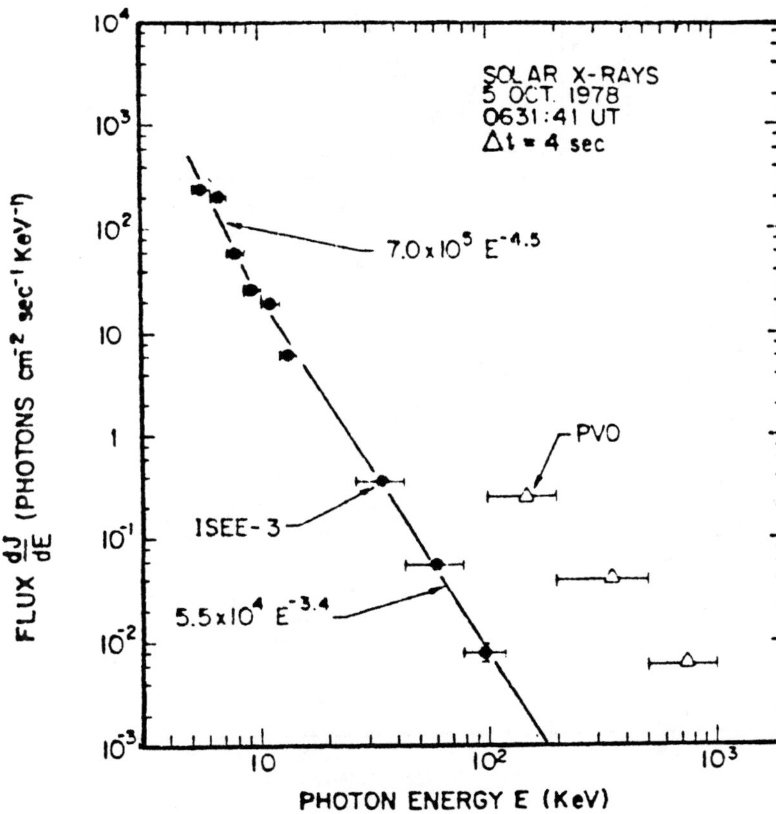

Fig. 5. Observations of an impulsive X-ray burst on 5 October, 1978 attributed to a large solar flare located ~15° behind the East limb of the Sun. X-ray spectra observed at the time of maximum with ISEE-3 (occulting height ~25000 km) and PVO (total flux) (from Kane, 1983).

flux produced at each height in the structure depends on the electron distribution at the same height (as in thin target models), while the spectrum of the total X-ray flux (integrated over the whole structure) is the spectrum of the thick target hard X-ray emission. The predicted height distribution, polarization and directivity of the X-ray emission are functions of the degree of anisotropy of the initial distribution in pitch angles as well as of the magnetic field convergence. Very recently, Takakura (1986) solved a time dependant Fokker–Planck treatment of the electron population evolving in similar media. His general results are consistent with those of a more simple approach which consists in the use of a continuity equation describing the mean evolution of the electron distribution due to collisional energy losses and scattering (Craig *et al.*, 1985; Vilmer *et al.*, 1986a, b). In such a situation and during the electron injection, the X-ray flux progressively hardens and increases with depth, this hardening depending on time

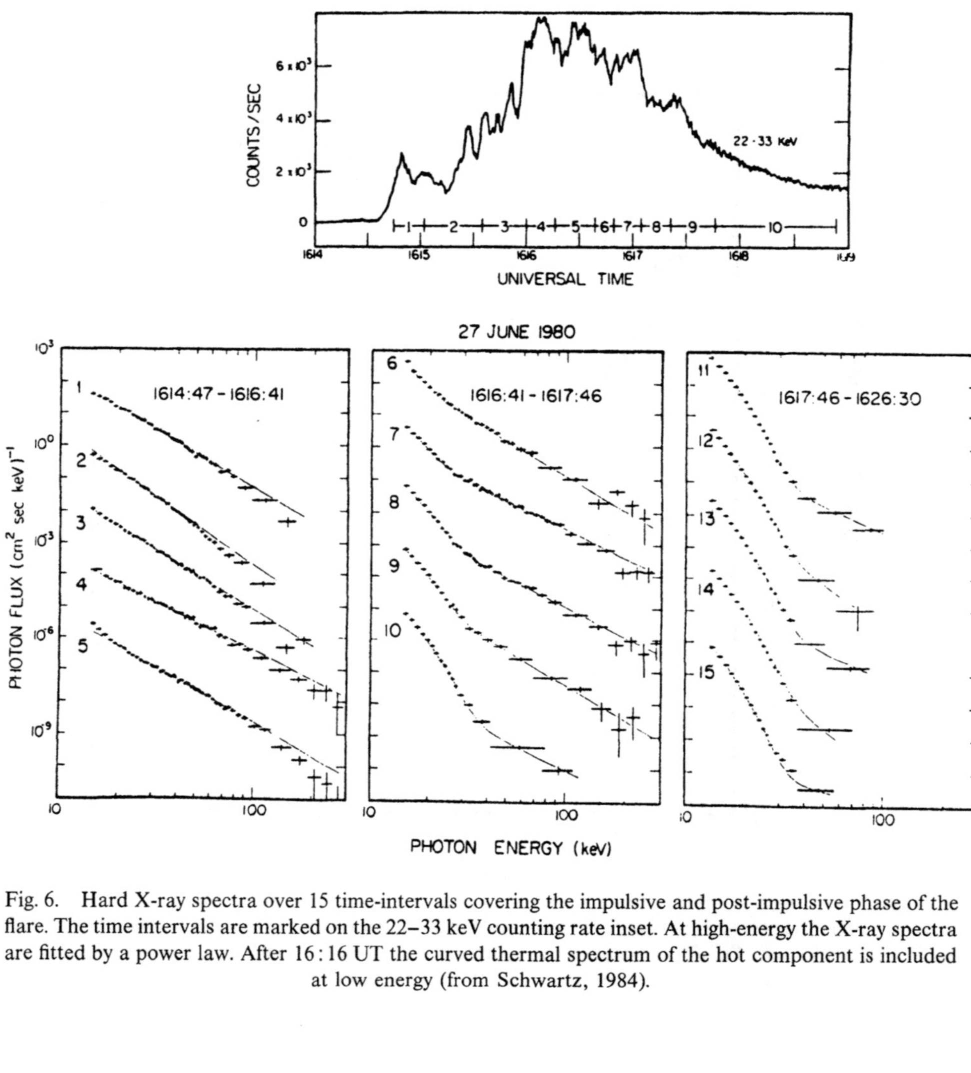

Fig. 6. Hard X-ray spectra over 15 time-intervals covering the impulsive and post-impulsive phase of the flare. The time intervals are marked on the 22–33 keV counting rate inset. At high-energy the X-ray spectra are fitted by a power law. After 16 : 16 UT the curved thermal spectrum of the hot component is included at low energy (from Schwartz, 1984).

and degree of anisotropy of the initial angular distribution (isotropic or beamed, see Figure 4). Indeed, the spectral hardness of the X-ray flux exhibits less variation in the case of an isotropic distribution (see also Figure 6 of Leach and Petrosian, 1983). The predicted gradual hardening of the X-ray flux is in agreement with the stereoscopic observations of Kane (1983) which show that, for coronally occulted events, the total X-ray flux is harder and more intense than the coronal part of the emission (Figure 5). However, even with the same occulting height, the spectral variations between the total flux and its coronal component vary from one flare to the other and the two fluxes may exhibit different temporal evolutions. The variations from one event to the other can easily be explained in the above model by variations from one flare to the other of the electron angular distribution and of the emission region characteristics. Moreover, HXIS observations (Haug and Elwert, 1985) have also shown for a limb flare that the coronal X-ray part of the event is softer than the footpoint emission, as predicted by the present models.

4. Discussion and Conclusions

As shown in the previous sections, the most probable hard X-ray emission mechanism is bremsstrahlung of energetic electrons interacting with the ambient medium, whether or not these electrons reflect the primary energy release or secondary effects. Thus, characteristics of the energetic electrons and of the emitting source can be inferred through models from X-ray observations. As concerns the long discussed controversy between a thermal or a non-thermal interpretation of X-ray bursts, it can be concluded that, although both interpretations explain some of the characteristics of X-ray emission, they both fail for different reasons. Thermal models, while potentially more energetically efficient, cannot easily explain, e.g., very energetic X-ray events, the good observed correlation between the (not necessarily cospatial) hard X-ray and microwave emissions and the association between hard X-rays and γ-rays (showing evidence for particle acceleration) observed for some events. On the other hand, although non-thermal interpretations of the X-ray flux succeed in reproducing different types of hard X-ray events and in accounting for the close temporal correlation between hard X-rays and broad band radio emission – such as the microwave radiation produced through electron gyrosynchrotron – (see, e.g., Klein *et al.*, 1986), these kinds of interpretations demand an embarrassingly high-acceleration efficiency. At the present time, X-ray observations do not generally allow a definitive statement on the thermal or non-thermal origin of the emission (see below). Both components are in fact present in solar flares in a proportion which may vary from one flare to the other and from one part of the event to the other. This is strongly suggested by observations obtained with excellent spectral resolution ($\sim$keV) by Lin *et al.* (1981) which show the gradual rise of a hot thermal X-ray component ($T \sim 3 \times 10^7$ K) – over a non-thermal one after the burst maximum (Figure 6).

Hard X-ray directivity should in principle provide valuable information on the characteristics of the X-ray emitting electron population. Indeed, as the differential electron bremsstrahlung cross section reaches a maximum in direction close to the electron velocity vector, different directivities are expected depending on the anisotropy of the radiating electrons. Contrary to thermal models, non-thermal models tend to give anisotropic distributions of the X-ray flux (see, e.g., Elwert and Haug, 1970, 1971), although at low-photon energies (less than ~ 300 keV), the intrinsic directivity may be substantially reduced by the photospheric Compton backscattering (see, e.g., Bai and Ramaty, 1978). However, the observation of an anisotropic distribution of the X-ray flux does not necessarily imply an intrinsic anisotropy of the accelerated electrons. Indeed, propagation effects in a non-uniform magnetic field may significantly alter the electron angular distribution from the energy release site to the emitting regions, thus changing the observed directivity (Leach and Petrosian, 1983; Petrosian, 1985). Measurements of the directivity can be obtained either directly by stereoscopic observations or statistically by center to limb variations of the hard X-ray events, once assuming an a priori luminosity distribution of bursts. The former observations (Kane *et al.*, 1980a) put severe limits on the emission anisotropy in the energy range 50–100 keV, thus favouring isotropic hard X-ray emission such as produced in thermal models. However, the results may be also compatible with a non-thermal anisotropic X-ray production, if the photospheric backscattering is taken into account. On the other hand, a recent statistical study carried out on $\sim$ a hundred events (Vestrand *et al.*, 1986) shows that, above 300 keV, the fraction of detected events close to the limb is higher than it is expected for isotropic X-ray emission. This effect is interpreted as an evidence of a small but real directionality of X-ray emission above 300 keV and as an indication of some anisotropy of the X-ray emitting electron distribution (Vestrand *et al.*, 1986). At higher energies (≥ 10 MeV), the effect is magnified since all the events are detected near the limb (Rieger *et al.*, 1983). It must be nevertheless reminded, that because of the small number of detected events (~ 10), this result may not be significant statistically.

Hard X-ray polarization measurements may also give valuable diagnostics on the nature of the X-ray emitting energetic electrons. Indeed, a thermal production of X-ray predicts polarizations of at most a few percent (Hénoux, 1975; Emslie and Brown, 1980), while non-thermal hard X-ray models predict high polarizations (10% or more) (see, e.g., Leach and Petrosian, 1983). However, even the recent measurements aboard the Space Shuttle (Tramiel *et al.*, 1984) are inconclusive. Indeed, the low-observed level of polarization at energies between 5 and 20 keV can either be produced by thermal bremsstrahlung or by electron beams once isotropization due to collisions or non converging magnetic fields is taken into account (Leach *et al.*, 1985). It must be noted that the present observations are restricted to very low X-ray energies where the thermal bremsstrahlung of the soft X-ray emitting plasma dominates and that polarization measurements obtained at higher energies could certainly bring more severe constraints on the anisotropy of the X-ray emitting electrons.

Finally, hard X-ray images should in principle allow one to determine which part of the emission is produced by a hot plasma or by a non-thermal electron population. Indeed, bremsstrahlung of non-thermal electron beams would rather produce compact emission at magnetic footpoints while bremsstrahlung of a hot plasma or of highly confined non-thermal electrons would favour large structureless X-ray sources near the top of magnetic loops. In fact, both X-ray sources have been observed with the detectors aboard SMM or Hinotori (e.g., Duijveman *et al.*, 1982; Tsuneta, 1983). However, at the present time, images are still restricted to energies less than 400 keV where the thermal bremsstrahlung of the soft X-ray emitting plasma may be important. In the future, higher resolution hard X-ray images obtained with better time resolution and extending to higher photon energies will certainly allow improved identification of the X-ray emitting electron population as well as the characteristics of the emitting source and the location of the energy release site. In particular, experiments such as the 'Pinhole Occulting Facility' (Tandberg-Hanssen *et al.*, 1983) providing high temporal, spatial and spectral resolution X-ray data should allow close examination of the dynamics of the emitting particles themselves once they are accelerated. As a conclusion, although X-ray emission is the most direct diagnostics of energetic electrons, hard X-ray observations by themselves do not give the only clues to the problems of particle acceleration in solar flares. To get a more complete understanding, it is in fact necessary to interprete quantitatively and simultaneously the different emissions produced by the particles as well as to study in more detail the interactions of the accelerated particles with the ambient plasma.

Acknowledgements

I would like to thank Drs K. L. Klein, A. L. MacKinnon, and G. Trottet for helpful discussions and comments on this paper. I am also grateful to the referee for valuable comments on the manuscript.

References

Bai, T. and Dennis, B. R.: 1985, *Astrophys. J.* **292**, 699.
Bai, T., Hudson, H. S., Pelling, R. M., Lin, R. P., Schwartz, R. A., and von Rosenvinge, T. T.: 1983, *Astrophys. J.* **267**, 433.
Bai, T. and Ramaty, R.: 1978, *Astrophys. J.* **219**, 705.
Bai, T. and Ramaty, R.: 1979, *Astrophys. J.* **227**, 1072.
Batchelor, D. A., Crannell, C. J., Wiehl, H. J., and Magun, A.: 1985, *Astrophys. J.* **295**, 258.
Brown, J. C.: 1971, *Solar Phys.* **18**, 489.
Brown, J. C.: 1972, *Solar Phys.* **26**, 441.
Brown, J. C.: 1973, *Solar Phys.* **31**, 143.
Brown, J. C.: 1974, in S. R. Kane (ed.), *Solar γ-, X- and EUV Radiation*, Joint AIS-COSPAR Symp. No. 68, D. Reidel Publ. Co., Dordrecht, Holland, 245.
Brown, J. C. and Bingham, R.: 1984, *Astron. Astrophys.* **131**, L11.
Brown, J. C. and Melrose, D. B.: 1977, *Solar Phys.* **52**, 117.
Brown, J. C., Melrose, D. B., and Spicer, D. S.: 1979, *Astrophys. J.* **228**, 592.
Brown, J. C. and Smith, D. F.: 1980, *Rep. Prog. Phys.* **43**, 125.
Chupp, E. L.: 1984, *Ann. Rev. Astron. Astrophys.* **22**, 359.
Craig, I. J. D., MacKinnon, A. L., and Vilmer, N.: 1985, *Astrophys. Space Sci.* **116**, 377.
Crannell, C. J., Frost, K. J., Mätzler, C., Ohki, K., and Saba, J. L.: 1978, *Astrophys. J.* **223**, 620.
Cromwell, D., McQuillan, P., and Brown, J. C.: 1985, *Proceedings of the SMM Rapid Fluctuations Workshop*, Lanham, MD, U.S.A., p. 12.
Duijveman, A., Hoyng, P., and Machado, M. E.: 1982, *Solar Phys.* **58**, 127.

Elliot, H.: 1964, *Planetary Space Sci.* **12**, 657.
Elwert, G. and Haug, E.: 1970, *Solar Phys.* **15**, 234.
Elwert, G. and Haug, E.: 1971, *Solar Phys.* **20**, 413.
Emslie, A. G.: 1980, *Astrophys. J.* **235**, 1055.
Emslie, A. G.: 1981, *Astrophys. J.* **245**, 711.
Emslie, A. G.: 1985, *Solar Phys.* **98**, 281.
Emslie, A. G. and Brown, J. C.: 1980, *Astrophys. J.* **237**, 1015.
Emslie, A. G. and Brown, J. C.: 1985, *Astrophys. J.* **295**, 648.
Emslie, A. G. and Smith, D. F.: 1984, *Astrophys. J.* **279**, 882.
Emslie, A. G. and Vlahos, L.: 1980, *Astrophys. J.* **242**, 359.
Haines, M. G. and Marsh, F.: 1982, *J. Plasma Phys.* **27**, 427.
Haug, E. and Elwert, G.: 1985, *Solar Phys.* **99**, 219.
Heristchi, D.: 1986, *Astrophys. J.* **311**, 474.
Hénoux, J. C.: 1975, *Solar Phys.* **42**, 219.
Hernandez, A. M., Machado, M. E., Vilmer, N., and Trottet, G.: 1986, *Astron. Astrophys.* **167**, 77.
Heyvaerts, J.: 1981, in E. R. Priest (ed.), *Solar Flare Magnetohydrodynamics*, Gordon and Breach Science Publishers, New York, Chapter 8, p. 429.
Kane, S. R.: 1983, *Solar Phys.* **86**, 355.
Kane, S. R., Anderson, K. A., Evans, W. D., Klebesadel, R. W., and Laros, J. G.: 1980a, *Astrophys. J.* **239**, L85.
Kane, S. R., Crannell, C. J., Datlowe, D. W., Feldman, U., Gabriel, A., Hudson, H. S., Kundu, M. R., Mätzler, C., Neidig, D., Petrosian, V., and Sheely, N. R., Jr.: 1980b, in P. Sturrock (ed.), *Solar Flares*, Univ. of Colorado Press, Boulder, Chapter 5, p. 187.
Kaufmann, P., Correia, E., Costa, J. E. R., and Zodi Vaz, A. M.: 1986, *Astron. Astrophys.* **157**, 11.
Kawamura, K., Omadaka, T., and Suzuki, I.: 1981, *Solar Phys.* **71**, 55.
Kennel, C. F.: 1969, *Rev. Geophys.* **7**, 397.
Klein, K. L.: 1987, *Astron. Astrophys.*, in press.
Klein, K. L., Trottet, G., and Magun, A.: 1986, *Solar Phys.* **104**, 243.
Knight, J. W. and Sturrock, P. A.: 1977, *Astrophys. J.* **218**, 306.
Korchak, A. A.: 1971, *Solar Phys.* **18**, 284.
Kovalev, V. A. and Korolev, O. S.: 1981, *Soviet Astron.* **25**, 215.
Kundu, M. R.: 1984, *Adv. Space Res.* **4**, 157.
Leach, J. and Petrosian, V.: 1981, *Astrophys. J.* **251**, 781.
Leach, J. and Petrosian, V.: 1983, *Astrophys. J.* **269**, 715.
Leach, J., Emslie, A. G., and Petrosian, V.: 1985, *Solar Phys.* **96**, 331.
Lin, R. P., Schwartz, R. A., Pelling, R. M., and Hurley, K. C.: 1981, *Astrophys. J.* **251**, L109.
MacKinnon, A. L.: 1985, *Solar Phys.* **98**, 293.
MacKinnon, A. L.: 1986, *Astron. Astrophys.* **163**, 239.
MacKinnon, A. L., Brown, J. C., Trottet, G., and Vilmer, N.: 1983, *Astron. Astrophys.* **119**, 297.
Mätzler, C., Bai, T., Crannell, C. J., and Frost, K. J.: 1978, *Astrophys. J.* **223**, 1058.
McClements, K. G.: 1987, *Astron. Astrophys.* **175**, 255.
McClements, K. G. and Brown, J. C.: 1986, *Astron. Astrophys.* **165**, 235.
McClements, K. G., Emslie, A. G., and Brown, J. C.: 1985, *Proceedings of the SMM Rapid Fluctuations Workshop*, Lanham, MD, U.S.A., p. 96.
Melrose, D. B. and Brown, J. C.: 1976, *Monthly Notices Roy. Astron. Soc.* **176**, 15.
Melrose, D. B. and Dulk, G. A.: 1982, *Astrophys. J.* **259**, 844.
Orwig, L. E. and Woodgate, B. E.: 1985, *Proceedings of the SMM Rapid Fluctuations Workshop*, Lanham, U.S.A.
Peterson, L. E. and Winckler, J. R.: 1959, *J. Geophys. Res.* **64**, 697.
Petrosian, V.: 1985, *Astrophys. J.* **299**, 987.
Pick, M.: 1986, *Solar Flares and Coronal Physics Using P/OF as a Research Tool*, NASA Conference Publ. 2421, p. 159.
Rieger, E., Reppin, C., Kanbach, G., Forrest, D., Chupp, E., and Share, G.: 1983, *18th ICRC* **10**, 338.
Rowland, H. L. and Vlahos, L.: 1985, *Astron. Astrophys.* **142**, 219.
Schwartz, R. A.: 1984, Ph.D. Thesis, University of California, Berkeley, CA, U.S.A.

Simnett, G. M.: 1986, *Proc. of the SMA Workshop*, Irkutsk USSR, June 1985, VNU, Science Press, Utrecht, The Netherlands.
Simnett, G. M. and Strong, K. T.: 1984, *Astrophys. J.* **284**, 839.
Smith, D. F.: 1986, *Astrophys. J.* **302**, 836.
Smith, D. F. and Auer, L. H.: 1980, *Astrophys. J.* **238**, 1126.
Smith, D. F. and Lilliequist, C. G.: 1979, *Astrophys. J.* **232**, 582.
Spicer, D. S. and Sudan, R. N.: 1984, *Astrophys. J.* **280**, 448.
Švestka, Z.: 1976, *Solar Flares*, D. Reidel Publ. Co., Dordrecht, Holland.
Takakura, T.: 1986, *Solar Phys.* **104**, 363.
Takakura, T. and Kai, K.: 1966, *Publ. Astron. Japan* **18**, 57.
Tandberg-Hanssen, E. A., Hudson, H. S., Dabbs, J. R., and Baity, W. A.: 1983, NASA, Technical Paper 2168.
Tramiel, L. J., Chanan, G. A., and Novick, R.: 1984, *Astrophys. J.* **280**, 440.
Trottet, G.: 1986, *Solar Phys.* **104**, 145.
Trottet, G. and Vilmer, N.: 1983, *XVIIIth International Cosmic Ray Conference Proceeding*, SP 3-15.
Trottet, G. and Vilmer, N.: 1984, *Adv. Space Res.* **4**, 153.
Tsunuta, S.: 1983, *Proceedings of the Japan–France Seminar, Paris*, p. 243.
Vestrand, W. T., Forrest, D. J., Chupp, E. L., Rieger, E., and Share, G.: 1986, *Adv. Space Res.* **6**, No. 6, 123.
Vilmer, N., Kane, S. R., and Trottet, G.: 1982, *Astron. Astrophys.* **108**, 306.
Vilmer, N., MacKinnon, A. L., and Trottet, G.: 1986a, *Astrophys. Space Sci.* **123**, 87.
Vilmer, N., Trottet, G., and MacKinnon, A. L.: 1986b, *Astron. Astrophys.* **156**, 64.
Vlahos, L. and Papadopoulos, K.: 1979, *Astrophys. J.* **233**, 717.
Vlahos, L., Machado, M. E., Ramaty, R., Murphy, R. J., and 23 co-authors: 1984, in M. Kundu and B. Woodgate (eds.), *Energetic Phenomena on the Sun*, NASA CP 2439, p. 2–1.
Vlahos, L., Sharma, R. R., and Papadopoulos, K.: 1979, *Astrophys. J.* **233**, 717.

HARD X-RAY AND RADIO EMISSION AT THE ONSET OF GREAT SOLAR FLARES*

K.-L. KLEIN[1], M. PICK[1], A. MAGUN[2], and B. R. DENNIS[3]

Abstract. A study of the onset phase of ten great hard X-ray bursts is presented. It is shown from hard X-ray and radio observations in different wavelength ranges that the energization of the electrons proceeds on a global time-scale of some tens of seconds. In nine of the bursts two phases of emission can be distinguished during the onset phase: the pre-flash phase, during which emission up to an energy limit ranging from some tens of keV to 200 keV is observed, followed ten to some tens of seconds later by the flash phase, where the count rate in all detector channels rises simultaneously to within some seconds. For two of the events strong γ-ray line emission is observed and is shown to start close to the onset of the flash phase.

1. Introduction

Hard X-ray and γ-ray observations from SMM have shown that energetic electrons and ions are accelerated from the early stage of solar flares. There is nevertheless still controversy as to the exact timing of the acceleration of low-energy electrons on the one hand, and of high-energy electrons and ions on the other hand. Furthermore, the relative proportions of particle acceleration and heating in the primary energization process are still subject to discussion.

Hard X-ray and radio emissions allow us to probe the evolution of the energetic electrons during a flare, while 4–8 MeV γ-ray line emission reveals energetic ions. These radiations are emitted through different mechanisms in different sources of the solar atmosphere. Any common signature in the time histories of the emissions can hence be attributed to the radiating particles. The difficulties of inferring properties of the particles from such observations arise because the radiative signature is the product of several complex processes:

– the acceleration and injection into atmospheric structures where the particles radiate;

– the dynamical evolution of the energetic particles in these structures through interaction with the ambient medium (e.g., Vilmer, 1986, and references therein);

– the radiation processes themselves which in different spectral bands are sensitive to different energy ranges of the particles.

The most direct diagnostic of energetic electrons during flares is their hard X-ray emission (20–500 keV), because it does not suffer from self-absorption and gives a better energy discrimination than any other radiative signature. Microwaves are in general self-absorbed at low frequencies. They do not allow for a sharp energy discrimnation, but as a rule the electron energies involved grow with the observing frequency. Deci-

* Proceedings of the Second CESRA Workshop on Particle Acceleration and Trapping in Solar Flares, held at Aubigny-sur-Nère (France), 23–26 June, 1986.

[1] Observatoire de Paris, Section de Meudon, UA 324, 92195 Meudon, France.

[2] Institut für Angewandte Physik, Sidlerstr. 5, 3012 Bern, Switzerland.

[3] Code 602.6, NASA – Goddard Space Flight Center, Greenbelt, Maryland, 20771, U.S.A.

Solar Physics **111** (1987) 225–233.

metre-to-metre wave emission cannot yet be associated with a particular range of energies.

Studies of the timing of these radiations have so far focussed mostly on the instant of maximum of the event. This time depends very much on the dynamical evolution of the particles. Time delays of the maxima of radiative signatures resulting from different electron energies or from different particle species hence are not necessarily a property of the energization process. On the other hand hard X-ray and microwave emission is the immediate response of a sufficiently dense and magnetized plasma to the injection of a sufficiently high number of energetic electrons. The starting time of these emissions is thus expected to be far less affected by the interaction processes. It is clear, however, that the detection thresholds of the instruments and their dependences on the energy of the radiating particles introduce problems into the interpretation. This makes the use of detectors with different characteristics necessary.

Studies on the onset phase of the radiation from energetic electrons in solar flares are still rare. Forrest and Chupp (1983) concluded for two events that the hard X-ray and γ-ray observations were consistent, within the limits of detector sensitivity, with a simultaneous start of photon emission from 40 keV to 8 MeV. Machado *et al.* (1986), on the other hand, noted an excess of hard X-rays below 100 keV and a nonthermal broadening of the Ca XIX line before the onset of the rapidly evolving burst. Benz *et al.* (1983) considered hard X-rays and radio waves at the very beginning of the impulsive phase. In some events they found evidence for two components of the impulsive phase: a weak hard X-ray emission lasting up to one minute, followed by a steep rise within some seconds to the peak count rate. They termed these phases respectively pre-flash and flash phase. Subsequently Raoult *et al.* (1985) showed for a sample of energetic hard X-ray bursts a systematic evolution from the pre-flash phase, associated with metric type III bursts, to the more energetic flash phase with associated continua in the decimetre to metre waveband (type V burst). Raoult *et al.* also suggested that the mixed type III/V activity observed during the flash phase appears to be a signature of flares where particles are accelerated to high energies and where occasionally γ-ray lines are observed. However, neither of the last two papers considered the dependence of these evolutions on the energy of the radiating electrons.

This paper attempts an observational analysis of the energetic particle populations during the onset and early rise of the X-ray emission above 20 keV with emphasis on the energy-dependence of this evolution. We start from an analysis of the hard X-ray emission, aiming at the precise definition of the starting time of the burst in each detector channel. The study is based on observations made with the Hard X-Ray Burst Spectrometer (HXRBS) on the Solar Maximum Mission (SMM) satellite (Orwig *et al.*, 1980). The results will be compared for some events with observations from ISEE-3 (courtesy S. R. Kane), with the simultaneous radio signature of energetic electrons and with the gamma-ray line emission.

2. Method of Analysis and Event Selection

We have analyzed the starting time of ten hard X-ray bursts observed with HXRBS. These events had peak count rates above 7000 s^{-1} integrated over all detector channels, corresponding to photon energies between about 25 keV and 500 keV (HXRBS event listing: Dennis *et al.*, 1985). Properties of the hard X-ray, gamma-ray and radio emission of these events, compiled from the HXRBS event listing, Bai and Dennis (1985), Raoult *et al.* (1985), from observations with the Gamma-Ray Spectrometer (GRS) on SMM (Chupp, 1986, personal communication), and from *Solar Geophysical Data*, are listed in Table I.

TABLE I

X-ray, γ-ray, and radio emission of the sample of events

Flare		Hard X-ray peak		4–8 MeV	*m*-wave
		Count rate	Spectral index		
3 June, 1982	11:42 UT	2.1×10^5	2	yes	III GG/V; II
5 Feb., 1982	09:03 UT	7.2×10^3	3.5	yes	III/V
26 Apr., 1981	11:40 UT	7.8×10^3	3.4	yes	III G; IV
26 Feb., 1981	14:23 UT	2.3×10^4	3.2	yes	III GG
9 Oct., 1980	11:23 UT	2.7×10^4	4.5	no	no report
29 June, 1980	10:41 UT	9.3×10^3	3.8	yes	III GG/V; II
21 June, 1980	01:12 UT	1.4×10^5	2.0	yes	III/V; IV
7 June, 1980	03:11 UT	3.9×10^4	2.8	yes	III CC/V; II
29 Mar., 1980	09:17 UT	1.9×10^4	3.5	no	III GG/V
29 Mar., 1980	09:53 UT	8.0×10^3	4.2	no	III G/V

In order to define a starting time of the burst, we computed in each channel the background count rate (mean value) and the noise (standard deviation σ, derived from the measured fluctuations in the background) in the minutes before burst onset. The starting time was defined by the instant after which for the first time three successive count rate values exceeded the mean background rate by more than $n\sigma$. We considered separately the cases $n = 3$ and $n = 5$ in order to have an approximate measure of the uncertainty of our results. In all channels count rates integrated during 1.024 s were used.

3. Results

Figure 1 shows the results for the strongest event of our sample, the neutron flare of 1982, June 3. Crosses give the photon energy where the count rate exceeds the threshold value, as a function of time: horizontal error bars delimit the time interval between the 3σ and 5σ levels' being exceeded (if no error bars are plotted, these levels are exceeded simultaneously), vertical error bars represent the width of the detector channels. For the

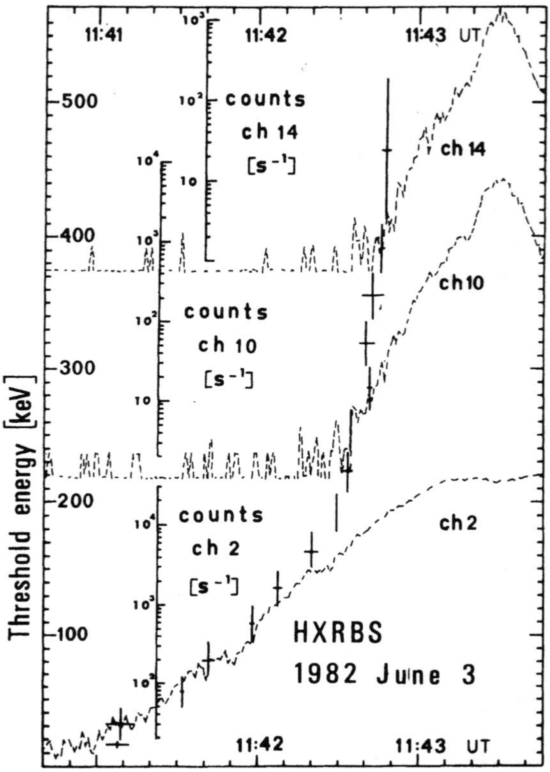

Fig. 1. Time history of the hard X-ray emission at the onset of the 1982, June 3 flare. Crosses denote the photon energy where the count rate exceeds the noise level. The dashed curves give the count rate in three HXRBS detector channels on a logarithmic scale: channel 2 (37–62 keV, maximum count rate $5 \times 10^4\ s^{-1}$), channel 10 (262–294 keV, $6 \times 10^3\ s^{-1}$), channel 14 (400–438 keV, $2 \times 10^3\ s^{-1}$).

purpose of illustration we have also plotted – on a logarithmic scale – the count rate time histories in three HXRBS channels. Two observations can be made from this figure:

(1) The onset of the burst occurs later at higher energies. Apparent delays of several seconds between successive detector channels are observed in the early onset phase.

(2) About 80 s after the start of the low-energy emission the delays shorten. The threshold is exceeded nearly simultaneously in the high energy channels.

The existence of a delay between burst onsets at high and low photon energies is expected when the noise level decreases more slowly with photon energy than the count rate of the burst emission. This is indeed the case, and the duration of the delay can thus not be assessed by our method. A possible reason for the shortening of the delays at higher photon energies could be a downward break of the noise spectrum at these energies. However, by inspection of the noise spectra determined from the pre-burst level in the detector channels it was found that this instrumental effect cannot explain the observed evolution.

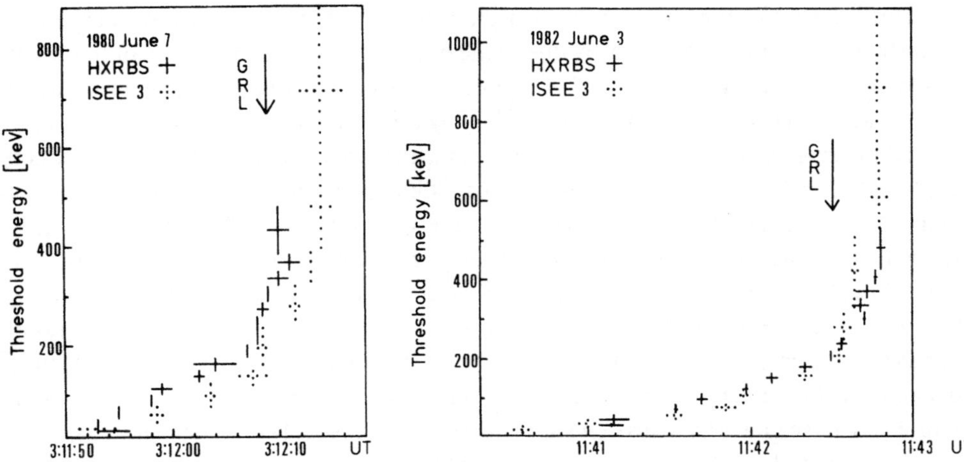

Fig. 2. Time histories of the threshold energy of two flares observed with HXRBS (full crosses) and with the spectrometer on ISEE-3 (dotted crosses). The arrow labelled 'GRL' indicates the starting time of the excess emission in the 4–8 MeV photon energy range of the SMM gamma-ray spectrometer.

Figures 2(a) and 2(b) compare the results obtained with HXRBS and with ISEE-3 for the 1980, June 7 event and the 1982, June 3 burst, respectively. Despite differences introduced by the detector characteristics, the conclusions from Figure 1 apply to both events and to the observations of both instruments. The temporal evolutions of the threshold energies observed with HXRBS and ISEE-3 are seen to be similar to within some seconds. The time scales of the evolutions are different in these events. As a working definition which does not yet allude to specific physical processes, we refer to the two successive parts of the rising hard X-ray emission as pre-flash and flash phase, respectively.

Both events exhibit excess emission in the 4–8 MeV photon energy range (Forrest and Chupp, 1983; Chupp, personal communication) which is intense enough to identify the time of its onset: the emission starts, as indicated by the arrows in Figures 2(a) and 2(b), close to the onset of the flash phase of hard X-rays.

In most of our events the section termed flash phase, which is the first observable signature of the hard X-ray burst in the high-energy channels, is also distinguished in channels of lower energy. However, its start is masked by the pre-flash emission and cannot be analyzed by the method of Section 2. We attempted to get an indication of the starting time in the low-energy channels by linearly extrapolating the rising branch of the count rate time history back to zero. This was possible for four events of our sample. The starting time of the flash phase found this way in the low-energy channels agreed with that in the high-energy channels obtained with the method of Section 2 within an uncertainty of $\pm$ 3 s. As the extrapolation yields only a crude estimate of the starting time, this difference cannot be considered as significant.

The analysis of the ten hard X-ray bursts of Table I is summarized as follows: in all of them the start of the emission at high photon energies lags that at low energies by some tens of seconds. This is partly due to the different energy-dependence of the count rate and the noise level. In nine of these events evidence is found for the speed up of this evolution. This phenomenon is described by the term flash phase in contrast to the preceding pre-flash emission. In the two events which are common to the study of Benz *et al.* (1983) and the present one, these terms, although inferred from different types of analysis, denote the same time periods. The energy above which flash phase emission is seen as the first detectable signature of the burst ranges between some tens of keV and 200 keV. The duration of the pre-flash phase is some seconds to 80 s. However, the two phases are not sharply delimited in time: when it exists, the flash phase emission appears as a gradual acceleration of the pre-flash phenomenon. The above figures depend on the detection threshold and should only be considered as an indication. During the flash phase the emission starts nearly simultaneously in all detector channels.

These results have been obtained with HXRBS and have been confirmed for four events of our sample with the spectrometer aboard ISEE-3. In the following we discuss in some detail the spectral evolutions of hard X-rays and radio waves at the onset of

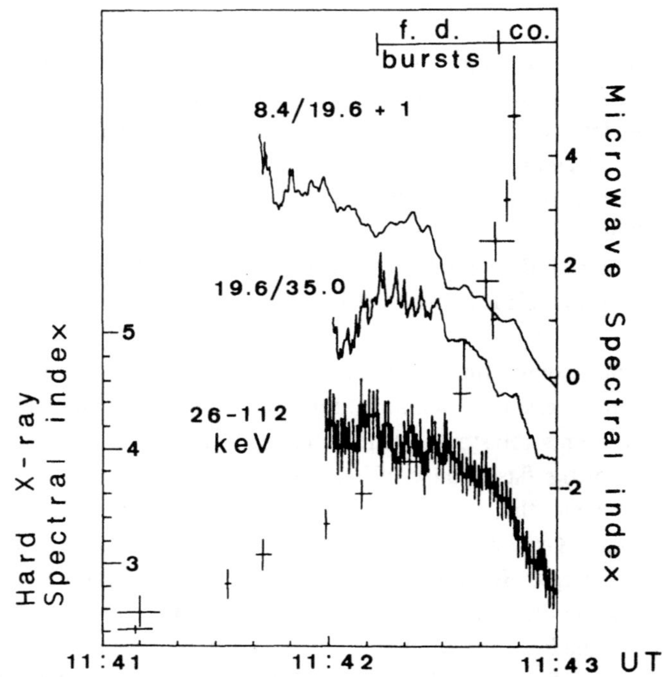

Fig. 3. Temporal evolution of the hard X-ray and radio spectra at the onset of the 1982, June 3 burst. The solid curves show, from *bottom* to *top*: (1) the power-law spectral index fitted to channels 2 to 4 (26–112 keV) of HXRBS; (2) the spectral indices of the microwave flux, derived respectively from observations at 19.6 and 35 GHz and at 8.4 and 19.6 GHz. In order to avoid superposition of the curves, the measured index between 8.4 and 19.6 GHz has been incremented by 1. The periods of fast drift bursts and continuum emission in the dm–m waveband are indicated by horizontal bars at the top of the figure. The time history of the threshold energies is plotted by the crosses (cf. Figures 1 and 2).

the 1982, June 3 burst, in order to separate as far as possible the properties of the radiating electrons from the influence of the detector response.

Figure 3 compares the radiative signatures of the flare in different spectral ranges. The solid line at the bottom shows the evolution of the power-law spectral index obtained from a fit to photon energies between 26 and 112 keV. These low energies were chosen so as to derive spectra during and after the pre-flash phase which are not contaminated by the poor count rate statistics of the high-energy channels. For comparison, the temporal evolution of the threshold energies is again indicated by the crosses. It is clear that a reliable spectral fit is only possible well after the onset of the event. In the early pre-flash phase before 11:42 UT count rate statistics are too poor. It is seen that during the late pre-flash phase the spectral slope is roughly constant within the limits of the statistical error, with a power-law index of 4.2 ± 0.4. As the flash phase emission rises, the photon spectrum starts a continuous hardening. The two upper curves in Figure 3 show the evolution of the microwave spectral index α, derived from observations at 8.4 and 19.6 GHz and 19.6 and 35 GHz, respectively (flux $\sim \nu^{-\alpha}$, where ν is the frequency; observations from Berne University). At the onset of the hard X-ray emission at low energies the flux is more than 5 times lower at 19.6 GHz than at 8.4 GHz. As the hard X-ray emission exceeds the detection theshold at successively higher photon energies, the high frequency radiation increases with respect to the low frequencies (i.e. α tends towards 0) and finally exceeds the latter ($\alpha < 0$) after 11:42:50 UT. The microwave spectral index follows closely the evolution of the hard X-ray spectrum. A detailed correspondence between the spectral indices in hard X-rays and microwaves is not expected, because self-absorption by the energetic electrons influences the microwave spectral index after 11:42:40 UT. Nevertheless, the speed up of the evolution is seen, especially at the highest frequencies, to be simultaneous with the hardening of the hard X-ray spectrum in the flash phase. Assuming that the microwaves are emitted through the gyrosynchrotron process, this evolution is due either to an increase of the magnetic field or to the presence of more numerous or more energetic electrons in the microwave source. The common evolution of the hard X-ray and microwave spectra, where the former is insensitive to the magnetic field, suggests that the observations should be interpreted in terms of the energy spectrum of the radiating electrons, which evolves on a time scale of some tens of seconds. The indication of the flash phase in the microwave spectral index was not found in the data of 1982, February 5 and 1981, February 26, although these events exhibit the two phases in their hard X-ray emission. It should be noted, however, that gyrosynchrotron radiation from energetic electrons is intrinsically wide-band, so that the microwave emission is far less indicative of a precise energy range than the hard X-rays. In the decimetre to metre waveband, changes of the spectrum and of the source configuration are observed during the onset phase with the Nançay radioheliograph and radiospectrograph. As indicated at the top of Figure 3, the emission starts with a weak fast drift burst at 11:42:13 UT. At 11:42:19 UT a series of fast drift bursts begins around 200 MHz. Their starting frequency drifts to higher values. At 11:42:45 UT a continuum (type V) becomes visible. The continuum and the superposed fast drift bursts extend very rapidly to high frequencies: at 11:43 UT the type III/V emission covers the whole observed band between 150 and 470 MHz. At 169 MHz a new radio source appears at 11 : 43 : 00 UT. The spread of the low-frequency emission to decimetric waves and the sudden appearance of a new source at metre waves accompanies the transition from pre-flash to flash phase emission as observed in the hard X-rays. Benz *et al.* (1983) and Raoult *et al.* (1985) have shown that this is a general feature of the low-frequency spectrum during the buildup of the impulsive phase.

The flare without an observable distinction between a pre-flash and flash phase occurred on 1981, April 26 at 11:40 UT. This event is part of a sequence of bursts between 10 : 30 UT and 13 : 00 UT observed with ISEE-3. During this period the hard X-ray emission up to some tens of keV is permanently above the background level. A reliable onset time cannot be given.

4. Summary and Conclusions

The present study of ten great hard X-ray bursts has shown that in their onset phase the number and energy of the energetic electrons increases on a global time scale of tens of seconds: The start of the hard X-ray emission in the highest detector channels is delayed or, equivalently, the photon spectrum hardens. In the four cases where the simultaneous microwave radiation has been analyzed, the bulk of the emission was found to shift towards higher frequencies during this period. In nine out of our sample of ten hard X-ray bursts the evolution speeds up after some tens of seconds. This is described by a phenomenological distinction of two sections of the impulsive phase: the pre-flash phase, observed at energies below some tens to 200 keV, varying from event to event, and the flash phase. During the latter the emission rises simultaneously in all detector channels within an uncertainty of some seconds.

In the two events where we have precise onset times measured with GRS, the rise of emission in the 4 to 8 MeV band occurs close to the onset of the flash phase of energetic electrons. Forrest and Chupp (1983) concluded that there were simultaneous starting times of the 40 keV to 8 MeV emission in the 1980, June 7 and June 21 events, observed with GRS. This is valid for the flash phase. The results obtained here for two events confirm this finding by a different approach. The pre-flash emission was not considered by Forrest and Chupp.

In conclusion, we have shown that observations in hard X-rays, microwaves and decimetric to metric radio waves can probe the evolution of the energetic electron population at the onset of solar flares, despite the problem of weak signals at these early times. This problem affects the quantitative interpretation of the observed spectral evolutions of hard X-rays and microwaves. However, the reality of these delays or, equivalently, of the hardening of the electron spectrum at the onset of the event, is suggested by the joint hard X-ray and microwave observations. The observations are consistent with two alternative interpretations:

– the evolution of a single mechanism of acceleration/injection during the onset of the flares;

– the transition between different processes, e.g., between emission from a hot plasma during the pre-flash phase and from accelerated electrons in the flash phase.

Further progress in interpreting the observations will be made through simulation of the HXRBS detector response to different model spectra.

Acknowledgements

The authors are indebted to Sharad Kane for having made ISEE-3 observations available. K.-L.K. acknowledges helpful discussions with D. Forrest, J. Brown, A. MacKinnon, G. Simnett, and G. Trottet. One of the authors (K.-L.K.) has benefitted from an ESA fellowship.

References

Bai, T. and Dennis, B.: 1985, *Astrophys. J.* **292**, 699.
Benz, A. O., Barrow, C. H., Dennis, B. R., Pick, M., Raoult, A., and Simnett, G.: 1983, *Solar Phys.* **83**, 267.
Dennis, B. R., Orwig, L. E., Kiplinger, A. L., Gibson, B. R., Kennard, G. S., and Tolbert, A. K.: 1985, NASA TM 86236.
Forrest, D. J. and Chupp, E. L.: 1983, *Nature* **305**, 291.
Machado, M. E., Orwig, L. E., and Antonucci, E.: 1986, *Adv. Space Res.* **6**, No. 6, 101.
Orwig, L. E., Frost, K. J., and Dennis, B. R.: 1980, *Solar Phys.* **65**, 25.
Raoult, A., Pick, M., Dennis, B. R., and Kane, S. R.: 1985, *Astrophys. J.* **299**, 1027.
Vilmer, N.: 1987, *Solar Phys.* **111**, 207 (this issue).

ANNOUNCEMENT

Particle Acceleration and Trapping in Solar Flares

Editors: Gérard Trottet and Monique Pick

Please note that a hardbound edition of this special issue of *Solar Physics*, Vol. 111, No. 1 (1987), is available from the publishers.

ISBN 90-277-2609-4 Prices: Dfl. 130,– / $64.00 / £40.25

GPSR Compliance

The European Union's (EU) General Product Safety Regulation (GPSR) is a set of rules that requires consumer products to be safe and our obligations to ensure this.

If you have any concerns about our products, you can contact us on ProductSafety@springernature.com

In case Publisher is established outside the EU, the EU authorized representative is:

Springer Nature Customer Service Center GmbH
Europaplatz 3
69115 Heidelberg, Germany

Batch number: 10614744

Printed by Printforce, the Netherlands